Principles and Applications of Atomic Absorption Spectroscopy

Principles and Applications of Atomic Absorption Spectroscopy

Edited by **Eddie Dillon**

New York

Published by NY Research Press,
23 West, 55th Street, Suite 816,
New York, NY 10019, USA
www.nyresearchpress.com

Principles and Applications of Atomic Absorption Spectroscopy
Edited by Eddie Dillon

International Standard Book Number: 978-1-63238-369-3 (Hardback)

Printed in the United States of America.

Contents

Preface

I am honored to present to you this unique book which encompasses the most up-to-date data in the field. I was extremely pleased to get this opportunity of editing the work of experts from across the globe. I have also written papers in this field and researched the various aspects revolving around the progress of the discipline. I have tried to unify my knowledge along with that of stalwarts from every corner of the world, to produce a text which not only benefits the readers but also facilitates the growth of the field.

The principles as well as applications of Atomic Absorption Spectroscopy (AAS) are elucidated in this book. Atomic absorption spectroscopy is one of the methods of analyzing qualitative and quantitative components in numerous fields like biomaterials, forensic science and industrial residues. The objective of this book is to encompass crucial spheres needed for helping scholars to comprehend current developments in this sphere. The book emphasizes on particular spheres in certain cases. It talks about history, basic principles, instrumentation and sample preparation, in-depth analysis with discussions on elemental profiling, functions, biochemistry and potential toxicity of metals, along with comparative techniques. This book also overviews the importance of sample preparation methods with emphasis on other details. Parts of the book also focus on ways to differentiate the components. The intervention results between various components have also been elucidated. This book will prove to be useful to everyone working in the related fields of AAS.

Finally, I would like to thank all the contributing authors for their valuable time and contributions. This book would not have been possible without their efforts. I would also like to thank my friends and family for their constant support.

Editor

Atomic Absorption Spectrometry (AAS)

R. García and A. P. Báez

Centro de Ciencias de la Atmósfera, Universidad Nacional Autónoma de México,
Ciudad Universitaria, Mexico City
Mexico

1. Introduction

Atomic Absorption Spectrometry (AAS) is a technique for measuring quantities of chemical elements present in environmental samples by measuring the absorbed radiation by the chemical element of interest. This is done by reading the spectra produced when the sample is excited by radiation. The atoms absorb ultraviolet or visible light and make transitions to higher energy levels. Atomic absorption methods measure the amount of energy in the form of photons of light that are absorbed by the sample. A detector measures the wavelengths of light transmitted by the sample, and compares them to the wavelengths which originally passed through the sample. A signal processor then integrates the changes in wavelength absorbed, which appear in the readout as peaks of energy absorption at discrete wavelengths. The energy required for an electron to leave an atom is known as ionization energy and is specific to each chemical element. When an electron moves from one energy level to another within the atom, a photon is emitted with energy E. Atoms of an element emit a characteristic spectral line. Every atom has its own distinct pattern of wavelengths at which it will absorb energy, due to the unique configuration of electrons in its outer shell. This enables the qualitative analysis of a sample. The concentration is calculated based on the Beer-Lambert law. Absorbance is directly proportional to the concentration of the analyte absorbed for the existing set of conditions. The concentration is usually determined from a calibration curve, obtained using standards of known concentration. However, applying the Beer-Lambert law directly in AAS is difficult due to: variations in atomization efficiency from the sample matrix, non-uniformity of concentration and path length of analyte atoms (in graphite furnace AA).

The chemical methods used are based on matter interactions, i.e. chemical reactions. For a long period of time these methods were essentially empirical, involving, in most cases, great experimental skills. In analytical chemistry, AAS is a technique used mostly for determining the concentration of a particular metal element within a sample. AAS can be used to analyze the concentration of over 62 different metals in a solution.

Although AAS dates to the nineteenth century, the modern form of this technique was largely developed during the 1950s by Alan Walsh and a team of Australian chemists working at the CSIRO (Commonwealth Science and Industry Research Organization) Division of Chemical Physics in Melbourne, Australia. Typically, the technique makes use of a flame to atomize the sample, but other atomizers, such as a graphite furnace, are also used. Three steps are involved in turning a liquid sample into an atomic gas:

1. Desolvation – the liquid solvent is evaporated, and the dry sample remains;
2. Vaporization – the solid sample vaporizes to a gas; and
3. Volatilization – the compounds that compose the sample are broken into free atoms.

To measure how much of a given element is present in a sample, one must first establish a basis for comparison using known quantities of that element to produce a calibration curve. To generate this curve, a specific wavelength is selected, and the detector is set to measure only the energy transmitted at that wavelength. As the concentration of the target atom in the sample increases, the absorption will also increase proportionally. A series of samples containing known concentrations of the compound of interest are analyzed, and the corresponding absorbance, which is the inverse percentage of light transmitted, is recorded. The measured absorption at each concentration is then plotted, so that a straight line can then be drawn between the resulting points. From this line, the concentration of the substance under investigation is extrapolated from the substance's absorbance. The use of special light sources and the selection of specific wavelengths allow for the quantitative determination of individual components in a multielement mixture.

2. Basic principle

The selectivity in AAS is very important, since each element has a different set of energy levels and gives rise to very narrow absorption lines. Hence, the selection of the monochromator is vital to obtain a linear calibration curve (Beers' Law), the bandwidth of the absorbing species must be broader than that of the light source; which is difficult to achieve with ordinary monochromators. The monochromator is a very important part of an AA spectrometer because it is used to separate the thousands of lines generated by all of the elements in a sample.

Without a good monochromator, detection limits are severely compromised. A monochromator is used to select the specific wavelength of light that is absorbed by the sample and to exclude other wavelengths. The selection of the specific wavelength of light allows for the determination of the specific element of interest when it is in the presence of other elements. The light selected by the monochromator is directed onto a detector, typically a photomultiplier tube, whose function is to convert the light signal into an electrical signal proportional to the light intensity. The challenge of requiring the bandwidth of the absorbing species to be broader than that of the light source is solved with radiation sources with very narrow lines.

The study of trace metals in wet and dry precipitation has increased in recent decades because trace metals have adverse environmental and human health effects. Some metals, such as Pb, Cd and Hg, accumulate in the biosphere and can be toxic to living systems. Anthropogenic activities have substantially increased trace metal concentrations in the atmosphere. In addition, acid precipitation promotes the dissolution of many trace metals, which enhances their bioavailability. In recent decades, heavy metal concentrations have increased not only in the atmosphere but also in pluvial precipitation. Metals, such as Pb, Cd, As, and Hg, are known to accumulate in the biosphere and to be dangerous for living organisms, even at very low levels. Many human activities play a major role in global and regional trace element budgets. Additionally, when present above certain concentration levels, trace metals are potentially toxic to marine and terrestrial life. Thus, biogeochemical perturbations are a matter of crucial interest in science.

The atmospheric input of metals exhibits strong temporal and spatial variability due to short atmospheric residence times and meteorological factors. As in oceanic chemistry, the impact of trace metals in atmospheric deposition cannot be determined from a simple consideration of global mass balance; rather, accurate data on net air or sea fluxes for specific regions are needed. Particles in urban areas represent one of the most significant atmospheric pollution problems, and are responsible for decreased visibility and other effects on public health, particularly when their aerodynamic diameters are smaller than 10 μm, because these small particles can penetrate deep into the human respiratory tract. There have been many studies measuring concentrations of toxic metals such as Ag, As, Cd, Cr, Cu, Hg, Ni, Pb in rainwater and their deposition into surface waters and on soils. Natural sources of aerosols include terrestrial dust, marine aerosols, volcanic emissions and forest fires. Anthropogenic particles, on the other hand, are created by industrial processes, fossil fuel combustion, automobile mufflers, worn engine parts, and corrosion of metallic parts. The presence of metals in atmospheric particles are directly associated with health risks of these metals. Anthropogenic sources have substantially increased trace metal concentrations in atmospheric deposition.

The instrument used for atomic absorption spectrometry can have either of two atomizers. One attachment is a flame burner, which uses acetylene and air fuels. The second attachment consists of a graphite furnace that is used for trace metal analysis. Figure 1 depicts a diagram of an atomic absorption spectrometer.

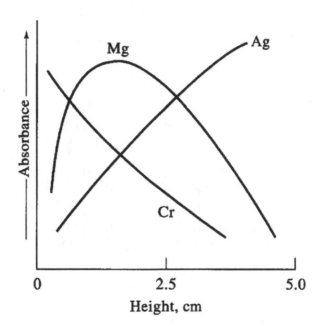

Fig. 1. The spectral, or wavelength, range captures the dispersion of the grating across the linear array.

Flame and furnace spectroscopy has been used for years for the analysis of metals. Today these procedures are used more than ever in materials and environmental applications. This is due to the need for lower detection limits and for trace analysis in a wide range of samples. Because of the scientific advances of Inductively Coupled Plasma Optical Emission

Spectroscopy (ICP-OES), Inductively Coupled Plasma Mass Spectrometry (ICP-MS), have left Atomic Absorption (AA) behind. This technique, however, is excellent and has a larger specificity that ICP does not have. Figure 2 shows a diagram of an atomic absorption spectrometer with a graphite furnace.

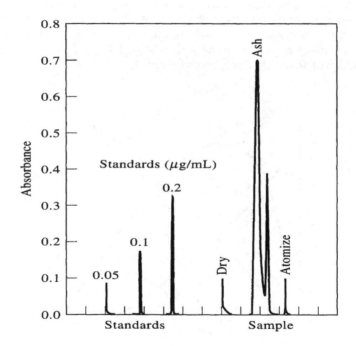

Fig. 2. Flame relatively insensitive - incomplete volatilization, short time in beam (Figure 1)

In this section, the authors will attempt to demonstrate that AAS is a reliable chemical technique to analyze almost any type of material. The chapter describes the basic principles of atomic absorption spectroscopy in the analysis of trace metals, such as Ag, As, Cd, Cr, Cu, and Hg, in environmental samples. For example, the study of trace metals in wet and dry precipitation has increased in recent decades because trace metals have adverse environmental and human health effects. Anthropogenic activities have substantially increased trace metal concentrations in the atmosphere. In recent decades, heavy metal concentrations have increased not only in the atmosphere but also in pluvial precipitation. Many human activities play a major role in global and regional trace element budgets. Additionally, when present above certain concentration levels, trace metals are potentially toxic to marine and terrestrial life. Thus, biogeochemical perturbations are a matter of crucial interest in science.

The atmospheric input of metals exhibits strong temporal and spatial variability due to short atmospheric residence times and meteorological factors. As in oceanic chemistry, the impact of trace metals in atmospheric deposition cannot be determined from a simple consideration of global mass balance; rather, accurate data on net air or sea fluxes for specific regions are needed.

Particles in urban areas represent one of the most significant atmospheric pollution problems, and are responsible for decreased visibility and other effects on public health, particularly when their aerodynamic diameters are smaller than 10 µm, because these small particles can penetrate deep into the human respiratory tract. There have been many studies measuring concentrations of toxic metals such as Ag, As, Cd, Cr, Cu, Hg, Ni, Pb in rainwater and their deposition into surface waters and on soils. Natural sources of aerosols include terrestrial dust, marine aerosols, volcanic emissions and forest fires. Anthropogenic particles, on the other hand, are created by industrial processes, fossil fuel combustion, automobile mufflers, worn engine parts, and corrosion of metallic parts. The presence of metals in atmospheric particles and the associated health risks of these metals. Anthropogenic sources have substantially increased trace metal concentrations in atmospheric deposition. In addition, acid precipitation favors the dissolution of many trace metals, which enhances their bioavailability. Trace metals from the atmosphere are deposited by rain, snow and dry fallout. The predominant processes of deposition by rain are rainout and washout (scavenging). Generally, in over 80 % of wet precipitation, heavy metals are dissolved in rainwater and can thus reach and be taken up by the vegetation blanket and soils. Light of a specific wavelength, selected appropriately for the element being analyzed, is given off when the metal is ionized in the flame; the absorption of this light by the element of interest is proportional to the concentration of that element. Quantification is achieved by preparing standards of the element.

- AAS intrinsically more sensitive than Atomic Emission Spectrometry (AES)
- Similar atomization techniques to AES
- Addition of radiation source
- High temperature for atomization necessary
- Flame and electrothermal atomization
- Very high temperature for excitation not necessary; generally no plasma/arc/spark in AAS

3. Flame (AAS)

Flame atomic absorption methods are referred to as direct aspiration determinations. They are normally completed as single element analyses and are relatively free of interelement spectral interferences. For some elements, the temperature or type of flame used is critical. If flame and analytical conditions are not properly used, chemical and ionization interferences can occur. Different flames can be achieved using different mixtures of gases, depending on the desired temperature and burning velocity. Some elements can only be converted to atoms at high temperatures. Even at high temperatures, if excess oxygen is present, some metals form oxides that do not redissociate into atoms. To inhibit their formation, conditions of the flame may be modified to achieve a reducing, nonoxidizing flame. Some aspects are discussed below:

- Simplest atomization of gas/solution/solid
- Laminar flow burner - stable "sheet" of flame
- Flame atomization best for reproducibility (precision) (<1%)
- Relatively insensitive – incomplete volatilization, short time in beam

Usually the measurement of metals can be done by an appropriate selection of one of the methods given below. During the analysis of the data, the interference and background should be considered. Three pattern-matching methods are most often employed: using calibration curves generated by a series of patterns, the method of standard additions and the internal standard method. The use of calibration curves from a series of patterns is possibly the most widely used method. It involves measuring the sample of interest in a series of samples of known concentration and all prepared under the same conditions.

1. Calibration Curve Method: Prepare standard solutions of at least three different concentrations, measure the absorbance of these standard solutions, and prepare a calibration curve from the values obtained. Then measure the absorbance of the test solution adjusted in concentration to a measurable range, and determine the concentration of the element from the calibration curve.

2. Standard Addition Method: To equal volumes of more than two different test solutions are used, then the standard solution is added so that the solutions contain stepwise increasing amounts of the element, and add the solvent to make a definite volume. Measure the absorbance for each solution, and plot the concentration of the added standard element on the abscissa and the absorbance on the ordinate. Extend the calibration curve obtained by linking the plotted points, and determine the concentration of the element from the distance between the origin and the intersection of the calibration curve on the abscissa. This method is applicable only in the case that the calibration curve drawn as directed in (1) above passes through the origin.

3. Internal Standard Method: Prepare several solutions containing a constant amount of the internal standard element. Using these solutions, measure the absorbance of the standard element and the absorbance of the internal standard element at the analytical wavelength of each element under the same measuring conditions, and obtain the ratios of each absorbance of the standard object element to the absorbance of the internal standard element. Prepare a calibration curve by plotting the concentrations of the standard element on the abscissa and the ratios of absorbance on the ordinate. Then prepare the test solutions, adding the same amount of the internal standard element as in the standard solution. Proceed under the same conditions as for preparing the calibration curve, obtain the ratio of the absorbance of the element to that of the internal standard element, and determine the concentration of the element from the calibration curve.

The calibration curve is always represented with the response of the instrument on the vertical (y) and concentrations on the horizontal axis (x). The value of C can limit the accuracy of the measurements for various concentrations; however there are statistical methods which can be used. The calibration method is generally used where there is a linear relationship between the analytical signal (y) and concentration (x), taking precautions to ensure that linearity in the experimental response is maintained over a wide range of concentrations. In these cases the way to proceed is to obtain the regression line of y on x (ie, the best-fit straight line through the points of the calibration graph, which can be obtained, for example, by the method of least squares) and used to estimate the concentration of unknown samples by interpolation, as well as to estimate the detection limit of the analytical procedure.

- Atomic absorption spectrometers allow operators to adjust sample temperatures. Some spectrometers are self-calibrating, position lights automatically, or compensate for stray light emissions and various types of spectral interference that bias analytical results. Intrinsically safe (IS) instruments do not release sufficient electrical or thermal energy to ignite hazardous atmospheric mixtures. In flame AAS, the measured absorbance depends on the absorptivity of the element in question, the concentration of that element in the flame (and therefore on its solution concentration), and the optical path length through the flame. The light from the flame region is collected on the left-side of this particular instrument. Under certain circumstances, it is desireable to reduce the optical path length. To this end, the burner head can also be rotated about its vertical axis. Different burner heads (having different slot sizes) are used for acetylene/air and acetylene/nitrous oxide flames, since the latter is a much hotter flame. Flow controls also allow both the total gas flow-rate and fuel-to-oxidant ratio to be varied. Gas flow affects the aspiration rate and aerosol drop-size distribution, while the fuel-oxidant ratio can influence chemical processes in the flame that would otherwise reduce the sensitivity of the measurement. Following you can describe some features of this technique:
- Primary combustion zone - initial decomposition, molecular fragments, cool
- Interzonal region – hottest, most atomic fragments, used for emission/fluorescence
- Secondary combustion zone – cooler, conversion of atoms to stable molecules, oxides
- element rapidly oxidizes – largest atom near burner
- element oxidizes poorly – largest atom away from burner
- most sensitive part of flame for AAS varies with analyte (Figure 3)

The measurement procedure is as follows. A small quantity of the extracted sample is injected into a flame where the ions are reduced to elements and vaporized. The elements present in the sample absorb light at specific wavelengths in the visible or the ultraviolet spectrum. A light beam with a single specific wavelength for the element being measured is directed through the flame to be detected by a monochrometer.

The light absorbed by the flame containing the extract is compared with the absorption from known standards to quantify the elemental concentrations. One of the disadvantages of this method is that only one element can be quantified at a time. AAS requires an individual analysis for each element, and sometimes a large filter or several filters are needed to obtain concentrations for a large variety of elements. Samples having high concentrations of elements beyond the linear range of the instrument should be diluted prior to the analysis. Báez et al., 2007 and García et al., 2009 characterized atmospheric aerosols, metals and ions that play an important role in the content of chemical species and of many elements in atmospheric ecosystem interfaces. Sodium, K^+, Ca^{2+} and Mg^{2+} were analyzed with a double beam atomic absorption spectrophotometer. Deuterium and hollow cathode lamps (Photron Super lamp) were used for background correction and analysis.

Calibration standards were prepared under a laminar flux hood, using certified standards for each ion (High-Purity Standards traceable from the National Institute of Standards and Technology, NIST).

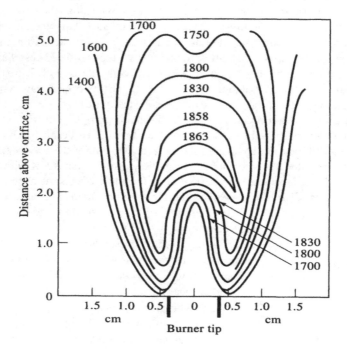

Fig. 3. Spectrometers element rapidly oxidizes - largest atom near burner

A calibration curve was used to determine the unknown concentration of an element in a solution. The instrument was calibrated using several solutions of known concentrations. The absorbance of each known solution was measured and then a calibration curve of concentration vs. absorbance was plotted.

The sample solution was fed into the instrument, and the absorbance of the element in this solution was measured. The unknown concentration of the element was then calculated from the calibration curve. The concentration of the analyte element was considered to be proportional to the ground state atom population in the flame; any factor that affected the ground state atom population was classified as interference.

Factors that might affected the ability of the instrument to read this parameter were also classified as interference.

The different types of interference that were encountered in atomic absorption spectroscopy were:

- Absorption of source radiation: an element other than the one of interest may absorb the wavelength being used.
- Ionization interference: the formation of ions rather than atoms causes lower absorption of radiation. This problem is overcome by adding ionization suppressors.
- Self absorption: atoms of the same kind as those that are absorbing radiation will absorb more at the center of the line than at the edges, thus resulting in a change of shape and intensity of the line.
- Background absorption of source radiation: This is caused by the presence of a particle from incomplete atomization. This problem is overcome by increasing the flame temperature.
- Rate of aspiration, nebulization, or transport of the sample (e.g. viscosity, surface tension, vapor pressure , and density) .

4. Atomic Absorption Spectroscopy with graphite furnace (GFAA)

The GFAA and flame AAS measurement principle is the same. The difference between these two techniques is the way the sample is introduced into the instrument. In GFAA analysis, an electrothermal graphite furnace is used instead. The sample is heated stepwise (up to 3000°C) to dry. The advantage of the graphite furnace is that the detection limit is about two orders of magnitude better than that of AAS. The analysis of different species of a given element is important because different oxidation states of the same element may present different toxicities and, consequently, different risks. Therefore, sequential extraction procedures for the separation and further analysis of a species have been developed for several metals. Sammut *et al.* (2010) developed a sequential extraction procedure for separating inorganic species of selenium found in the particulate matter of emissions in working areas. Essentially, the method is based on sequential dissolutions of the sampled matrix. Se(IV) and Se(0) can be independently analyzed by GFAA, whereas only total selenium can be detected when analyzed by AAS. This research group also proposed the analysis of beryllium species using GFAA and ICP-AES because metallic beryllium and beryllium oxide in workplaces are associated with different toxicities. These examples highlight the necessity of improving analytical techniques for measuring not only the total concentration but also the different species of a metal dispersed in contaminated air.

The graphite furnace has several advantages over a flame furnance. First it accepts solutions, slurries, or solid samples. Second, it is a much more efficient atomizer than a flame furnance and it can directly accept very small absolute quantities of sample. It also provides a reducing environment for easily oxidized elements. Samples are placed directly into the graphite furnace and the furnace is electrically heated in several steps to dry the sample, ash organic matter, and vaporize the analyte atoms. It accommodates smaller samples but it is a difficult operation, since the high energy that is provided to atomize the sample particles into ground state atoms might also excite the atomized particles into a higher energy level, thus lowering the precision.

GFAA has been the most common instrument used for Pb analysis, and in countries where this element is a criterion for pollution standards, AAS is generally the technique used in the reference methods to quantify it. Garcia *et al.*, 2008 determined Pb in total suspended particles in southwest Mexico City, using a nitric acid digestion–based AAS method. An AAS method was used in conjunction with voltammetry for Pb and Cd analysis to evaluate a rapid digestion technique using a microwave oven, comparing this with traditional methods after establishing the optimal efficiency of digestion in terms of power setting, time and the use of different acids (Senaratne and Shooter, 2004).

For low level determination of volatile elements such as As, Ge, Hg, Sb, and Se, hydride generation coupled with AAS provides lower detection limits (milligram–microgram range). The three-step sample preparation for graphite furnaces is as follows:

1. Dry - evaporation of solvents (10–100 s)
2. Ash - removal of volatile hydroxides, sulfates, carbonates (10–100 s)
3. Fire/Atomize - atomization of remaining analyte (1 s)

4.1 Specific sample considerations

GFAA has been used primarily for the analysis of low concentrations of metals in water samples. GFAA can also be used to determine the concentrations of metals in soil, but sample preparation for metals in soil is somewhat extensive and may require the use of a mobile laboratory. The more sophisticated GFAA instruments have a number of lamps and, therefore, are capable of simultaneous and automatic analyses of more than one element.

*   Plants: Solid samples must be in liquid form to be aspirated by the instrument. Therefore, solid material must be liquefied by means of some form of extraction or digestion protocol. Procedures have been devised that make the total amount of an element in the sample available for assay or that use some particular properties to extract the portion of the element that exists in some chemical forms but not in others. For example, the plant dry ash/double acid extraction method determines the total element content of the sample.

*   Soil: For ecological purposes, there is more interest in measuring extractable or labile soil constituents than in determining total element content. Certain partitions of the total soil content of a given element are operationally defined by an extraction procedure, and arguments are usually offered that suggest that such partitions correspond to different levels of biological availability or activity. The HCl/H_2SO_4 double acid extraction method, also referred to as North Carolina and Mehlich-1, is widely used to determine bioavailable Ca, K, Mg, Mn, P, and Zn in sandy acid soils characteristic of the eastern and southeastern United States.

*   Water: Aquatic samples, of course, need no liquefaction step, but researchers must still decide which analyte partition (dissolved, suspended, total) is of interest. Various treatments for each sample partition are detailed in the U.S. EPA's discussion of the content partitioning of water samples.

The advantages of using atomic absorption spectroscopy are, in our opinion:

-   Greater sensitivity and detection limits than other methods
-   Direct analysis of some types of liquid samples

- Low spectral interference
- Very small sample size

Finally, it is worth mentioning that in the last two decades, quantification and characterization of different matrices and materials by atomic absorption spectroscopy has been significant. The development of continuous measurement techniques for certain chemical species is important for governments, mainly in areas where the epidemiological studies are still being processed.

Element	Mean	Min	Max	25th Percentile	75th Percentile	Geometric Mean	Harmonic Mean	Std. Dev.
Al	1536.4	97.65	8939.1	570.04	2027.9	1047.8	691.56	1543.6
Cd	2.64	0.09	8.05	1.06	3.99	1.85	0.99	1.75
Cr	23.79	14.08	44.02	18.24	27.20	22.93	22.11	6.62
Fe	1098.6	27.88	4963.8	263.19	1593.8	623.6	315.00	1102.2
Mn	19.23	2.66	45.18	12.11	25.46	14.32	14.32	9.41
Ni	23.03	8.78	52.93	18.03	26.49	20.77	20.77	7.57
Pb	50.78	8.36	128.07	29.48	68.01	35.94	35.94	28.16
V	16.22	2.15	44.52	8.40	23.41	9.93	9.93	10.02

N=100 (number of samples)

Table 1. Descriptive Statistics: Mean, Minimum (Min), Maximum (Max), 25th and 75th percentiles, Geometric Mean, Harmonic Mean and Standard deviation (Std. Dev.), concentrations of metals in PST collected 2003–2004

5. References

A. Báez, R. Belmont, R.M. García, H. Padilla, M.C.Torres. (2007). "Chemical composition of rainwater collected at a southwest site of Mexico City, Mexico". Rev. Atmospheric Research. 86: 61-75.

Sammut, M. L., Noack, Y., Rose, J., Hazemann, J. L., Proux, O., Depoux, M., et al. (2010). Speciation of Cd and Pb in dust emitted from sinter plant. Chemosphere, *78*, 445.

Senaratne I. and Shooter D. (2004). Elemental compositi of brown haze in Auckland, New Zealand. Atmospheri 3059.

R. García, M. C. Torres and A. Báez. (2008). "Determination of trace elements in total suspended particles at the Southwest of Mexico City from 2003 to 2004". Rev. Chemistry and Ecology. 24(2), pág. 157-167.

R. García, R. Belmont, H. Padilla, M. C. Torres and A. Báez. (2009). "Trace metals and inorganic ions measurements in rain from Mexico City and a nearby rural area". Rev. Chemistry and Ecology. 25(2), pp. 71-86.

Microextraction Techniques as a Sample Preparation Step for Metal Analysis

Pourya Biparva[1] and Amir Abbas Matin[2]
[1]Department of Nanotechnology, Islamic Azad University, Langaroud Branch, Langaroud,
[2]Research Department of Analytical Chemistry, Iranian Academic Center for Education,
Culture & Research (ACECR), Urmia
Iran

1. Introduction

Analytical methods consist of several steps including; sampling, sample preparation, analysis, calculations and statistical evaluation of the results. Each step has a direct impact on accuracy, precision and sensitivity of the method. Among theses steps, sample preparation is the most time consuming step. Result of studies showed that more than 60% of analysis time is spent for sample preparation. Sample preparation follows two main aims; sample clean-up and concentration. Sample Clean-up is carried out for isolating the target analytes from matrix components which interfere on determination and concentration is done for enrichment of the analytes in sample because despite advances in analytical instrumentation, sensitivities are limited.

Characteristics of an ideal sample preparation technique are listed as below:
- Minimum loss of the sample and maximum recovery of the analyte
- Elimination of accompanying compounds with high yield
- Simple, fast and cheap method
- Capable with analytical instruments
- In agreement with green chemistry

In the case of atomic absorption spectrometry (AAS) which is the subject of this book, there are two priciple systems which are familiar with readers of this book, flame and electrothermal AAS. In continue of our discussions about microextraction techniques for metal analysis by AAS, we will emphasis on reduced volumes of extracting phases in the microlitre scale. It is clear that due to consumption of large volumes (in the mililitre scale) of the sample in flame AAS, coupling of microextraction techniques with flame AAS is difficult. But in the case of electrothermal AAS, this is so easy. Because volume of the samples introduced to graphite furnaces are very low and in microlitre scale. So a review on literature show that most of the microextraction methods are capable with electrothermal AAS not with flame AAS.

1.1 Liquid-Liquid Extraction (LLE)

Liquid-Liquid Extraction (LLE) is a versatile classical sample preparation technique. LLE is based on establishment of distribution equilibrium of the analytes between two immiscible phases, an aqueous and an organic phase. Apparatus for LLE is a separating funnel. If distribution equilibrium constant is enough large, a quantitative extraction of the analytes can be occurred in one step. But most of the LLEs are multistep.

LLE commonly is used for extraction of organic and inorganic compounds. In the case of metal analysis, extraction of them as ammonium pyrrolidin dithiocarbamat (APDC) complexes using methyl isobuthyl ketone (MIBK) as extraction solvent is known as Standard method. Other examples for application of LLE in metal analysis are briefly described here; Extraction of As (inorganic and organic) in urine and water samples as their iodide salts extracted in chloroform and re-extracted in dilute dichromate solution for total As determination by electrothermal atomic absorption spectrometry (ET-AAS) was reported (Fitchett et al., 1975). Another method for extraction of As (inorganic and organic) is based on extraction to toluene and back extraction with cobalt nitrate solution (Lauwerys et al. 1979). A simple LLE method for extraction of methylmercury was introduced by using toluene as extraction solvent before analysis by ET-AAS (Saber-Tehrani et al. 2007). Ease of operation and simplicity of the method are advantages of LLE. But important disadvantages such as consumption of large volumes of expensive and toxic solvents, emulsion formation at the interface of the two phases and difficult phase separations and finally low concentration factor lead the analytical chemists to introduce alternative methods for LLE by decreasing volume of the extracting solvent at micro liter scale known as liquid phase microextraction (LPME) which introduced in the late 1990s and early 2000s.

1.2 Solid Phase Extraction (SPE)

One of the alternative methods for LLE is solid phase extraction (SPE) which was introduced in early 1970 and developed during 1980-90. SPE process is based on distribution of analytes between solid sorbent packed in a cartridge and liquid sample which moves through the solid phase. Solid phase usually consists of small porous particles of silica with or without bonded organic phase, organic polymers and ion exchangers. Mechanisms of extractions are based on adsorption, partitioning or ion exchange according to kind of solid phase. SPE is used for extraction of both of the organic and inorganic compounds. A wide group of chemicals which their SPE procedures were reported in the literature is metal ions.

Main SPE methods for metal ions are summarized in Table 1 (Fritz, 1999).

Increasing development of SPE during 1990s continues until recent years by introducing novel solid sorbents such as molecularly imprinted polymers (Lucci et al., 2010) and nanostructured materials (Faraji et al., 2010). SPE has many attractive features in comparison with classical solvent extraction methods. However, it has its limitations. Some of the main limitations of SPE are listed below:

1. Clogging the pores of the solid phase by large biomolecules, oily materials and fine solids in the sample.
2. Despite, decrease in solvent consumption in SPE in comparison with LLE, SPE needs at least 100 μL of the solvent.
3. It is time consuming method due to several steps of operation including; conditioning, sample loading and elution

Despite popularity of the SPE, miniaturization of it caused to introduction of solid phase microextraction (SPME) (Arthur et al. 1990).

Metal ion	Extracted as	Solid Phase
Mo(VI), W(VI), Ta(V), Nb(V), Ti(IV), V(IV), V(V)	Hydrogen Peroxide complexes	Cation- exchanger
+2, +3 and +4 cations	Bromide complexes	Strong-acid cation-exchanger
Al(III), Mo(VI), Nb(V), Sn(IV), Ta(V), Ti(IV), U(VI), W(VI), Zr(IV)	Fluoride complexes	Cation- exchanger
Fe(III)	Chloride complexe	Haloport-F
Mo(VI)	Chloride complexe	Granular polymer impregnated with MIBK
Au(III)	Chloride complexe	Amberchrome resin
Cd(II), Co(II), Cu(II), Fe(II), Ni(II), Pb(II)	pyrrolidin dithiocarbamat complexes	C18
Cu(II), Fe(III), U(VI), Al(III)	Chelate with solid phase	Hydroxamic Acid resins (Chelating resin)
Ag(I), Au(III), Bi(III), Cd(II), Cu(II), Fe(III), Hg(II), Pb(II), Sb(III), Sn(IV), U(VI), Zn(II)	Chelate with solid phase	Thioglycolate resins (Chelating resin)
U(VI)	Chelate with solid phase	Cellulose phosphate ion exchanger (Chelating resin)

Table 1. SPE techniques for metal ion analysis.

2. Microextraction techniques

In order to get rid of limitations of classical sample preparation methods and reach to an ideal method, miniaturization of extraction techniques is necessary. The first step in miniaturization is reducing volume of extraction solvent in LLE (Liquid Phase Microextraction). Different ways of this miniaturization causes various modes of LPME like single drop microextraction, dispersive liquid-liquid microextraction, hollow fiber based supported liquid membrane microextraction and liquid phase microextraction based on solidification of floating organic drop. Miniaturization was also applied to SPE. Product of this process is known as solid phase microextraction. Review of scientific literature shows that design; development and applications of microextraction techniques are growing rapidly. Popularity and applicability of microextraction techniques requires discussing these subjects in books.

2.1 Solid Phase Microextraction (SPME)

Solid phase microextraction as a solvent free alternative method for conventional sample preparation methods was introduced (Arthur et. al., 1990). In SPME, a small volume of extraction phase (usually less than 1 μL) coated on fused silica support is mounted in a modified Hamilton 7000 series syringe. Extraction phase could be a high molecular weight polymeric liquid or a solid porous sorbent with high surface area. Fig. 1 illustrates the structure of a SPME device manufactured by Supelco Company. A stainless steel tube was replaced with inner wire of syringe needle and fiber was installed on inner tube. With pulling the syringe plunger in, the fiber is protected in the needle and with pulling out; the fiber is exposed to the sample. SPME process is carried out in three modes. Headspace mode which fiber is exposed into the headspace of the sample suitable for volatile analytes, direct mode which fiber is immersed directly in the sample suitable for nonvolatile analytes and direct mode with membrane protection suitable for biological or dirty samples.

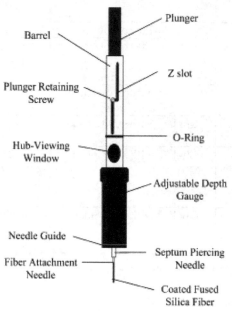

Fig. 1. SPME device manufactured by Supelco Co. (Mester et. al., 2005)

2.1.1 Theoretical aspects of SPME

SPME is based on distribution of the analytes between the sample and the fiber. Based on nature of fiber coating, mechanism of microextraction differs. Mechanism of microextraction for polymeric liquid phases and solid sorbents are partitioning and adsorption, respectively. Amount of extracted analytes depend on distribution constant (D) of them.

$$D = K_d \frac{V_2}{V_1} \tag{1}$$

Fraction of extracted analyte from aqueous sample to the fiber is calculated from Eq. 2

$$F_{ex} = \frac{D}{1+D} = \frac{K_d V_2}{K_d V_2 + V_1} \tag{2}$$

2.1.2 Factors aaffecting SPME

In order to reach equilibrium conditions in SPME, factors such as nature of fiber coating, time and temperature of extraction, time and temperature of desorption, salt addition, sample agitation and solution pH must be optimized. One of the most advantages of SPME is ability for hyphenation with various analytical instruments. The most capable instrument with SPME is gas chromatograph. But HPLC, CE and AAS had been coupled with SPME. According to subject of this book, we will focus on coupling of SPME with AAS. Direct coupling of SPME with AAS can be carried out using hydride generation apparatus, including quartz tube equipped with electric heater as flow cell. So due to this system carrier gas flow is another parameter which needs to investigate.

2.1.2.1 Fiber coating selection

The first step in SPME is selection of appropriate fiber. SPME fiber coatings with different natures including polar, non-polar and semi polar are available. According to chemical nature of the analytes, the best fiber must be selected. Polydimethylsiloxane (PDMS) is the most useful coating for SPME fibers which is commercially available. Now a days, in addition to commercial fibers, various coating compositions are made at laboratories. Therefore analysts can select appropriate available fibers or design novel coatings suitable for their aims.

2.1.2.2 Microextraction temperature

Temperature has a major effect on efficiency of SPME. Increasing the temperature causes an increase in distribution coefficient of the analytes between the sample and the fiber. In the case of headspace SPME, it causes increase in distribution coefficients between the sample and the headspace and between headspace and the fiber. But temperature is a limiting parameter, because increasing the temperature more than a certain value causes a significant decrease in distribution coefficient of the analytes between the sample (direct SPME) or headspace and the fiber and results decrease in amount of extracted analytes.

2.1.2.3 Microextraction time

Exposure time of the fiber in the headspace of samples is usually kept long enough to achieve equilibrium between the headspace and the adsorbent in order to maximize the extraction efficiency.

2.1.2.4 Desorption temperature and time

In order to transfer the analytes from the fiber to the analytical instruments like GC and AAS, thermal desorption is the best method. So desorption temperature and time of fiber duration in the desorption chamber (injection port and heated quartz tube for GC and AAs, respectively) must be studied.

2.1.2.5 Sample aagitation

Various modes of agitation could be applied to the sample for faster achieving the equilibrium. Magnetic stirring and sonication are usual methods for transporting the analytes from bulk of the sample solution to the surface of the fiber. Power and time of sonication or rate of magnetic stirrer must be adjusted at different levels for selection of optimum condition.

2.1.2.6 Salting out effect

Addition of an inorganic salt has often been used in order to enhance the activity coefficients of volatile components in aqueous solutions, increasing the concentration in the headspace vapor. The salts are also added to equalize the activity coefficients of analytes in different matrices (Zuba et al., 2001). For this purpose, microextraction processes are carried out in presence of various salt concentrations and also in salt less solution. Results demonstrates role of salt addition in microextraction of target analytes.

2.1.3 Application of SPME for extraction of metallic analytes before atomic absorption spectrometric determination

SPME can be coupled with AAS easily via heated quartz tube. Quartz tube flow cells equipped with electric heater usually used for mercury determination and hydride generation techniques. Quartz tube AAS (QT-AAS) not only is used for direct coupling of SPME with AAS but also it can be used as gas chromatographic detector (SPME-GC-QT-AAS). This technique is mostly reported for determination of organometallic compounds or derivatized metals as organometallics. A method for determination of organomercury species based on solid phase microextraction after hydride generation using KBH_4 and determination using GC-QT-AAS was reported. As mentioned above, in the case of mercury hydrides commercial fibers are not suitable and researchers had to design a novel fiber based on acid treated fused silica. Suitable capillary column for separating mercury hydrides is CPL-SIL 5CB (10 m × 0.25 mm) under 40 °C isothermal condition (He, et al., 1998). This method was applied for determination of methylmercury in biological samples and sediments. Another similar method was reported for determination of methyl, ethyl and phenyl mercury species in soil samples using above mentioned method (He, et al., 1999). Recently successful efforts were done for direct coupling of SPME with QT-AAS (Fragueiro et al., 2004). In this method direct coupling between headspace SPME and QT-AAS was evaluated for speciation of methylmercury in seafood after volatilization of its hydride or chloride derivates. The best limit of detection (LOD) for methyl mercury obtained by using PDMS/DVB fiber coating as 0.06 ng mL[-1].

2.2 Single Drop Microextraction (SDME)

The LPME is a miniaturized version of the LLE in which the extraction solvents, volume reduced to about 1-10 µL. Various techniques are known as LPME. Single drop microextraction (SDME) is one of these methods in which extraction solvent is a single drop. Jeannot and cantwell was reported the SDME for first time by suspending an 8 µL organic solvent drop at the end of Teflon rod immersed in the stirring sample. After extraction, solvent drop was removed from the end of the Teflon rod using a micro syringe and injected to analytical instrument (Jeannot and cantwell, 1996). Fig. 2 shows the schematic SDME system.

Some modifications were made by He and Lee on primary reported method. In newer version of SDME, teflon rod was replaced by a micro syringe (Fig. 3). A one µL immiscible extracting solvent drop is exposed into the sample (liquid or gaseous) from a micro syringe. After establishment of distribution equilibrium the organic drop is retracted back into the micro syringe and is injected to the analytical instrument for determination of the analytes (He and Lee, 1997). SDME is suitable for coupling with various instrumental methods such as gas and liquid chromatography (GC & HPLC), atomic absorption spectrometry (AAS) and inductively coupled plasma atomic emission spectrometry (ICP).

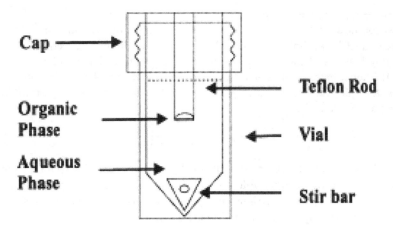

Fig. 2. Illustration of SDME method reported by Jeannot and cantwell (Jeannot et. al., 1996)

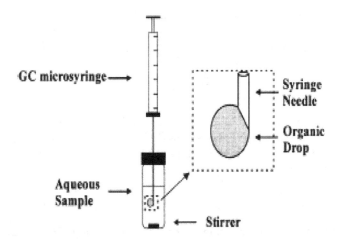

Fig. 3. Schematic diagram of SDME using micro syringe (He et. al., 1997)

Some of the advantages of the SDME are:
- SDME is a cheap technique
- SDME needs simple equipment
- Its operation is easy
- Use of minimum amounts of solvents which introduces the technique as a green approach to sample preparation
- In situ derivatization of the analytes is possible

In addition to above mentioned advantages, instability of the drop, small surface of the drop and slow kinetics of extraction are disadvantages of the method (Dadfarnia and Haji Shabani, 2010). Like SPME, SDME can be operated in various modes known as direct immersed SDME (DI-SDME), headspace SDME (HS-SDME) and three phases SDME.

2.2.1 Theoretical aspects of SDME

Extraction of the analytes via SDME process is an equilibrium phenomenon. Microextraction conditions must be optimized for establishment of thermodynamic equilibrium of analyte partitioning between the sample and extracting phase. In order to better understanding of microextraction process a model was introduced. This model is based on dynamic mass balances for the analytes in each phase or both. Mass balance equation for SDME is presented as equation (3).

$$C_A^{aq} V^{aq} + C_A^o V^o = C_{A,o}^{aq} V^{aq} \tag{3}$$

Where C_A^{aq} and C_A^o are the concentrations of the analyte in the sample and the microdrop respectively; V^{aq} is the sample volume and V^o is the microdrop volume, and $C_{A,o}^{aq}$ is the initial concentration of the analyte in the aqueous sample. The dynamic mass balance of the analyte in the microdrop is given by following equation:

$$\frac{d(C_A^o V^o)}{dt} = k_{tot}^o A_i [K_A C_A^{aq} - C_A^o] \tag{4}$$

Where A_i is the interfacial area (the surface area of the microdrop). K_A is the equilibrium partition coefficient and k_{tot}^o is the total mass transfer coefficient of the analyte with respect to the organic phase. If the two-film theory is considered, k_{tot}^o is given by:

$$\frac{1}{k_{tot}^o} = \frac{1}{k^o} + \frac{K_A}{k^{aq}} \tag{5}$$

Where k^o and k^{aq} are the mass transfer coefficients for the analyte in the film of the organic and the aqueous phases respectively. If we consider the volume of microdrop constant, then A_i is also constant. So we can obtain Eq. 4 as result of combination of Eq. 1 and 2. Eq. 6 presents C_A^o as a function of time.

$$C_A^o(t) = C_A^{o,aq} [1 - e^{-\lambda t}] \tag{6}$$

Where $C_A^{o,aq}$ is the analyte concentration in the microdrop at equilibrium and λ is the rate constant.

2.2.2 Factors affecting SDME

In order to reach equilibrium status, all the factors affecting microextraction process must be optimized. Kind and amount of organic solvent drop, extraction time and temperature, salting out and agitation rate are important factors.

2.2.2.1 Kind and volume of extraction solvent

In SDME solvent selection is the most important parameter. Selectivity and extraction efficiency directly depends on solvent's nature. Solvents with different polarities must be examined for extraction of studied analytes. Volume of the selected solvent is also important. Solubility of the analytes and partitioning of them in extraction solvent depends on volume of solvent microdrop. Increase in solvent volume, increases the amount of extracted analytes but after reaching quantitative recoveries, increasing the solvent volume causes a significant decrease in concentration factor due to dilution of the analytes. On the other hand, hanging of large volumes of organic solvents made the microdrop unstable. So in SDME solvent volume is a limited parameter.

2.2.2.2 Extraction time

Exposure time of the microdrop to the samples is an important parameter in achieving distribution equilibrium of analytes between solvent drop and sample; it is a decisive factor for improving the extraction efficiency. So it is necessary to be optimized. But according to difficulty of microdrop expose to the sample in this technique, long extraction times are not preferred.

2.2.2.3 Extraction temperature

As mentioned before, SDME process is a thermodynamic equilibrium, so effect of temperature is not negligible. But extraction temperature in SDME, strongly limited by solvent's boiling point.

2.2.2.4 Salt addition

Addition of salt to the sample increases the ionic strength of the solution. Depending on the solubility of the target analytes, extraction is usually enhanced with increased salt concentration (salting out effect). But in the case of SDME, presence of salt changes the physical properties of the extraction film and reduces the diffusion rates of the analytes into the solvent drop. So, salt addition has a negative effect on method efficiency.

2.2.2.5 pH Adjustment

Effect of pH on extraction efficiency, depends on analyte nature. Extraction of analytes with weak acidity and basicity is strongly pH sensitive. Solution pH easily adjusted by appropriate buffers according to the pKa values of the analytes.

2.2.2.6 Sample agitation

Similar to the other extraction methods, sample agitation has a significant effect on enhancement of extraction yield. Increase in agitation rate, decreases the extraction time due to faster establishment of the distribution equilibrium. But despite positive effect of higher agitation rates, it is so critical parameter in SDME because instability of microdrop on syringe tip.

2.2.3 Application of the SDME for extraction of metal ions before atomic absorption spectrometric determination

Due to low volumes of extraction solvents in SDME, electrothermal atomic absorption spectrometry is suitable for determination of metals after extraction with SDME technique. The first coupling of SDME with ET-AAS was reported for extraction of As in aqeous samples. Total arsenic species were converted to As (III) using $NaBH_4$ and extracted by a 4 μL organic drop consisting pyridine and benzyl alcohol containig silver diethyldithiocarbamate (AgDDC) as complexing agent with arsine (Chamsaz et al., 2003). Palladium is a good sorbent for arsenic and act as matrix modifier in ET-AAS. So a green headspace SDME method for extraction of As (III) and total As was presentd using a 3 μL aqeous drop of Pd (30 mg L^{-1}) as extraction solvent prior to determination by ET-AAS

(Fragueiro et al, 2004). This method is faster than peviously reported method by Chamsaz et al, but both of the concentration factor and detection limit of previouse method is better.

Another green method using aqeous micro drop of Pd (II) and Pt (IV) was reported for determination methylmercury in fish samples. Volatilization of methylmercury was performed using hydride generation reaction and after extraction analysis was done by ET-AAS (Gil et al., 2005). Microextraction of Se using SDME prior to ET-AAS analysis was reported by photogeneration of volatile hydride and alkyl Selenium derinatives. Aqeous microdrop containing Pd (II) was used as extracting phase (Figueroa et al., 2005). A series of SDME methods were reported for lead determination in various samples by ET-AAS. A microdrop of benzene containing 1-phenyl-3-methyl-4-benzoyl-5-pyrazolone (PMBP) was used for extraction of Pb (II) from biological samples and its determination by ET-AAS (Liang et al., 2008). Dithizone is another extractant in SDME which was used for microextraction of lead from water samples prior to assay by ET-AAS (Liu and Fan, 2007). Application of ionic liquids as novel extractants in SDME was reported for determination of manganese and lead. Complex of manganese with 1-(2-thiazolylazo)-2-naphtol (TAN) was extracted into a microdrop of $[C_4MIM][PF_6]$ (Manzoori et al., 2009). The same group extracted lead ions after complexation with ammonium pyrrolidine dithiocarbamate (APDC) into the same ionic liquid (Manzoori et al., 2009). Simulteneous direct SDME of Cd and Pb after complexation with dithizone in aqeous samples into toluene microdrop followed by determination with ET-AAS was reported (Jiang and Hu et al., 2008). Ag, Tl, Cr and Sb are another metals which the SDME-ET-AAS method was reported for them.

2.3 Dispersive liquid-liquid microextraction (DLLME)

Dispersive liquid-liquid microextraction (DLLME) is a newer mode of LPME (Rezaee, et al., 2006). DLLME is a ternary solvent system consisting of aqeous sample solution, extraction solvent and disperser solvent. Extraction solvent must be immisible with aqeous sample solution and disperser solvent must solouble in both of the extraction solvent and aqeous sample solution. In DLLME, 5-10 mL of sample solution is placed in a test tube with conical bottom, then optimized volumes of extraction solvent (μL) and disperser solvent (mL) are mixed and mixture of these solvents is injected rapidly in the sample solution. After injection of solvents mixture a stable cloudy suspension consisting of fine droplets of extraction solvent is appered which can be easily separated by centrifugation (Fig. 4). Aqeous phase is discarded and sedimented organic phase is transfered to an analytical instrument using a micro syringe. Early works on DLLME were based on denser solvents than water (mostly chlorinated solvents) but later DLLME methods based on lighter extraction solvents than water were reported (Farajzadeh, et al., 2009).

Some of the advantages of DLLME are listed below:

- Simple, fast and cheap
- High preconcentration factor
- High recovery of the analytes
- Small volumes of the sample are needed
- Minimum volumes of the organic solvents (μL) are used
- Easy coupling with most of the instrumental methods

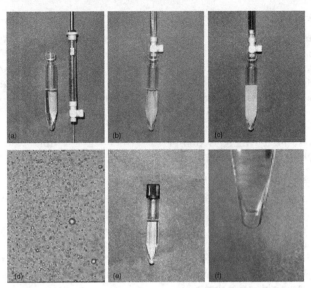

Fig. 4. Shematic illustration of a DLLME process: a) prior to injection of extraction and disperser solvents b) starting the injection c) end of injection, cloudy suspension was made d) fine droplets of solvents in solution bulk e) after centrifugation f) conical bottom of the test tube with sedimented phase (Rezaee et. al., 2006)

2.3.1 Theoritical aspects of DLLME

Inorder to design an accurate DLLME process, some terms must be identified:

Preconcentration or enrichment factor is the ratio of the analyte concentration in sedimented phase (C_{sed}) to analyte concentration in the sample (C_0)

$$EF = \frac{C_{sed}}{C_0} \tag{7}$$

Extraction recovery of the analyte (%) is the percentage of extracted analyte in sedimented phase, where n_0 is the amount of analyte in the sample prior to extraction and n_{sed} amount of analyte in sedimented phase.

$$ER = \frac{n_{sed}}{n_0} \times 100 = \frac{C_{sed} \times V_{sed}}{C_0 \times V_0} \times 100 \tag{8}$$

$$ER = (\frac{V_{sed}}{V_{aq}})EF \times 100 \tag{9}$$

2.3.2 Factors affecting DLLME

Like every other methods several experimental factors affect DLLME process. For development of a sensitive and accurate DLLME method these parameters including Kind and volume of extraction solvent, kind and volume of disperser solvent, extraction temperature, solution pH and salting out must be optimized.

2.3.2.1 Kind and volume of extraction solvent

Its clear that extraction solvent is the most important parameter in a LPME method. As we disscussed above, extraction solvent must be immicible with water and must be a good solvent for the analytes. Soloubility of the extraction solvent in water has an inverse relation with stability of cloudy suspension. Volume of extraction solvent is important too, because it is obvious that with increasing extraction solvent volume, amount of extracted analytes increase. But when we reach to quantitative recoveries, inrease of extraction solvent volume causes a significant decrease in an anlytical signal due to dillution and reducing enrichment factor. Extraction solvent can be a denser or lighter solvent than water. Usuall denser solvents are tetrachloroethylene, trichloroethylene, carbon tetrachloride and etc.

2.3.2.2 Kind and volume of disperser solvent

Disperser solvents usually are polar solvents such as aceton, acetonitril and methanol. Disperser solvent acts as a bridge between two immicible phases. The role of an ideal disperser solvent is making more fine droplets of extraction solvent and dispesion of extraction phase in sample bulk. Amount of disperser solvent is also important. Increasing in disperser solvent volume, increases the volume of sedimented phase and naturally lowers the analytical signal due to dillution. So, optimization of disperser solvent volume is as important as extraction solvent volume.

2.3.2.3 Extraction temperature and time

Temperature is an important parameter in all of the equilibrium systems. But in DLLME temperature is a critical parameter. Because boiling point of the extraction and disperser solvents are limited the process. Extraction time is contact time between the sample and extraction solvent. It is clear that extraction time directy depends on extraction efficiency. But in the case of DLLME, time is not so important parameter due to high surface contact between the sample and fine droplets of extraction solvent. So in most of the reports, time has no effect on extraction efficiency.

2.3.2.4 Salting out

Despite other extraction methods, in DLLME salting out has dual effect. Similar to other extraction techniques addition of an inorganic salt, causes an increase in amount of extracted analytes. But in DLLME salt addition causes a significant increase in volume of sedimented phase. So effect of salt addition on extraction efficiency are controlled by these two factors.

2.3.2.5 Complexation of metal ions prior to extraction

Similar to SDME, extraction of metal ions with organic solvent needs to convert them to suitable form. Appropriate chealating agents must be selected and complexation conditions such as solution pH, concentration of ligand and etc, must be adjusted. It is clear that extraction of metal ions after conversion of them to organometallic compounds is possible.

2.3.3 Application of DLLME for extraction of metal ions before determination by AAS

A simple and powerful microextraction technique was used for determination of selenium in water samples using dispersive liquid–liquid microextraction (DLLME) followed by ET-AAS (Bidari et al., 2007). In this study, complex of Se and APDC extracted by a mixture of ethanol (disperser solvent) and carbon tetrachloride (extraction solvent) from water samples. The concentration of enriched analyte in the sedimented phase was determined by iridium-modified pyrolitic tube graphite furnace atomic absorption spectrometry. Determination of trace levels of lead is possible with DLLME followed by ET-AAS (Liang and sang, 2008). In the proposed approach, 1-phenyl-3-methyl-4-benzoyl-5-pyrazolone (PMBP) was used as a chelating agent, and carbon tetrachloride and ethanol were selected as extraction and dispersive solvents. Another simple DLLME-Flame AAS method was reported for determination Pd(II) as complex with thioridazine HCl (Ahmadzadeh Kokya and Farhadi, 2009). Ethanol as disperser solvent and chloroform as extraction solvent were used in this study.

Extraction of Co(II) as its complex with Br-TAO via a DLLME method was reported (Baliza et al., 2009). The procedure is based on a ternary system of solvents, where appropriate amounts of the extraction solvent, disperser solvent and the chelating agent Br-TAO are directly injected into an aqueous solution containing Co(II). A cloudy mixture is formed and the ions are extracted in the fine droplets of the extraction solvent. After extraction, the phase separation is performed with a rapid centrifugation, and cobalt is determined in the enriched phase by FAAS.

Use of ionic liquids is a novel development in DLLME. A new ionic liquid based DLLME method was developed for preconcentration and determination of Pb (II) and Cd (II) in aqueous samples containing very high salt concentrations (Yousefi and Shemirani, 2010). This is believed to arise from dissolving of the ionic liquids in aqueous samples with high salt content. In this method, the robustness of microextraction system against high salt concentration (up to 40%, w/v) is increased by introducing a common ion of the ionic liquid into the sample solution. The proposed method was applied satisfactorily to the preconcentration of lead and cadmium in saline samples. After preconcentration, the settled IL-phase was dissolved in 100μL ethanol and introduced to Flame-AAS.

2.4 Liquid phase microextraction based on solidification of floating organic drop (LPME-SFO)

Liquid phase microextraction based on solidification of floating organic drop (LPME-SFO) is one the newest versions of LPME (Khalili-Zanjani et al., 2007) in which the extraction solvent with lower density than water, low toxicity and proper melting point near room temperature (in the range of 10–30 °C) was used. In this method, small volume of an extraction solvent is floated on the surface of aqueous solution. The aqueous sample solution is agitated for an optimized time. After the extraction, tube containing the sample is transferred in the ice bath and the floated extractant droplet solidified at low temperature. The solidified organic solvent can be separated from the sample and melted quickly at room temperature, which is then determined by appropriate analytical methods (Fig. 5).

In comparison with other methods, LPME-SFO has some advantages which is listed below:
- Simplicity of operation
- Small amount of low toxic solvent used
- Good repeatability
- Low cost
- High preconcentration factors
- More suitability for the analysis of complex matrix samples.

LPME-SFO is only applicable for the analytes with high or moderate lipophilic property and can not be used to those neutral analytes with high hydrophilic property and this is the main disadvantages of the method.

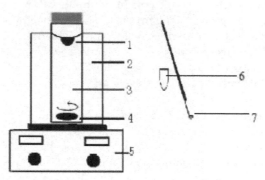

Fig. 5.1 Floated organic solvent, 2) Wter bath, 3) Sample, 4) Stirring bar, 5) Heater-Stirrer, 6) Conical vial, 7) Spatula (Ying-Ying et. al., 2010)

2.4.1 Thepritical aspects of LPME-SFO

LPME-SFO is based on the distribution of the analytes between floated extraction solvent and the aqueous sample matrix. As mentioned previously, the organic extraction solvent of LPME-SFO must have the melting point near room temperature (in the range of 10–30 °C) and lower density than water. It should be readily solidified at low temperatures, and thus its droplet can be collected easily. Maximum sensitivity and precision were obtained by stirring the sample solution until the equilibrium was obtained.

Distribution coefficient (K) is defined as the ratio between the analyte concentration in extraction solvent and sample solution. The enrichment factor (ER) and extraction recovery (ER) are calculated as follows:

$$EF = \frac{C_{o,F}}{C_{aq}} \tag{10}$$

Where $C_{o,F}$ and C_{aq} are the analyte concentration in the organic solvent, and the initial concentration of the analyte in the aqueous sample, respectively.

$$ER(\%) = 100 \times \frac{C_{o,F} V_{o,F}}{C_{aq} V_{aq}} \tag{11}$$

Where $V_{o,F}$ and V_{aq} are the volume of the organic phase and the volume of the aqueous sample, respectively.

2.4.2 Factors affecting LPME-SFO

The partitioning of the analytes between two phases (extraction solvent and the aqueous solution) was affected by various parameters, such as the type and volume of the organic solvent, aqueous sample volumes, the extraction time, the stirring rate, and salt addition. In order to design an accurate method, every parameter must be optimized.

2.4.2.1 Selection of extraction solvent

The selection of an appropriate extraction solvent is of major importance for the optimization of the LPME-SFO process. The selected extraction solvent must satisfy several requirements. First, it should be immiscible with water, have low volatility, low density, low melting point near room temperature and be able to extract the desired analytes. According to these considerations, limited number of the organic extraction solvents commonly used in LPME-SFO which are listed below. 1-Undecanol (13-15 °C), 1-Dodecanol (22-24 °C), 2-Dodecanol (17-18 °C), 1-Bromo hexadecane (17-18 °C), n-Hexadecane (18 °C) and 1,10-Dichlorodecane (14-16 °C).

2.4.2.2 Effect of extraction temperature

Generally, increasing sample solution temperature has a positive effect on extraction efficiency. Based on the extraction kinetics, higher temperatures would facilitate the diffusion and mass transfer of the analytes from sample solution to the organic solvent, and the time required to reach the equilibrium would be decreased. However, at high temperatures, the over-pressurization of the sample vial could make the extraction system unstable. On the other hand in LPME boiling point of the solvents is a limiting factor.

2.4.2.3 Effect of organic solvent volume

Similar to other LPME methods, increasing organic solvent volume increases extraction recovery. But after reaching quantitative recoveries, increasing solvent volume causes a significant decrease in enrichment factor. Usually in LPME-SOF, 5–100 μL of extraction solvent is selected.

2.4.2.4 Effect of stirring rate

For the SFO-LLME, sample agitation is an important parameter that influences the extraction efficiency. According to the film theory of convective-diffusive mass transfer for LPME system, high stirring speed can decrease the thickness of the diffusion film in the aqueous phase, so the aqueous phase mass-transfer coefficient will be increased with increased stirring speed (rpm). But incresing stirring rate must be controlled, because it may be cause to sputtering of the solvent drops and influence the extraction efficiency.

2.4.2.5 Effect of extraction time

LPME is not an exhaustive extraction, so extraction time has great effect on extraction efficiency, precision, sensitivity, and the repeatability of the LPME-SFO. It is necessary to choose an appropriate extraction time to guarantee the equilibrium between aqueous and organic phases and the maximum extraction of analytes.

2.4.2.6 Salting out

As described before, salt addition can improve the extraction efficiency of the analytes due to salting out effect. But higher salt concentration can be affect physical properties of the extraction film. This causes reducing in the diffusion rates of the analytes into the organic phase. Therefore, the amount of salt should be optimized in LPME-SFO.

2.4.3 Application of LPME-SFO for extraction of metal ions before determination by AAS

LPME-SFO can be used for extraction of metal ions from aqeous solution in combination with ET-AAS. A LPME-SFO technique was used for the extraction of lead in water samples (Dadfarnia et al., 2008) 1-undecanol containing dithizone as the chelating agent was used as floated organic drop for extraction of lead ions. After stirring the sample for a certain time, the sample vial was cooled in an ice bath for 5 min. The solidified extract was transferred into a conical vial where it melted immediately, and then 10 µL of it was analyzed by ET-AAS. Another LPME-SFO method was designed for extraction of Co(II) and Ni (II) as complexes with 1-(2-Pyridylazo)-2-naphthol (PAN) as chelating agent (Bidabadi et al., 2009). A highly efficient LPME-SFO method for the determination of arsenic by electrothermal atomic absorption spectrometry (ETAAS) was reported (Ghambarian et al., 2010). In this method extraction of As(III) as its complex with APDC was carried out in ppb level. Another literature reported the same method for the determination of trace lead and cadmium in water samples (Rivas et al., 2010).

When LPME-SFO is coupled with flame atomic absorption spectrometry (FAAS), direct injection analysis cannot be performed, since the volume of the extraction solvent is too little for Flame-AAS. For this purpose a LPME-SFO method was proposed for extraction of cadmium ions in different water samples for determination by Flame-AAS (Dadfarnia et al., 2009). In the method, the extraction solution was first diluted with ethanol to 250 µL and then 100 µL of it was analyzed by flow injection flame atomic absorption spectrometry (FI-FAAS).

2.5 Hollow fibre based liquid phase microextraction

An alternative concept for LPME was developed using hollow fiber membranes (Pedersen-Bjergaard, et al., 1999). This technique is based on the use of hollow fibers, typically made of polypropylene. This form of LPME consists of a donor phase (the sample), an acceptor phase (in the lumen of the hollow fiber) and the hollow fiber between them. Pores of the hollow fiber membrane are impregnated with organic solvent and the system is called as supporting liquid membrane. This configuration can be used in two modes; two phase and three phase. In two phase system donor phase is aqueous; acceptor phase is organic with oraganic solvent in hollow fiber pores. But in three phase system, donor phase and acceptor phase are both aqueous with hollow fiber impregnated with organic solvent (Fig. 6)

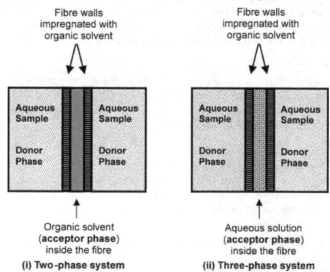

Fig. 6. Illustration of Hollow fiber based LPME (i) two phase (ii) three phase system (Psillakis et. al., 2003)

2.5.1 Theoretical aspects of hollow fiber based LPME

Two phase extraction mode: Extraction through hollow fiber membranes by passive diffusion of analytes from donor phase (the sample) into acceptor phase (extractant) is based on a partitioning equilibrium. Partition coefficient is presented by:

$$K_{a/d} = C_{eq,a} / C_{eq,d} \qquad (12)$$

Where $C_{eq,a}$ and $C_{eq,d}$ are analyte concentration at equilibrium in acceptor and donor phases, respectively. Extraction recovery and enrichment factor are given by:

$$R = (100K_{a/d}V_a) / (K_{a/d}V_a + V_d) \qquad (13)$$

$$E = (V_d R) / (100V_a) \qquad (14)$$

Where V_a and V_d are volume of acceptor and donor phase solutions.

Three phase extraction mode: Similar to two phase mode, this system is based on passive diffusion of analytes through the membrane, too. But two partitioning equilibrium system must be established. The first equilibrium between donor phase and organic phase (impregnated in membrane pores) and second one between organic phase and donor phase.

$$K_{org/d} = C_{eq,org} / C_{eq,d} \qquad (15)$$

$$K_{a/org} = C_{eq,a} / C_{eq,org} \qquad (16)$$

$$K_{a/d} = C_{eq,a} / C_{eq,d} = K_{org/d}.K_{a/org} \qquad (17)$$

Where $C_{eq,org}$ is the analyte concentration in organic phase at equilibrium also $K_{a/d}$ as partition coefficient of the analytes between acceptor and donor phases is the main force for performing the microextraction. Extraction recovery (R) in three phase system can be given as:

$$R = (100K_{a/d}.V_a)(K_{a/d}.V_a + K_{org/d}.V_{org} + V_d) \qquad (18)$$

2.5.2 Factors affecting hollow fiber based LPME

Several parameters affects hollow fiber based LPME. Kind of hollow fiber, nature of organic phase, volume of donor, acceptor and organic phases, pH of donor and acceptor phases, salt addition and sample agitation must be optimized.

2.5.2.1 Kind of hollow fiber

The hollow fibers used in LPME should be compatible with the organic solvent. Polypropylene hollow fibers are the most popular fibers for this purpose. Common dimensions of employed hollow fibers are as listed below. Inner diameter 600 µm, wall thickness 200 µm, nominal and maximum pore sizes are 0.2 and 0.64 µm, respectively. Length of the hollow fiber is based on optimum volume of acceptor phase according to volume of fibers lumen.

2.5.2.2 Organic solvent

An important step in method development for different modes of hollow fiber based LPME is organic solvent selection. Water solubility and polarity are the main characteristics of the organic solvent which must be considered. To prevent dissolution of it in aqueous phase, it should have minimum solubility. Low volatility is also another important characteristic of the selected organic solvent which will restrict solvent evaporation during extraction. Solubility of the analytes in organic solvents and high partition coefficients of the analytes between two phases are other principle considerations in solvent selection.

2.5.2.3 Agitation of the sample

In order to improve extraction kinetics, use of factors increasing convection of the analytes from solution bulk into the extracting phase is usual. In this method sample and extracting phase are not in direct contact with each other and sample agitation facilitates transfer of the analytes through the supported liquid membrane. Stirring, vibration and sonication are popular agitation techniques. Suitability of the selected agitation method must be considered. Rate and power of agitation must be optimized to prevent air bubble formation around the membrane or accelerate solvent evaporation which causes repeatability and imprecision problems in analysis.

2.5.2.4 Salt addition

In both of the two phase and the three phase LPME methods, salting out effect must be investigated by adding optimized amounts of selected salt to donor phase. Positive effect of salt addition on extraction efficiency depends on nature of the analyte and ionic strength of the sample solution.

2.5.2.5 Volumes of donor and acceptor solutions

It's clear that, the main aim of all preconcentration techniques is reaching to higher enrichment factors. This is available with increasing the ratio of donor phase volume to acceptor phase volume. In this case kind of analytical instrument which LPME is coupled is important. Because various volumes are required for analyte introduction. In the case of the atomic absorption spectrometry (AAS) it depends on the mode of the instrument. For Flame-AAS appropriate minimized volume of the analyte solution is about one mL which is large enough in comparison to the volume of membrane lumen. But electrothermal-AAS is more compatible with this technique due to lower volumes (μL scale) of the samples need to be introduce to graphite furnace.

2.5.2.6 Adjustment of pH

pH of donor and acceptor phases must be adjusted according to the nature of the analytes. In both of the two phase and three phase modes of this technique extraction efficiency has great dependence to solution pH. If the analytes are dissociable in different pHs (weak acids and bases) effect of pH on extraction efficiency must be investigated.

2.5.2.7 Extraction time

In all of the extraction methods, extraction time is an important parameter. Because sample preparation is rate determining step of an analytical method. Mass-transfer of the analytes between two phases is a time consuming process. As discussed before sample agitation is one of the methods which is applied for faster extraction and lower extraction times.

2.5.3 Application of the hollow fiber based LPME for extraction of the metal ions before atomic absorption spectrometric determination

One decade after introduction of hollow fiber based LPME for enrichment of the organic compounds, the first application of this technique for extraction of inorganic compounds was reported (Xia, et al., 2006). In this study extraction of Se (IV) and Se (VI) from water samples before determination by ICP-MS was investigated. The first report on use of hollow fiber based LPME before AAS is about determination of Cd (II) in sea water using electrothermal AAS (Peng, et al., 2007). In this three phase extraction system dithizone/oleic acid in 1-octanol – HNO_3 was used as extraction phase 0.8 ng L^{-1} and 387 are LOD and preconcentration factor of the method, respectively. Other two phase and three phase methods for extraction of organomercury and As before electrothermal AAS were reported and summarized in a review article (Dadfarnia, et al., 2010).

2.6 Cloud point extractin (CPE)

Cloud point extraction (CPE), known also as phase separation extraction and surfactant (or micelle)-mediated phase separation is based on the phase behavior of non-ionic surfactants in aqueous solutions, which exhibit phase separation after an increase in temperature or the addition of a salting-out agent. Separation and preconcentration based on cloud point extraction (CPE) are becoming an important and practical application of surfactants in analytical chemistry. The technique is based on the property of most nonionic surfactants in aqueous solutions to form micelles and to separate into a surfactant-rich phase of a small volume and a diluted aqueous phase when heated to a temperature known as the cloud point temperature.

The small volume of the surfactant-rich phase obtained with this methodology permits the design of extraction schemes that are simple, cheap, highly efficient, fast and environmental friendly in comparison with classical methods. CPE might be an interesting and efficient alternative, once it eliminates or reduces consumption of organic solvents significantly. Trace elements can be extracted to the surfactant-rich phase usually after formation of a hydrophobic complex with an appropriate chelating agent (ghaedi et al., 2009).

2.6.1 Surfactants and micelles

Surfactants are compounds that lower the surface tension of a liquid, the interfacial tension between two liquids or between a liquid and a solid. The term surfactant is a blend of surface active agent (Rosen, 2010). Surfactants are amphiphilic molecules, one of whose parts (the head) is polar or hydrophilic in nature and the other (the tail) hydrophobic (Fig. A). This latter part is generally a hydrocarbon chain with different numbers of carbon atoms and may be linear or branched. It may also contain aromatic rings. Therefore, a surfactant molecule contains both a water insoluble (oil soluble component) and a water soluble component. Surfactant molecules migrate to the water surface, where the insoluble hydrophobic groups may extend out of the bulk water phase, either into the air (Fig. B) or if water is mixed with an oil, into the oil phase, while the water soluble head group remains in the water phase. Surfactants in dilute aqueous solutions arrange on the surface. With increasing the concentration of the surfactant the solution surface becomes completely loaded with surfactant and any further additions must arrange as micelles (Fig. C). Therefore, when the surfactant concentration is increased above a certain threshold, called the critical micellar concentration (CMC), the surfactant molecules become dynamically associated to form molecular aggregates of colloidal size. These aggregates containing 60 to 100 monomers are at equilibrium with surfactant molecules in solution with concentration near to CMC.

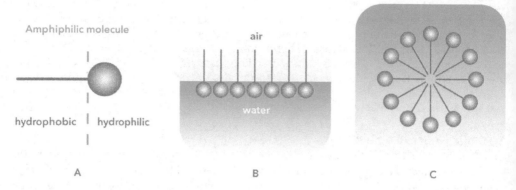

Fig. 7. Schematic representation of the surfactant and formation of a micelle from its monomers beyond its critical micellar concentration (CMC).

2.6.2 Phase separation in CPE
When a micellar solution of a non-ionic surfactant is heated, it becomes turbid over a narrow temperature range, which is referred to as its cloud-point temperature (Hinze & Pramauro, 1993). Above the cloud-point temperature, the system initially in an isotropic phase is separated into two isotropic phases, one of them surfactant-rich phase which is separated from the bulk aqueous solution; and the other aqueous phase, in which the surfactant concentration will be approximately equal to the critical micelle concentration. The phenomenon is reversible and upon cooling, a single phase is again obtained. The mechanism by which separation occurs is poorly understood but some authors have explained the cloud point phenomenon on the basis of the dehydration process that occurs in the external layer of the micelles of non-ionic surfactants when temperature is increased (Hinze, 1987).

2.6.3 Experimental procedure of CPE in metal analysis
The extraction process of CPE for metal ions is very simple and is shown in Fig. 7. First, a few ml of the surfactant or a concentrated surfactant solution is added to some tens to hundreds of millilitres of an aqueous sample containing the metal ions. The final surfactant concentration must exceed its CMC in order to ensure formation of micelle aggregates. The chelating agent is added, where necessary, along with the surfactant, dissolved in an organic solvent or directly to the water, depending on its solubility. Next, the solution is heated above the cloud point and separation of the phases usually takes place after centrifugation. Any analyte solubilized in the hydrophobic core of the micelles, will separate and become concentrated in the small volume of the surfactant-rich phase. After cooling in an ice bath, the surfactant-rich phase became viscous and retain at the bottom of the tube. The supernatant aqueous phases can readily be discarded by inverting the tube.
Finally, a volume (microliter) of nitric acid in aqueous or organic solvent can be added to the surfactant-rich phase to reduce its viscosity and to facilitate sample handling prior to AAS assay.

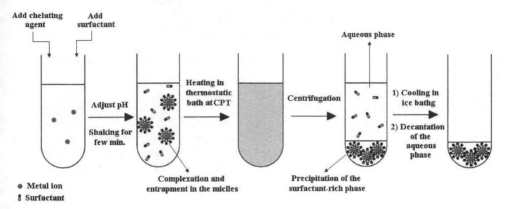

Fig. 8. Experimental schemes for cloud point extraction of metal ions prior to atomic absorption spectrometry.

2.6.4 Calculations

The CPE of metal ion from micellar solutions was evaluated in terms of extraction recovery (ER), distribution coefficient (D), selectivity $S_{X/A}$ and the concentration factor CF that are defined as follows:

The distribution coefficient is a parameter used to describe the degree of analyte partitioning from the aqueous to the surfactant-rich phase, is given by

$$D = \frac{[A]_S}{[A]_W} \tag{19}$$

where $[A]_S$ and $[A]_W$ are the final analyte concentrations in the surfactant-rich phase and in the aqueous phase, respectively. $[A]_W$ can be calculated from the mass balance equation:

$$[A]_O V_O = [A]_W V_W + [A]_S V_S \tag{20}$$

where $[A]_O$ refers to the analyte concentration in the original aqueous solution prior to the extraction step, and V_O is the volume of original aqueous solution. V_W and V_S are the volumes of aqueous solution and surfactant rich phase obtained after the extraction step, respectively.

The extraction recovery (percent of analyte extracted) can be expressed as:

$$ER = \frac{[A]_O - [A]_W}{[A]_O} \times 100 \tag{21}$$

Selectivity of analyte in the CPE process is described as:

$$S_{X/A} = \frac{D_X}{D_A} \tag{22}$$

Where D_X and D_A are distribution coefficient of foreign ion and analyte, respectively. Finally the concentration factor is given by:

$$CF = \frac{m_S}{m_W} = \frac{[A]_S V_S}{[A]_W V_W} = \frac{[A]_O V_O - [A]_W V_W}{[A]_W V_W} \tag{23}$$

Where m_S and m_W are total mass of analyte in the surfactant rich phase and aqeous phase, respectively.

2.6.5 Factors affecting the CPE efficiency

In CPE, extraction needs to be carried out under optimal conditions in order to maximize the preconcentration factor and extraction recovery. It is well known that the extraction/preconcentration process can be altered by pH, complexing agent, the types and concentration of surfactant, ionic strength (additive), equilibration temperature and incubation time (Quina & Hinze, 1999).

2.6.5.1 Effect of solution pH

The separation of metal ions by CPE involves prior formation of a complex with sufficient hydrophobicity to be extracted into the small volume of surfactant-rich phase. The formation of metal-ligand comples and its chemical stability are two important factors for CPE. The pH, which plays a unique role on formation of the complex and subsequent extraction, is proved to be a main parameter for CPE. Extraction yield depends on the pH at which complex formation is carried out (Biparva & Hadjmohammadi, 2007).

2.6.5.2 Effect of the complexing agent

Generally, CPE of a metal ion is taken place via a complex formation of the analyte with a lipophilic ligand. These complexes interact with the micellar aggregate and can be extracted from the aqueous solution into the surfactant-rich phase. The selectivity and efficiency of the method depend directly on the hydrophobicity of the ligand and the complex formed, the apparent equilibrium constants in the micellar medium, the kinetics of the complex formation, and the transfere of the complex between the phases (Constantine, 2002; Carabias-Martinez, et al., 2000).

2.6.5.3 Effect of surfactant type and concentration

Depending on the nature of the hydrophilic group, surfactants are classified as non-ionic, zwitterionic, cationic, and anionic. Up to now, non-ionic, zwitterionic and anionic surfactants are most widely used for CPE of metal ions. However, it is very important to select an appropriate surfactant for a successful CPE analysis since it can directly affect the extraction and preconcentration, and accuracy of the final analytical results. Reports showed that Triton X-114 and PONPE-7.5 (cloud point temperature, near room temperature) (Paleologos et al., 2000, 2001) are proper surfactants to perform CPE for trace elements because of its commercial availability in a high purified homogeneous form, low toxicity and cost. Also, low cloud point temperature (23–26 °C) and high density of the surfactant-rich phase facilitates phase separation by centrifugation. A successful cloud point extraction should maximize the extraction efficiency by minimizing the phase volume ratio $(\frac{V_S}{V_W})$. This shows that the smaller surfactant concentration ptovides higher preconcentration factor; but when the volume of surfactant-rich phase is small, the extraction process becomes more difficult and the accuracy and reproducibility probably suffer (Eiguren Fernandez et al., 1999; Moreno Cordero et al., 1993). However, since the volume of the surfactant- rich phase must be manageable, a compromise must be reached so that the surfactant concentration will allow a high phase ratio and a manageable surfactant-rich phase.

2.6.5.4 Effect of Ionic strength

The cloud point of micellar solutions can be controlled by addition of salts, alcohols, non-ionic surfactants and some organic compounds (salting-out effects). To date, most of the studies conducted have shown that ionic strength has no appreciable effect on the extraction efficiency. An increase in the ionic strength in the CPE does not seriously alter the efficiency of extraction of the chemical forms. Moreover, the addition of a salt can markedly facilitate the phase separation process. As demonstrated with some non-ionic surfactant systems, it alters the density of the bulk aqueous phase.

2.6.5.5 Effects of equilibration temperature and incubation time

Optimal incubation time and equilibration temperature are necessary to complete the reaction, and to achieve easy phase separation and preconcentration as efficient as possible. The greatest analyte preconcentration factors are thus expected under conditions where the CPE is conducted using equilibration temperature that are well above the cloud point temperature of the surfactant. It was desirable to employ the shortest equilibration time and the lowest possible equilibration temperature, which compromise completion of the reaction and efficient separation of phases.

2.6.6 Applications of CPE in atomic absorption spectrometry

The use of CPE coupled with atomic absorption spectrometry (AAS) offers a conventional alternative to more traditional extraction systems and permits the design of extraction schemes that are simple, cheap, of high efficiency, reducing in extraction time and environmentally clean methodology due to low consumption of a solvent, apart from the results that comparable to those obtained with other separation procedures. CPE technique has been successfully employed for the preconcentration of micro amounts of several metals in different matrices, as prior step before their determinations by Atomic Absorption Spectrometry.

Watanabe et al. used this unconventional liquid–liquid separation to extract metal ions for the first time (Watanabe & Tanaka, 1978). They extracted Ni using Triton X-100 but this surfactant has a relatively high cloud point, around 70 °C. Later, to extract Zn, they used PONPE 7.5, which requires a lower temperature (around 5 °C) for the phase separation. This surfactant and Triton X-114 which has a very convenient cloud point were used in other studies of metal ion extraction. Cu (Kulichenko et al., 2003), Mn (Doroschuk et al., 2004), Co (Nascentes & Arruda, 2003), Cd and Ni (Manzoori & Karim-Nezhad, 2004), Ag and Au (Mesguita da Silva et al., 1998) were determined by FAAS. CPE was also applied for extraction of Fe(III) (Ohashi et al., 2005) and As(III) and As(V) (Shemirani et al., 2005) before determination by ET-AAS. Cd, Cu, Pb, and Zn (Chen & Teo, 2001) were simultaneous extracted as complex with 1-(2-thiazolylazo)-2-naphthol (TAN) using TritonX-114 prior to determination by FAAS. Several ligands such as 1-(2-pyridylaso)-2-naphthol (PAN), 2-(2-thiazoylazo)-4-methylphenol (TAC), dialkyldithiophosphates (DDTP), 4-(2-pyridylazo) resorcinol (PAR), 2-(5-bromo-2-pyridylazo)-5-diethilaminophenol (Br-PADAP) have been used in CPE of metal ions.

3. Conclusion

Analysis of metallic analytes is so important in environmental, biological and food samples. Due to complex matrix of real samples and trace concentration of them use of sample preparation methods is necessary. Development of efficient sample preparation methods in agreement with green chemistry is one of the most exciting research fields for analytical chemists. Growing number of scientific publishing in this area shows emerging needs for newer methods with higher concentration factors, higher recoveries, more cheap and simple and environmental friendly methods. All of the techniques discussed in this chapter have some advantages and some limitations. An analyst must be able to choose the best method according to his problem. Kind of analyte and sample, complexity of the sample, range of concentration which analyte exist in the sample are some of the parameters which must be considerd.

4. Acknowledgment

Financial supports of Iranian Academic Center for Education, Culture and Research (ACECR) from our research projects gratefully acknowledged. Also the authors express their deep sense of gratitude to the publisher for their efforts in order to development of science and technology.

5. References

Ahmadzadeh Kokya, T. & Farhadi, K. (2009). Optimization of Dispersive liquid–liquid Microextraction for the Selective Determination of Trace Amounts of Palladium by Flame Atomic Absorption Spectroscopy. *Journal of Hazardous Materials*, Vol 169, pp. 726-733

Arthur, C. L. & Pawliszyn, J. (1990). Solid Phase Microextraction with Thermal Desorption. *Analytical Chemistry*, Vol 62, pp. 2145-2148.

Baliza, P. X., Teixeira, L. S. G. & Lemos, V. A. (2009). A Procedure for Determination of Cobalt in Water Samples after Dispersive liquid–liquid Microextraction. *Microchemical Journal*, Vol 93, pp. 220-224.

Bidabadi, M. S., Dadfarnia, S. & Shabani, A. M. H. (2009). Solidified Floating Organic Drop Microextraction (SFODME) for Simultaneous Separation/Preconcentration and Determination of Cobalt and Nickel by Graphite Furnace Atomic Absorption Spectrometry (GFAAS). *Journal of Hazardous Materials*, Vol 166, pp. 291-296.

Bidari, A., Zeini Jahromi, E., Assadi, Y. & Milani Hosseini, M. R. (2007). Monitoring of Selenium in Water Samples using Dispersive liquid–liquid Microextraction Followed by Iridium-modified Tube Graphite Furnace Atomic Absorption Spectrometry. *Microchemical Journal*, Vol 87, pp. 6-12.

Biparva, P. & Hadjmohammadi, M. R. (2007). Cloud Point Extraction Using NDTT Reagent for Preconcentration and Determination of Copper in Some Environmental Water Samples by Flame Atomic Absorption Spectroscopy. *Acta chimica Slovenica*, Vol 54, pp. 805–810.

Chamsaz, M., Arbab-Zavar, M. H. & Nazari, S. (2003). Determination of Arsenic by Electrothermal Atomic Absorption Spectrometry using Headspace Liquid Phase Microextraction after in situ Hydride Generation. *Journal of Analytical Atomic Spectrometry*, Vol 18, pp. 1279-1282.

Carabias-Martinez, R., Rodriguez-Gonzalo, E., Moreno-Cordero, B., Perez-Pavon, J. L., Garcia Pinto, C. & Fernandez Laespada, C. E. (2000). Surfactant Cloud Point Extraction and Preconcentration of Organic Compounds prior to Chromatography and Capillary Electrophoresis. *Journal of Chromatography A*, Vol 902, pp. 251-265.

Chen, J. & Teo, K. C. (2001). Determination of Cadmium, Copper, Lead and Zinc in Water Samples by Flame Atomic Absorption Spectrometry after Cloud Point Extraction. *Analytica Chimica Acta*, Vol 450, pp. 215-222.

Constantine, D.S. (2002). Micelle-mediated Extraction as a Tool for Separation and Preconcentration in Metal analysis. *Trends in Analytical Chemistry*, Vol 21, pp. 343-355.

Dadfarnia, S. & Shabani, A. M. H. (2010). Recent Developments in Liquid Phase Microextraction for Determination of Metals-A Review. *Analytica Chimica Acta*, Vol 658, pp. 107-119.

Dadfarnia, S., Salmanzadeh, A. M. & Shabani, A. M. H. (2008). A Novel Separation/Preconcentration System based on Solidification of Floating Organic Drop Microextraction for Determination of Lead by Graphite Furnace Atomic Absorption Spectrometry. *Analytica Chimica Acta*, Vol 623, pp. 163-167.

Dadfarnia, S., Shabani, A. M. H. & Kamranzadeh, E. (2009). Separation/Preconcentration and Determination of Cadmium Ions by Solidification of Floating Organic Drop Microextraction and FI-AAS. *Talanta*, Vol 79, pp. 1061-1065.

Doroschuk,V. O., Lelyushok, S. O., Ishchenko, V. B. & Kulichenko, S.A. (2004). Flame Atomic Absorption Determination of Manganese(II) in Natural Water after Cloud Point Extraction. *Talanta*, Vol 64, pp. 853-856.

Eiguren Fernandez, A., Sosa Ferrera, Z. & Santana Rodriguez, J. J. (1999). Application of Cloud-point Methodology to the Determination of Polychlorinated Dibenzofurans in Sea Water by High-Performance Liquid Chromatography. *Analyst.* Vol 124, pp. 487-491.

Faraji, M., Yamin, Y., Saleh, A., Rezaee, M., Ghambarian, M. & Hassani, R. (2010). A Nanoparticle-based Solid-Phase Extraction Procedure Followed by Flow Injection Inductively Coupled Plasma-Optical Emission Spectrometry to Determine some Heavy Metal Ions in Water Samples. *Analytica Chimica Acta*, Vol 659, pp. 172-177.

Farajzadeh, M. A., Seyedi, S. E., Shalamzari, M. S. & Bamorovat, M. (2009). Dispersive Liquid–Liquid Microextraction using Extraction Solvent Lighter than Water. *Separation Science*, Vol 32, pp. 3191-3200.

Figueroa, R., Garcia, M., Lavilla, I. & Bendicho, C. (2005). Photoassisted Vapor Generation in the Presence of Organic Acids for Ultrasensitive Determination of Se by Electrothermal-Atomic Absorption Spectrometry Following Headspace Single-Drop Microextraction. *Spectrochimica Acta Part B: Atomic Spectroscopy*, Vol 60, pp. 1556-1563.

Fitchett, A. W., Hunter Daughtrey, E., & Mushak, P. (1975). Quantitative Measurements of Inorganic and Organic Arsenic by Flameless Atomic Absorption Spectrometry. *Analytica Chimica Acta*, Vol 79, pp. 93-99.

Fragueiro, S., Lavilla, I. & Bendicho, C. (2004). Headspace Sequestration of Arsine onto a Pd(II)-containing Aqueous Drop as a Preconcentration Method for Electrothermal Atomic Absorption Spectrometry. *Spectrochimica Acta Part B: Atomic Spectroscopy*, Vol 59, pp. 851-855.

Fragueiro, S., Lavilla, I. & Bendicho, C. (2004). Direct Coupling of Solid Phase Microextraction and Quartz Tube-Atomic Absorption Spectrometry for Selective and Sensitive Determination of Methylmercury in Seafood: an Assessment of Chloride and Hydride Generation. *Journal of Analytical Atomic Spectrometry.* Vol 19, pp. 250-254.

Fritz, J. S. *Analytical Solid Phase Extraction,* WILEY-VCH, ISBN 0-471-24667-0, New York, USA

Ghaedi, M., Niknam, K., Niknam, E. & Soylak, M. (2009). Application of Cloud Point Extraction for Copper, Nickel, Zinc and Iron Ions in Environmental Samples. *Journal of the Chinese Chemical Society,* Vol 56, pp. 981-986.

Ghambarian, M., Khalili-Zanjani, M. R., Yamini, Y., Esrafili, A. & Yazdanfar, N. (2010). Preconcentration and Speciation of Arsenic in Water Specimens by the Combination of Solidification of Floating Drop Microextraction and Electrothermal Atomic Absorption. *Talanta,* Vol 81, 197-201.

Gil, S., Fragueiro, S., Lavilla, I. & Bendicho, C. (2005). Determination of Methylmercury by Electrothermal Atomic Absorption Spectrometry using Headspace Single-Drop Microextraction with in situ Hydride Generation. *Spectrochimica Acta Part B: Atomic Spectroscopy,* Vol 60, pp. 145-150.

He, B., Jiang, G. B. & Ni, Z. (1998). Determination of Methylmercury in Biological Samples and Sediments by Capillary Gas Chromatography Coupled with Atomic Absorption Spectrometry after Hydride Derivatization and Solid Phase Microextraction. *Journal of Analytical Atomic Spectrometry.* Vol 13, pp. 1141-1144.

He, B. & Jiang, G. B. (1999). Analysis of Organomercuric Species in Soils from Orchards and Wheat Fields by Capillary Gas Chromatography on-line Coupled with Atomic Absorption Spectrometry after in situ Hydride Generation and Headspace Solid Phase Microextraction. *Fresenius Journal of Analytical Chemistry.* Vol 365, pp. 615-618.

He, Y. & Lee, H. K. (1997). Liquid-Phase Microextraction in a Single Drop of Organic Solvent by Using a Conventional Microsyringe. *Analytical Chemistry.* Vol 69, pp. 4634-4640.

Hinze, W. L. *Ordered Media in Chemical Separations,* ACS Symposium Series, Vol 342, 1987.

Hinze, W. L. & Pramauro, E. (1993). A Critical Review of Surfactant-Mediated Phase Separations (Cloud-Point Extractions): Theory and Applications. *Critical Reviews in Analytical Chemistry.* Vol 24, pp. 133-177.

Jeannot, M. A. & Cantwell, F. F. (1996). Solvent Microextraction into a Single Drop. *Analytical Chemistry.* Vol 68, pp. 2236-2240.

Jiang, H. & Hu, B. (2008). Determination of Trace Cd and Pb in Natural Waters by Direct Single Drop Microextraction Combined with Electrothermal Atomic Absorption Spectrometry. *Microchimica Acta,* Vol 161, pp. 101-107.

Khalili-Zanjani, M. R., Yamini, Y., Shariati, S. & Jonsson, J. A. (2007). A New Liquid-Phase Microextraction Method based on Solidification of Floating Organic Drop. *Analytica Chimica Acta,* Vol 585, pp. 286-293.

Kulichenko, S. A., Doroschuk,V. O. & Lelyushok, S. O. (2003). The Cloud Point Extraction of Copper(II) with Monocarboxylic Acids into Non-ionic Surfactant Phase. *Talanta,* Vol 59, pp. 767-773.

Lauwerys, R. R., Buchet, J. P. & Roels, H. (1979). The Determination of Trace Levels of Arsenic in Human Biological Materials. *Archieves of Toxicology,* Vol 41, pp. 239-247.

Lucci, P., Derrien, D., Alix, F., Perollier, C. & Bayoudh, S. (2010). Molecularly Imprinted Polymer Solid-Phase Extraction for Detection of Zearalenone in Cereal Sample Extracts. *Analytica Chimica Acta*, Vol 672, pp. 15-19.

Liang, P., Liu, R. & Cao, J. (2008). Single Drop Microextraction Combined with Graphite Furnace Atomic Absorption Spectrometry for Determination of Lead in Biological Samples. *Microchimica Acta*, Vol 160, pp. 135-139.

Liang, P. & Sang, H. (2008). Determination of Trace Lead in Biological and Water Samples with Dispersive Liquid–Liquid Microextraction Preconcentration. *Analytical Biochemistry*, Vol 380, pp. 21-25.

Liu, X. & Fan, Z. (2007). Determination of Trace Pb in Water Samples by Electrothermal Atomic Absorption Spectrometry After Single-Drop Microextraction. *Atomic Spectroscopy Journal*, Vol 6, pp. 215-219.

Manzoori, J. L., Amjadi, M. & Abulhassani, J. (2009). Ionic Liquid-Based Single Drop Microextraction Combined with Electrothermal Atomic Absorption Spectrometry for the Determination of Manganese in Water Samples. *Talanta*, Vol 77, pp. 1539-1544.

Manzoori, J. L., Amjadi, M. & Abulhassani, J. (2009). Ultra-trace Determination of Lead in Water and Food Samples by using Ionic Liquid-Based Single Drop Microextraction-Electrothermal Atomic Absorption Spectrometry. *Analytica Chimica Acta*, Vol 644, pp. 48-52.

Manzoori, J. L. & Karim-Nezhd, G. (2004). Development of a Cloud Point Extraction and Preconcentration Method for Cd and Ni prior to Flame Atomic Absorption Spectrometric Determination. *Analytica Chimica Acta*, Vol 521, pp. 173-177.

Mesguita da Silva, M. A., Azzolin Frescura, V. L., Nome Aguilera, F. J. & Curtius, A. J. J. (1998). Determination of Ag and Au in Geological Samples by Flame Atomic Absorption Spectrometry after Cloud Point Extraction. *Anal. At. Spectrom.* Vol 13, pp. 1369-1373.

Mester, Z. & Sturgeon, R. (2005). Trace element speciation using solid phase microextraction. *Spectrochim. Acta. Part B*, Vol 60, 1243-1269.

Moreno Cordero, B., Perez Pavon, J. L., Garcia Pinto, C. & Fernandez Laespada, E. (1993). Cloud Point Methodology: A New Approach for Preconcentration and Separation in Hydrodynamic Systems of Analysis. *Talanta*, Vol 40, pp. 1703-1710.

Nascentes, C. C. & Arruda, M. A. Z. (2003). Cloud Point Formation Based on Mixed Micelles in the Presence of Electrolytes for Cobalt Extraction and Preconcentration. *Talanta*, Vol 61, pp. 759-768.

Ohashi, A., Ito, H., Kanai, C., Imura, H. & Ohashi, K. (2005). Cloud Point Extraction of Iron(III) and Vanadium(V) using 8-quinolinol Derivatives and Triton X-100 and Determination of 10^{-7} mol dm^{-3} level Iron(III) in River Water Reference by a Graphite Furnace Atomic Absorption Spectroscopy. *Talanta*, Vol 65, pp. 525-530.

Paleologos, E. K., Stalikas, C. D., Tzouwara-Karayanni, S. M., Pilidis, G. A. & Karayannis, M. I. (2000). Micelle-Mediated Methodology for Speciation of Chromium by Flame Atomic Absorption Spectrometry. *Journal of Analytical Atomic Spectrometry*, Vol 15, 287–291.

Paleologos, E. K., Stalikas, C. D. & Karayannis, M. I. (2001). An Optimised Single-Reagent Method for the Speciation of Chromium by Flame Atomic Absorption Spectrometry Based on Surfactant Micelle-Mediated Methodology. *Analyst.* Vol 126, pp. 389–393.

Pedersen-Bjergaard, S. & Rasmussen K. E. (1999). Liquid–Liquid–Liquid Microextraction for Sample Preparation of Biological Fluids Prior to Capillary Electrophoresis. *Anal. Chem*, Vol 71, 2650-2656.

Peng, F., Liu, J. F., He, B., Hu, X. L. & Jiang, G. B. (2007). Ultrasensitive Determination of Cadmium in Seawater by Hollow Fiber Supported Liquid Membrane Extraction Coupled with Graphite Furnace Atomic Absorption Spectrometry. *Spectrochim. Acta. Part B*, Vol 62, 499-503.

Psillakis, E. & Kalogerakis, N. (2003). Developments in liquid-phase microextraction. *Trends Anal. Chem*, Vol 22, 565-574.

Quina, F.H. & Hinze, W.L. (1999). Surfactant-Mediated Cloud Point Extractions: An Environmentally Benign Alternative Separation Approach. *Industrial & Engineering Chemistry Research*, Vol 38, 4150-4168.

Rezaee, M., Assadi, Y., Milani Hosseini, M. R., Aghaee, E., Ahmadi, F. & Berijani, S. (2006). Determination of Organic Compounds in Water using Dispersive Liquid–Liquid Microextraction. *Journal of Chromatography A*, Vol 1116, pp. 1-9.

Rivas, R. E., Garcia, I. L. and Cordoba, M. H. (2010). Microextraction Based on Solidification of a Floating Organic Drop Followed by Electrothermal Atomic Absorption Spectrometry for the Determination of Ultratraces of Lead and Cadmium in Waters. *Analytical Methods*, Vol 2, pp. 225-230.

Rosen, M. J. *Surfactants and Interfacial Phenomena* (3rd ed.) John Wiley & Sons, Hoboken, New Jersey, (2010).

Saber-Tehrani, M., Givianrad, M. H. & Hashemi-Moghaddam, H. (2007). Determination of Total and Methyl Mercury in Human Permanent Healthy Teeth by Electrothermal Atomic Absorption Spectrometry after Extraction in Organic Phase. *Talanta*, Vol 71, pp. 1319-1325.

Shemirani, F., Baghdadi, M. & Ramezani, M. (2005). Preconcentration and Determination of Ultra Trace Amounts of Arsenic(III) and Arsenic(V) in Tap Water and Total Arsenic in Biological Samples by Cloud Point Extraction and Electrothermal Atomic Absorption Spectrometry. *Talanta*, Vol 65, 882-887.

Watanabe, H. & Tanaka, H. (1978). A Non-Ionic Surfactant as a New Solvent for Liquid – Liquid Extraction of Zinc(II) with 1-(2-pyridylazo)-2-naphthol. *Talanta*, Vol 25, 585-589.

Xia, L., Hu, B., Jiang, Z., Wu, Y., Chen, R. & Li, L. (2006). Hollow Fiber Liquid Phase Microextraction Combined with Electrothermal Vaporization ICP-MS for the Speciation of Inorganic Selenium in Natural Waters. *J. Anal. At. Spectrom*, Vol 21, 362-365.

Ying-Ying, W., Guang-Ying, Z., Qing-Yun, C., Xiao-Huan, Z., Chun, W. & Zhi, W. (2010). Developments in liquid-phase microextraction method based on solidification of floating organic drop. *Chinese J. Anal. Chem*, Vol 38, 1517-1522.

Yousefi, S. R. & Shemirani, F. (2010). Development of a Robust Ionic Liquid-Based Dispersive Liquid–Liquid Microextraction Against High Concentration of Salt for Preconcentration of Trace Metals in Saline Aqueous Samples: Application to the Determination of Pb and Cd. *Analytica Chimica Acta*, Vol 669, pp. 25-31.

Zuba, D., Parczewski, A. & Rozanska, M. (2001). 39th Annual TIAFT meeting, Prague, Czech Republic.

3

State-of-the-Art and Trends in Atomic Absorption Spectrometry

Hélcio José Izário Filho[1], Rodrigo Fernando dos Santos Salazar[2],
Maria da Rosa Capri[1], Ângelo Capri Neto[1],
Marco Aurélio Kondracki de Alcântara[1] and André Luís de Castro Peixoto[1]

[1]Universidade de São Paulo,
[2]Universidade Federal de São Carlos
Brazil

1. Introduction

Atomic Absorption Spectrometry, or AAS, is an analytical technique commonly used for the quantitative and qualitative determination of elements in samples such as aqueous solutions, waters, sea-waters, metals and alloys, glass, drugs, food, environmental samples, industrial wastes, biological samples among others.

This technique is based on measuring the amount of electromagnetic energy of a particular wavelength (ultraviolet or visible region), which is absorbed as it passes through a cloud of atoms of a particular chemical element (the analyte) coming from samples and standards. An appropriate mathematical treatment allows relating the amount of absorbed energy to the number of absorbed atoms by providing a measurement of the element concentration in the sample. This technique is established, relatively quickly, economically affordable and allows to determine more than 60 chemical elements from a huge type of samples. It is used by most of research laboratories and industry quality control around the world.

The aim of this text is to present concisely this powerful technique providing basic information about fundamental concepts, instrumentation and application. Relevant and effective techniques will be shown including flame atomization, hydride generation, cold vapor and electrothermic atomization or graphite furnace. Fundamental and theoretical details will not be presented on mechanical components, electronic, software and data processing. They are deeply discussed in specific literature.

2. A brief history

Isaac Newton made important discoveries about solar radiation in the 17th Century. He used a glass prism to break up white light into its constituent spectral colors. He studied the separation of white light in details and, from that discovery, he created the reflecting telescope, the Newtonian telescope. He, then, published his corpuscular theory of light by contraposition of Huygens' undulatory theory.

Two-hundred years after, Wollaston observed dark lines in the solar spectrum known as Fraunhofer lines, after being extensively studied by this German physicist. In 1832, Brewsler concluded that the Fraunhofer lines were produced due to the presence of atomic vapors in the solar atmosphere that absorbed part of radiation emitted by Sun.

From 1859 to 1861, Robert Bunsen and Gustov Kirchoff demonstrated that each chemical element had a characteristic color or spectrum when heated to incandescence (Na yellow; K violet). Heating several elements during a flame test, they identified characteristic spectra of these elements and established a relation between emission spectrum and absorption spectrum. This explains the black lines in the solar spectrum: atoms in the solar corona absorbing part of energy emitted by Sun (continuous spectrum), originating the black lines observed. This also permits to identify absorbing atoms normally present in the corona, comparing the black lines to elements' emission spectrum produced in the laboratory.

At the beginning of the 20th century, the development of quantum mechanics theory provided mathematical patterns explaining the phenomena, which is the interaction of radiation with the matter. Considering this point, the necessary theoretical basis to develop a new technique of elementary analysis using the phenomenon of atomic absorption was established, but only in 1955, the first proposal of a practical instrument was introduced. At the beginning of the 60s, the first commercial equipment appeared and so far, equipments have been using the same basic components, although with more technologically involved.

Understanding this technique requires understanding the phenomenon of emission and absorption of radiant-energy through the matter or, more specifically, through atoms.

3. Basic principles

3.1 Wave-particle duality: Light

Electromagnetic radiation is a form of energy described by classical physics as a wave made up of mutually perpendicular, fluctuating electric and magnetic field that propagates at a constant speed. It is characterized by wavelength (λ), (distance between two adjacent waves) and by frequency v (number of waves per unit of time). The speed of light propagated in a vacuum is C=299.792.458 m s^{-1}.

This model explains the energy propagation, but cannot explain its interaction with the matter. That interaction can be explained if we treat energy as a particle, called photon. A photon is characterized by frequency and wavelength and can transfer energy amount E= hv/ λ, where h is Planck constant (4.135667516×10^{-15} eV s).

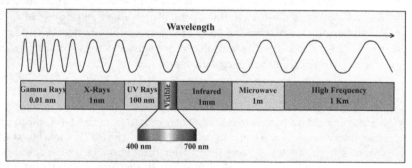

Fig. 1. Regions of the electromagnetic spectrum.

Electromagnetic spectrum is conveniently divided in wavelength rates with specific names (from gamma rays to radio waves) as shown in figure 1. That wavelength division does not have a physical meaning itself, being only a practical classification in accordance with the available technological equipment for its generation and detection. For example, the visible region of light between 400 and 700 nanometers (nm) is directly detected by the human eye and perceived as visible light (1 nm = 10^{-9} m).

3.2 The atom, energy of a quantum state and electronic transitions

Bohr model has an atom consisting of a nucleus containing protons and neutrons surrounded by a cloud of electrons in fact inhabit specific regions in space. This is known as an orbital. The further an orbital gets from the nucleus, the more they gain potential energy associated to a determined orbital.

Quantum Mechanics explains that orbitals have quantized energy levels and for moving an electron to another level, it has to receive or emit the exact amount of energy corresponding to the difference between the two electronic levels ($\Delta E = E_1 - E_0$).

The amount of energy required to move and electron from energy level E_0 to energy level E_1 can be provided by heat due to a collision with other particles or absorb the energy of a photon. In this case, the energy of a Photon ($E = h\nu = hc/\lambda$) should be equal to the difference between the orbitals (ΔE), this is, only a Photon of a particular wavelength is absorbed and can promote that transition. This phenomenon is known as atomic absorption.

A more stable electron configuration of an atom is the one with less energy, also known as ground state configuration. The difference of energy between the last full orbital and the next empty orbital of the atom in a ground state is of the same order of magnitude of photons with wavelengths between 200 and 800 nm, this means, photons in ultraviolet regions and visible light of electromagnetic spectrum.

Sodium atom in ground state, e.g., has an electronic configuration of $1s^2\, 2s^2\, 2p^6\, 3s^1$. The 3s electron can receive a photon with energy of 589.0 nm (E=2.2 eV) and passes to 3p orbital, which is an unstable state known as excited state. Being unstable, the excited atom loses its energy quickly (approximately in 10^{-8} s) and returns to ground state.

One way to lose excitation energy is by emitting a photon of 589.0 nm, a phenomenon known as atomic emission. A photon of 330.3 nm can also be absorbed by sodium. This is the difference of energy between 3s and a 4p orbital (3.6eV), but one photon of 400 nm cannot be absorbed because there are not two orbitals in a sodium atom with the same difference of energy. The return of electron from 4p orbital to ground state can also occur in two steps: first to 4s and then to 3s by emitting two photons with energies correspondent to the two transitions in a phenomenon known as atomic fluorescence.

The sodium atom can also receive enough energy to remove an electron, turning into sodium ion (Na^+), known as ionization. In this case, a change occurs in orbitals of different energy levels, so that the ion has a new set of transition being able to absorb or emit photons of wavelengths differently from metallic sodium.

Each chemical element has a unique electronic structure that differentiates from others. This implies in a possible and unique set of transitions, a set of characteristic absorption/emission lines that can be used for identification. Although the set of transitions is unique for each element, there may be a coincidence of spectrum in some lines of two or more elements, which means that different atoms can absorb or emit photons of same wavelength.

Even though the theoretical basis was established in the beginning of the twentieth century, only in the early 1950s an Australian physicist, Sir Alan Walsh, proposed the phenomenon of atomic absorption as an elementary analysis technique. However, the first scientific equipment was made in the early of 1960s. Despite unavailable technology by that time, two main problems of the technique were solved: to obtain a source that would emit radiant energy of specific wavelength for each element and to generate and to put atoms into a ground state from a sample so they can absorb energy from the source.

3.3 Basic instrumentation
Every atomic absorption spectrometer presents the same basic components, however, each manufacturer differentiates the configuration due to the analytical demand and according to technological advance. Figure 2 shows the main components of an atomic absorption spectrometer.

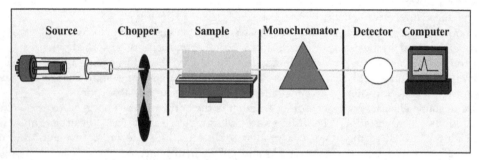

Fig. 2. Main components of an atomic absorption spectrometer.

The radiation source may be continuous, emitting from visible to infrared wavelengths or from lines that emit discreet lines, specifically from each chemical element. The modulator helps to differentiate radiation emitted by radiation lamp coming from the environment and, mainly, from the atomization system. The atomization system removes analyte atoms in solution and generates atomic vapor composed of atoms in ground state, putting them between the source and the detector to absorb the radiation emitted. The monochromator is responsible for the selection of photons due to the wavelength that will reach the detector. The detector transforms the energy of photons into a proportional electronic signal and amplifies it. The signal intensity obtained is treated by systems for data acquisition and processing.

At this point, it is important to stand out that the element cannot be directly determined by the atomic absorption spectrometer; even though the system provides that response. A spectrometer measures the amount of electromagnetic energy coming from the source before and after passing through a sample; this is, an indirect measurement of the absorption of energy by atoms present in the sample. To obtain a final result, a series of physical and chemical phenomenon should occur under controlled conditions and the measurement performed should be treated by an appropriate mathematical model. As each component of the equipment plays a part on this work, a more detailed study is essential to understand the performance of this analytical technique as well as its power and limitations.

3.3.1 Radiation source
Around 1952, when the development of the technique was starting, sources of continuous emissions were used. This means that it was used light in the visible and adjacent

ultraviolet ranges. First experiments showed the necessity of a spectrometer with 0.002 nm resolution for using this source, which was impossible with the equipment available by that time. A solution was found changing the emission sources of discrete lines. Then, the first sealed hollow cathode lamp (HCL) was manufactured (Welz et al., 2010), a high intensity atomic emission light source essential for developing the technique and widely used today.

3.3.2 Hollow cathode lamp (HCL)

HCL as shown in figure 3 consists of a sealed glass tube filled with inert gas, argon or neon, where an anode (positive pole) and a cathode (negative pole) are placed. The cathode contains an element whose spectrum is required basically in a pure source or an appropriate alloy. For elements that emit radiation in the ultraviolet region, the front window of emission is made of quartz because the glass of the tube absorbs UV radiation.

When an appropriate difference of power is applied, some atoms of Argon (filling gas) are ionized (Ar^+) and accelerated in the cathode direction where they are struck enough with force to eject atoms of the element of interest (M^0). These atoms are struck by other filling gas ions and pass from the ground state into the excited state (M^*), which will return rapidly to the ground state by emitting photon with a characteristic wavelength.

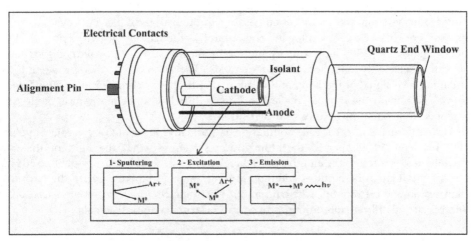

Fig. 3. Scheme of the hollow cathode lamp (HCL).

The HCL is a source of high intensity that emits simple spectra lines permitting the use of a low resolution monochromator. The disadvantage is the need of a different lamp for each element. The analysis turns slow and become expensive for multi-element analysis (determination of more than one element in the same sample).

To reduce these problems, there are multi-elemental lamps, whose cathodes are manufactured with a mixture of elements so they emit a characteristic spectrum for each one. The elements that form the cathode should be compatible. It is possible to form an alloy with them and the main lines of spectrum cannot overlap. The disadvantage is that the intensity of emission for each element is lower than a monoelementar lamp, which increases the signal-to-noise ratio and decreases the sensitivity of the analyses.

3.3.3 Electrodeless Discharge Lamp (EDL)

An electrodeless discharge lamp (EDL) is a discrete light source and was presented as an alternative to HCLs (hollow cathode lamps), but with high intensity emissions. They are very useful for more volatile elements with emission lines in UV region, where HCL has a lower performance.

EDLs consist of the element or a salt of the element sealed in a quartz bulb containing an inert gas. The capsule is placed inside a ceramic cylinder involved with a metallic wire coin. A radio-frequency power source distributes energy that passes through a conducting coil creating an oscillating electric field that ionizes gas inside the bulb. As the electric field is oscillating, the ions are accelerated so that they all have enough kinetic energy to excite the metal atoms inside the bulb when colliding with them. When they return to the ground state, the excited atoms release absorbed energy by emitting light with characteristic element spectrum.

Despite of HCLs efficiency, the cost-effective asset acquisition of EDLs is higher over its lifetime so it is normally used when HCLs are not enough to solve a specific analytical problem. Thus, it is more common to find a big quantity of HCLs in laboratories and none or only few units of EDLs.

3.3.4 Modulators

The atomic absorption spectrometric method is based on measuring the amount of radiation coming from the lamp before and after passing through the sample. However, there are other light sources as the environment and, mainly, atomization system captured in the monochromator. The other light sources interfere in the measurement by reducing signal-to-noise ratio, sensitivity and accuracy of analysis. The solution to this problem is the modulated drive signal of the lamp and combined with the synchronization of the detector to amplify selectively the signal of the lamp. The modulation can be mechanical or electrical and, practically, both of them have similar results.

With old equipments, mechanical modulation is normally performed by a rotating shield located between the lamp and the atomization system, so that the light of the lamp alternately passes through the atomizer or is blocked, at a general frequency of 50 to 60 Hz.

Electronic modulation is the most common among new equipments, where the current of the lamp is on and off rapidly. The equipment electronic circuits compensate variations of the lamp's current, filtrate interfering sources and measure analytical signals.

3.3.5 Atomization system

Although there are some different techniques used to work with solid and gaseous samples in most of the atomic absorption equipments, these samples are introduced in solution.

To determine the contamination level and amount of lead in fish, e.g., a representative sample of fish is collected, treated, and dissolved in acid or in an appropriate acid mixture. The resulting solution has its volume adjusted and finally taken to the equipment for respective analysis. The lead in the solution is used as an ionic form (Pb^{2+}) but the radiation emitted by the lamp is specific to metallic lead (Pb°), and is not absorbed by the ion lead.

To make a determination, the lead ion has to be separated by a solution, reduced to metallic lead (gaseous form) and placed in the optical path between the lamp and the monochromator. This all is done by the atomization system and mostly common by using a flame to provide necessary energy so that the atomization process occurs. When this kind of atomizer is used, the technique is known as Flame Atomic Absorption Spectrometry (FAAS).

Figure 4 shows an atomizer with a pre-mixture chamber, actually, most commonly used in FAAS. The solution containing the analyte is aspired by a nebulizer, mixed in a nebulization chamber with the gases. The obtained mixture is directed to the burner, where they are burned, occurring sample atomization and absorption.

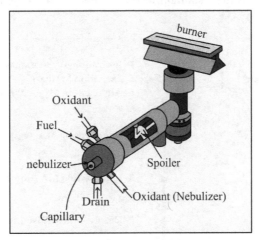

Fig. 4. Scheme of the pre-mixture atomizing.

The nebulizer has an orifice for gases and a central capillary tube where the solution to be analyzed is introduced. The gas, generally the oxidant, goes at a high speed and causes a pressure decrease in the capillary that pulls the solution (venture effect) and injects it in the nebulizer chamber as a thin aerosol. This kind of nebulizer typically has a flow rate measurement of 3 to 5 ml per minute and a constant and stable flow is obtained in less than 30 s after starting the aspiration.

The mixture, oxidant and fuel gas are homogenized in the nebulizer chamber with the aerosol coming from the nebulizer. It is coated with a resistant material to acid, alkali and various kinds of organic solvent. At the entrance to the chamber, in front of the nebulizer, there is a flow spoiler used to select aerosol droplets: the smaller ones are pulled by gas flow that passes around the spoiler and goes into the burner. This is necessary because small droplets improve the atomization and its homogenized size helps to produce a stable signal. The bigger ones, as they have more mass, are not deviated by the flow of gases, colliding with the spoiler, falling into the back of the chamber and are released by a drain. Approximately 95% of solution is discarded, being one the main restrictions to increase the technique sensitivity.

After passing through the chamber, solution and gases get to the burner. A burner is a metallic piece, generally a titanium based-alloy or another with high chemical and thermal resistance with a slit of 5 to 10 cm length and 1 to 2 mm width. Gases escape through that slit and are incinerated by forming an aligned thin layer of fire with optic axis of the equipment where the analyte atomization and selective radiation absorption emitted by the lamp occurs. The length of the slit defines the optical path, term used in spectrometry to describe the size of the sample cell, this is, the distance covered by radiation at the same means of the absorbent species. In this case, the flame is the sample cell with 5 or 10 cm of optical path, depending on the burner used. The amount of radiation absorbed is proportionally directed to the optical path, thus, the size of the burner influences the sensitivity and the detection limit that can be obtained with this technique.

Figure 5 shows several steps that occur during the atomization process of any element M, initially in solution of ionic M^+. The nebulizer and the chamber of nebulizer are responsible for transferring part of a flask solution to the burner, in continuous flow of very small droplets.

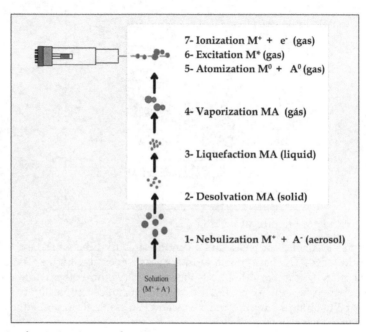

Fig. 5. Scheme of atomization mechanism.

When getting to the flame, the solvent dries fast and a solid MA compound is formed where A is an anion present in the solution. The heat is enough to liquefy, vaporize and break the bond of compound MA, releasing gaseous atomic specie M^0 in the ground state. This species absorbs a specific wavelength radiation emitted by the lamp. If the temperature were high enough, M^0 could be excited to M^* or lose an electron and passed to M^+, but these two species do do absorb the wavelength that is measured by decreasing the analytical signal.

The temperature is the variable that most influences the atomization process. The best way to control this variable is through a critical choice of fuel and oxidants gases, and varying the percentage to a fine adjustment among them. As examples of fuel gases, we can cite propane, hydrogen and acetylene. As examples of oxidant gases, we can cite are air, oxygen and nitrous oxide (Skoog et al., 2006).

Although this is a dynamic process, the solution is pulled with constant speed and, therefore, the population of M^0 species at the light path is also fairly constant over the time, producing a well stable absorption signal. The number of readings and its respective time can be freely adjusted by the operator (generally between 0.5 and 2s), however, it can be obtained quickly a very reliable and reproducible average.

A relative concentration on each species of elements M (M^+, MA, M^o, M^*, etc.) varies according to different factors, as the temperature and nature of the flame (oxidation or reduction), composition and speed of the solution aspiration, M ionizing power, stability of MA species and flame region observed. As only gaseous species M^o produces the desired analytical signal, all variables are optimized and their concentration can reach a maximum value exactly in the flame region aligned of the optical path.

Air-acetylene is one of the most used flames for elements that show fewer tendencies to form thermally stable compounds. Nitrous oxide-acetylene is used because of its high temperature and low and safe speed burn. For safety measures, the burner with a slit of 5 cm is used for nitrous oxide-acetylene. When air-acetylene mix is used, the preference is about 10 cm. Other compositions are rarelly used and only for very specific analyses.

Pneumatic atomizing systems for FAAS are simple, robust and reliable demanding little maintenance, however their technique sensitivity is limited because they discard most of the solution away and also for the little time of resistance of the absorption species in the optical path. This was noticeable since the first beginning of its development and, along the last five decades, several authors proposed alternative ways of atomization. Some of them were developed until the commercialization phase, initially, as basic equipment accessories, achieving then, the status of alternative analyses techniques.

3.3.6 Other kinds of atomizers

Using a same spectrometer, three more basic atomizers can be used to produce ground state atoms, which are used for specific techniques. These atomizers were adapted because of its limitation to flame AA sensitivity. Despite FAAS is a rapid and precise method of analysis, the need for trace metal analyses at μg L^{-1} level calls for a more sensitive technique. By examining the flame AA process, a number of areas limit the sensitivity of the technique: the concentration of analyte, nebulizer for flow-low rate liquid sample, and the major sample loss during the nebulizer process until reaching a burner, little time of the sample in the flame and, finally, chemical interferences occurred during the atomization process.

With exception of analyte concentration, these limitations occur practically with a large amount of atoms in an optical one-way for a longer period of time. This is the basic principle of three different atomization processes that, noticeably, receive the following analytical techniques denomination: Hydride Generation, Cold Vapor Mercury and Graphite Furnace Atomic Absorption.

Hydride Generation Technique – First, elements forming volatile hydrides (As, Pb, Sn, Bi, Sb, Te, Ge and Se) can react with reducing agents in mid-acid, generally with sodium borohydride. The gaseous hydrides are transported by the stripping argon flow directly to a cell on the central burner head (air/acetylene) and perfectly adjusted in the optical path of the atomic absorption spectrometer. In either case, they hydride gas is dissociated in the heated cell into free atoms, occurring the absorption process. The maximum absorption reading, or peak height, or the integrated peak area is taken as the analytical signal. For these elements, detection limits below μg L^{-1} range are achievable. The great advantage of this technique is the separation of specific elements as hydrides (pre-concentration stage), eliminating the interference from the matrix. Figure 6 shows a schematic hydride generator being atomized into the cell by the flame.

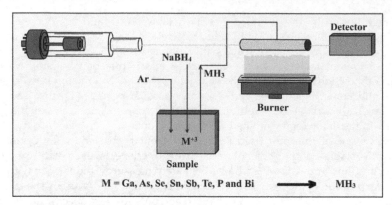

Fig. 6. Scheme of the hydride generation technique.

The major limitation to the hydride generation technique is that it is restricted primarily to the elements listed above, and results depend heavily on the valence state of the analyte, reaction time, gas pressures, acid concentration, and cell temperature. The formation of the analyze hydrides is also suppressed by a number of common matrix components, leaving the technique subject to chemical interference.

Cold Vapor Mercury Technique – This technique is only used to determine mercury due to its chemical characteristic reduced to atomic state by a sample reaction with a strong reducing agent as tin chloride or sodium borohydride in a closed reaction system. Similar to the previous system, the volatile free mercury is then driven from the reaction flask by air or argon bubbling through the solution, being carried in the light path of the AA spectrometer. Also, all of the mercury in the sample solution placed in the reaction flask is chemically atomized and transported to the sample cell for measuring the absorbance. The advantage of this technique is that sensibility can be increased by simply addition of the amount of the sample. The detection limit for mercury by this technique is approximately 0.02 μg L^{-1}.

Graphite Furnace Technique - So far the most advanced and widely used high sensitivity sampling technique for atomic absorption is the graphite furnace. In this technique, a tube of graphite is placed inside a furnace with rigorous control of temperature, where it can be cooled by current inert gas. The furnace is aligned to permit the passage of the lamp light beam. The importance of this technique requires a more detailed discussion in the following items.

3.3.7 Monochromator
Lamps used in atomic absorption, as previously mentioned, emit discrete spectral lines. The number of lines depends on different factors as the number of possible electronic transitions from the emitter element, applied current, presence of other emitter elements, and others. Besides that, the atomizing system is a powerful radiation emitter.

The mathematical description to relate radiation absorption with the concentration of analyte in the solution (Beer-Lambert's Law) is based on the measurements with monochromatic radiation, this means, a unique wavelength. On the other hand, the detectors are sensitive to photons and all the range of ultraviolet and visible radiation. Though, it is necessary to have a device between the source and the detector to select and isolate the appropriate wavelength.

The chosen wavelength is emitted and absorbed with greater intensity by the element and is related to the most probable electronic transition for the atom in ground state. The more transition probability, more frequently it will occur, so , more photons from the wavelength will be emitted. That wavelength or spectral line is known as resonance line. Due to its intensity, it provides a better signal-to-noise relation and the best sensitivity for the analysis. The monochromator is responsible for selecting the appropriate wavelength for analysis. It is a hermetical closed box with entrance and exit slits of 0.2 to 2nm, lenses and mirrors to focus the radiation and a dispersed element that can be a prism or a diffraction grating or a combination of them. It is filled up with inert gas to avoid ultraviolet radiation absorption by the air and normally has the walls painted in matte black to avoid reflection and scattered radiation. The polychromatic radiation enters through the entrance slit, it is separated and only the chosen wavelength of radiation reaches the exit slit and goes on to the detector.

In practice, the monochromators are not able to isolate a unique wavelength. So, a wavelengths range goes to the detector. The bandwidth depends on the resolution, which is the ability to distinguish between two near wavelengths. The more sophisticated an expensive the monochromator is, the better its resolution is. Most of the lamps used emit simple spectra and the monochromator can have a low resolution, around 0.1nm.

Figure 7 shows a design of a Czerny-Turner monochromator used in atomic absorption. In this project, the radiation enters through the entrance slit and is focused in the diffraction grating (dispersant element) by a spherical mirror. The grating is set up in a gyratory support that allows determining the angles of incidence and reflection and, after separated, the chosen wavelength is focused in the exit slit by another spherical mirror until reaching the detector. This is known as sequential monochromator, because it is possible to determine the selection of any wavelength by only revolving the diffraction grating.

In this case, the dispersed element is a diffraction grating. The grating is made on a substratum glass or quartz struck with grooves, cut with very precise angles and covered with a reflexive surface. When reaching this surface, the radiation is reflected at an angle that depends upon the wavelength of the incident ray. Different wavelengths are reflected at different angles, occurring dispersion of radiation. The more the number of grooves, the bigger the resolution of grating and separation are. Although this is not unique, the grating resolution is one of the main factors affecting the monochromators' performance. Monochromators used in atomic absorption equipments generally have grating up to 1000 slots per millimeter.

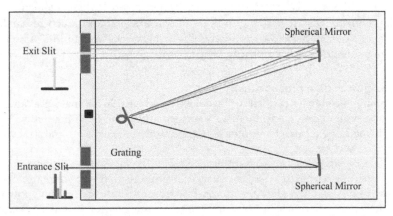

Fig. 7. Diagram of a discrete light diffraction in a Czerny-Turner monochromator.

Echelle monochromators, use a diffraction grating and a prism as dispersing elements that spread the spectrum in two dimensions. In the last decade, this kind of monochromator has been mainly chosen because they are well adapted to solid-state detectors are available today. These monochromators are called simultaneous or multi-channels and the measured wavelengths depend on the relative positions of dispersing elements and detector. In this case, it is not possible to determine any wavelengths for the determination, but only the ones that are pre-determined when the equipment is built. In the practice, this is not a very serious limitation, because the solid-state detectors are formed by millions of pixels and each one of them can work as an individual detector. This way, there are hundreds of available lines to be determined.

One of the advantages of these monochromators is that they do not have mobile parts that avoid mechanical defects such as misalignments and looseness occurring with sequential types. Besides that, they permit to determine several elements simultaneously. It does not have much utility when used with lines emission lamps, but there are equipments using continuous sources associated to monochromators of high resolution, which permit a simultaneous determination of over 60 elements in few seconds. These equipments can solve one of the main limitations of the technique with enough time to determine several elements in one same sample.

3.3.8 Detector

The detector is placed in front of the exit slit and receives the determined photons by the monochromator. It transforms lightning energy in an electrical signal, that is amplified and measured. Detectors mostly used in old equipments are photo-multiplier valves. With modern equipments, they are being substituted by a system based on solid-state detectors type CCD (charge coupled device). They are similar to digital cameras sensors with a set of sensitive pixels to electromagnetic radiation in the UV regions and visible that transforms the energy of photons coming from the monochromator into a digital signal.

Modern CCDs have from 1 to 50 millions of pixels (megapixels) working each one of them as individual detectors. When coupled to an Echelle monochromator of high resolution and to a continuous radiation source, they can transform the atomic absorption spectrometry in a simultaneous technique of analysis. This is a technological advance of great importance and, in practice, this means to pass the determination of one to tens elements every five minutes. Equipments with that configuration are still unusual and much research is being performed about this matter, however, many researchers believe than in a close future this will turn into a standard configuration.

3.3.9 Acquisition of data processing

In the beginning, the signal obtained in the detector was read in an analogical amperemeter. The value was written down and then, all the calculation was done manually. Nowadays, everything is made by a microcomputer equipped with a specific program for operational control and data processing.

Each manufacturer develops a program according to the characteristics and implemented accessories in the spectrometry, so that everything can be done by software, e.g., how to turn on, how to pre-heat and align the lamps, how to control the gases flow, how to change and align the atomization system, how to program and control automatic samplers and graphite furnace, among others.

In the acquisition and data processing part, the number of repetitions and the time of reading integration, number of samples and number and concentration of the analytical curve solutions, etc. can be programmed. Once the readings are obtained, the program draws the analytical curve, performs interpolation values of absorption and calculates the concentration of the sample solutions. If the mass of a solid sample and the dilutions done after solubilization are provided, the program will directly calculate the concentration of the analyte in the solid sample.

4. Sample preparation

Atomic absorption analyses are most commonly performed with standards and samples in solutions. Before introduction into the equipment, the solid samples need to be solubilized in a process generically known as sample preparation. This is a very slow process in the development of an analytical methodology. Considering all analytical operations, this is the most critical and most time-consuming. This analysis might present a large possibility of errors, which may result in a more expensive cost (Kingston & Jassie, 1988).

There are two large categories of analytical matrices: organic and inorganic samples. The inorganic samples incorporate soil, inorganic substance, sediment, clay and metal. The organic samples or carbon-based includes biological, polymeric, petrochemical and pharmaceutical samples. Organic samples can be solubilized in a dry or wet basis, and in a open or closed systems. These methods guarantee the elimination of organic material and the quantitative recovery of analyte in solution (Kingston & Jassie, 1988; Oliveira, 2003).

Decomposition methods via dry ashing consist of burning the sample at elevated temperatures (450-550°C), O_2 atmosphere to atmosphere pressure. This method allows to calcinate big amounts of samples and to dissolve the ash in a small volume of acid facilitating the sample analysis in case the element of interest comes in small quantity. Some problems may result in the loss of volatile elements due to high temperatures, reaction of the analyte with the crucible material or difficulty with the dissolution of the residue (Hoenig & Kersabiec, 1996; Watson, 1994).

Wet digestion in open systems consists of the oxidation by an acid or by a mixture of acids aided by heating. This is a technique that requires hard work and can evolve a great amount of toxic products (nitrogen oxides) (Hoenig & Kersabiec, 1996), it also increases the risk of constituent loss and contamination. These problems could be minimized if they were performed in closed places, digestion pumps of PTFE (Teflon), and heated by conduction or convection. In this case, the increase of temperature and pressure increases the speed of the reaction; however, it would take time-consuming or long hours for the digestion/cooling (Figueiredo et al., 1985). The use of digestion procedure in microwave systems is an alternative. Besides reducing the time of decomposition, it avoids contaminating the sample with maximal efficiency of solubilization (Kingston & Jassie, 1988). The microwave has been widely used in the solubilization of samples with high organic content (Binstock, 2000; Costa et al., 2001; Pouzar et al., 2001; Wasilewsk et al., 2002).

Some kinds of inorganic samples take much time to be solubilized via wet digestion methods, even using closed systems in microwaves-assisted. As for example, samples of silicate cement, refractory oxides, minerals and others. Here, a fusion method can be used. It consists of a fusion and reaction of the sample with high-temperature fluxes (300 to 1000°C), followed by the solubility product formed by an acid or a mixture of acids (Oliveira, 2003). Some common fondants used are lithium metaborate, sodium carbonate and potassium nitrate, among others. Additional quantity of fondants varies between 5 to 50 times of the sample mass and the high-temperature reaction transforms insoluble compounds in products that can be easily solubilized. As a reference example it can be mentioned silica (SiO_2, insoluble in acids except for HF) with sodium carbonate, forming soluble sodium silicate (Na_2SiO_3). The disadvantage of this method is the time and sample manipulation for its performance and the possible contamination of the samples by impurity fondants. This factor is critical for trace and ultra trace analyses.

After solubilization, other procedures are required for adequate samples and conditions of analysis as the separation of interferences or pre-concentration of elements of interest. The ideal solubilization method should be the simplest to get to dissolve the sample completely and, the most important, the final solution should contain all analytes of interest.

5. Interferences

Atomic absorption is a very specific technique with a few interferences. Such interferences can be classified into six categories: chemical interference, ionization interference, matrix interference, emission interference, spectral interference, background interference. Since they are very precise, there are available methods that can overcome most of the problems.

5.1 Chemical interference

If the sample in the flame can produce a thermical stable component of the analyte to be analyzed and if it does not dissociate completely in the flame, the population of analyte atoms that can absorb will be reduced. This will decrease the sensitivity analysis. This problem is normally solved in two ways: with temperature higher than the flame (where the compound is unstable and can be broken down), or adding a sequestering agent. This sequestering agent is a cation that competes with analyte of interest by formation of the stable compound. If the sequestering agent has a higher concentration, the analyte of interest is released to form the atom. A good example of a sequestering agent is the addition of lanthanum to determine Ca in samples that contain phosphate. A high concentration of lanthanum, which also forms a thermally stable compound with phosphate, releases calcium to the atomic state and allows it to be detected with good sensitivity, even when air-acetylene is used as a mix to produce flame. In case of using nitrous oxide-acetylene, this technique is not necessary since the flame is hotter and can break down the calcium phosphate.

5.2 Ionization interference

Ionization interference occurs when the flame temperature is too high for the analyte and, therefore, has enough energy to take it beyond the neutral atomic state and to produce ions. This also decreases the population of atoms that can absorb. A strategy to solve this problem is by adding ionization suppression to the sample, standards and white. The most common ones are K, Na, Rb and Cs, alkali metals that are easily ionizable and increase the electrons population in the flame, minimizing the formation of other cations. The analyte in the sample will not be ionized so easily. Another strategic is by working with cooler flames, which reduce ionization effects.

5.3 Matrix interference

Matrix interferences may increase or decrease the signal. They occur because the physicochemical characteristics of the sample matrix (viscosity, burn speed, surface tension) may differ substantially from the calibration curve standards. This may occur because of the difference on concentration of dissolved salts, acids and bases, because of the use of different solvents, or when the patterns and sample temperatures are very different. When it is not possible to reproduce the matrix or there are no standardized samples, it is recommended to use the matrix spike test.

5.4 Emission interference
If the emission intensity of the interest analyte is high, the analyte absorption may decrease due to its reduction of the ground state atoms population. The modulation of the lamp signal eliminates the problem of the detector to measure the emission signal mixed with signal coming from the lamp. The use of cooler flames may reduce the absorption.

5.5 Spectral interference
There are rare cases in which two elements present in the sample absorb the same spectral line. If the problem is detected, the simplest solution is to change the spectral line. It is very important to obtain previous information about the elements that may interfere in the line used and whether they are likely to be in the sample so that it can be confirmed and solved.

5.6 Background interference
This problem cannot be corrected by the matrix spike test and it occurs due to the light distribution caused by particles present in the flame, or for the light absorption caused by molecular fragments of materials coming from the sample matrix. To solve this problem, the background absorption must be determined and subtracted from the total. This determination is obtained since the analyte absorbs in a straight line and the background forms a wide absorption band. Therefore, an accessory lamp, as a deuterium lamp, that emits a continuous spectrum can be used, either the sample gets the light from the hollow cathode lamp or from the deuterium lamp. As the analyte absorbs very little from the continuous spectrum, the continuous absorption is all from the interference and can be subtracted from the total absorption measured by the hollow cathode lamp.

6. Quantitative analysis in absorption (flame atomization)

Quantitative analysis, as mentioned before, is possible due to the relation of the absorbance signal by the concentration of the analyte (Lambert-Beer Law). Basically, the atomic absorption spectrometry analysis consists of comparing absorbance signals of a group from patterns and samples.

In order to have a good instrumental performance, that is, to work with maximum analytical sensitivity, it is imperative the evaluation of some of the instrumental parameters. Every part of an atomic absorption equipment needs to be adjusted in order to have an optimization. Basically, these parameters according to figures of merit in AAS are: selection of the wavelength in relation to the concentration of analyte, adjustment of the lamp current, adjustment of oxidant and reducing gases flow, adjustment of the height of the burner and nebulizer pressure, due to the distance of the needle to the nebulizer chamber, selection of the monochromator exit slit and adjustment of the background effect, if necessary. In modern equipments, most of the adjustments are automatic.

Once adjusted all parameters, the determination starts reading the absorbance of the analytical blank, followed by the standards with increasing concentrations. Considering the obtained values, an analytical curve is constructed. If the correlation for this curve is significant, it is possible to start the aspiration of the sample solutions and to read the respectives absorbance values. By interpolation in the analytical curve, it is obtained the sample solution analyte concentration value. The concentration value of the original sample is calculated considering the amounts of the samples (mass or aliquot) and respective dilutions.

7. Introduction to graphite furnace atomic absorption

Differently from other spectrometric techniques, electrothermal atomization can get a total atomization of the sample inside the graphite tube, the signal observed depends on analyze mass rather than concentration, the term "characteristic mass (m_0)" is used as a measure of the sensitivity of the furnace. Characteristic mass is analogous to characteristic concentration for flame AA except that mass, rather than concentration, is related to absorbance. The characteristic mass of an analyte is defined as the mass of analyte in picograms required to produce a peak height signal of 0.0044 absorbance or an integrated peak area signal of 0.0044 absorbance-seconds (A.s), which corresponds to only 1% of the total light absorbed (log 100/99). This way, the detection limits in graphite furnace are usually defined in mass units (pg) instead of concentration units.

A basic graphite furnace atomizer is comprised of the following components: graphite tube, electrical contacts, enclosed water cooled housing and inert purge gas controls. Figure 8 shows the basic constituents of graphite furnace atomizer.

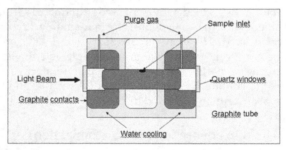

Fig. 8. Scheme of the longitudinally-heated graphite furnace atomizer.

The tube is held in place between two graphite contact cylinders, which provide electrical connection. An electrical potential applied to the contacts causes current to flow through the tube, leading to the heating of the tube and the sample. All this system is placed and aligned with the light beam.

Basically, GFAAS methodology consists of 5 steps that, evidently, can be in greater number according to the complexity of the sample. First, the sample (5 to 50 μL) is introduced by an automatic injector, and the analytical programming begins. The temperature of the furnace is slowly increased up to the temperature in which the solvent vaporizes, remaining at this temperature for a certain time. The elapsed time on each analytical phase is crucial, since loss of the analyte by evaporation may occur, or the physicochemical process of the sample may be incomplete, especially in this drying phase, in which the amount of volatile constituents present in the samples is bigger. After the drying phase, the temperature is gradually increased until it reaches the pyrolysis value (Ash), when fusion and evaporation of the more volatile constituents in relation to the volatilization temperature of the analyte of interest will occur, remaining for a while to be effective. Immediately and quickly, the temperature of the furnace reaches the temperature of the analyte atomization, differently from others in this phase, the inert gas, usually argon, is interrupted so that the maximum interaction of the atomic cloud with the electromagnetic lamp may occur. This also helps to maximize sample residence time in the tube and increase the measurement absorbance signal. Finally, the cleaning phase (Clean Out) occurs, by using higher temperatures in order to assure there is no remaining residual causing memory effect, which interferes with future atomizations. After all these steps, the system cooling takes place (Cool Down). Besides helping cooling the system and not interfering with the chemical properties of the sample inside the tube during the analysis phase, the inert gas is used to minimize graphite oxidation of the tube and furnace.

The figure 9 shows a typical programming in graphite furnace atomization, the characteristic effect inside the tube (lateral side) and the absorption measurement profile.

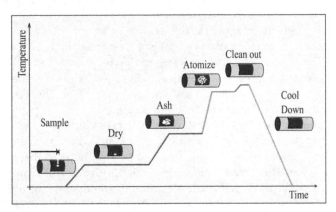

Fig. 9. Furnace thermal stages.

A significant characteristic of the graphite furnace is its high sensitivity. This technique has been implemented as an analytical routine in several kinds of samples, especially the ones related to environmental. Also, with the possibility of controlling the atmosphere and temperature inside the graphite tube, complex samples can be modified physically and chemically (the later with the addition of the so called chemical modifiers) during the analysis phases, enabling, therefore, a "cleaner" sample during the atomization phase and minimizing matrix interference. Due to advancements in technology and scientific researches, new materials and new modifiers have enabled better sensitivity and the increase of chemical elements can be quantified by this technique. They have made it possible to analyze very complex sample matrices, such as those frequently found in biological and geological samples. The microliter sample sizes used, offer additional benefits in which the amount of sample available for analysis is limited, as in many clinical analysis.

Opposite to the significant sensitivity, the signal handling requirements for graphite furnace atomic absorption are much more demanding than those for flame atomic absorption. Graphite furnace signals change rapidly, background levels are higher, and signal interpretation needs are more extensive. Only with the development of an analysis protocol known as the Stabilized Temperature Platform Furnace (STPF), the GFAAS implementation has been intensified due to the analytical efficiency acquired. This way, combining the stability in the analyte atomization provided by background correction (Zeeman) and the experimental conditions that minimize graphite furnace interferences, this development is mainly financed by the manufacturers of AA equipment and a large number of scientific work have been published.

These factors (STPF) contribute significantly to precision and detection limit performance: the L'vov platform (pyrolytic coating), whose function is to delay atomization and longitudinal heating until the furnace tube has reached steady state conditions; fast electronics to accurately follow the changing signal absorbance with symmetrical profiles; use of integrated peak absorbance signals to accurately quantitative the analyte regardless of changes in peak shape caused by the matrix; matrix modifiers to convert the analyte to less volatile compounds in the step atomization and which are atomized only after the furnace tube has reached a steady state temperature or to make the matrix components more volatile that are eliminated in the ash step, and maximum power heating to bring the furnace to a steady state temperature before the analyte is atomized.

The Zeeman Effect with the magnet on the analyte is an ideal performance to perform background correction for graphite furnace analyses. Correction occurs at precisely the same wavelength used for analysis, thus correcting properly even when the background is structured. The Zeeman Effect uses the identical light source with the same electrical source conditions to produce both the analytical and background signals. This corrects at the same time for source or detector drift and for non-atomic absorption.

In fact, with the condition of STPF applications, the diverse kinds and powerful background corrections and high sensitivity make electrothermal atomization be one of the most used among the atomic absorption techniques. Allied to other separation techniques (chromatography), continuous measurement systems (CMS) and solid sample, this latter in a vertiginous application in diverse matrices, GFAAS stands out not only in the use of analytical routine for diverse industrial segments, but also in scientific aspects and they will be referred in the application item of atomic absorption spectrometry.

7.1 Background correction

Radiation from lamp is attenuated by non-atomic source: molecular species, solid particles, absorption and scatter. The signal is added to atomic signal results in falsely high signal. This problem is more severe in graphite furnace than in FAAS.

Small analyte signals should not be measured in the presence of high background since it reduces energy at detector according to the high absorbance, reduces signal-to-noise-ratio and degraded precision and accuracy. Several techniques allow subtraction of background from total absorbance measurement.

The measurement of signal absorption in a determination by AAS is given according to the following sequence: first, the detector receives the signal: atomic + non-specific after the atomization; then, the same detector receives the signal only from the background correction: non-specific only; finally, measurements are timely separated in a few milliseconds, atomic absorption calculated: total absorbance - background absorbance = atomic absorbance.

Common background correction techniques are Deuterium for the atomization flame, Zeeman for the atomization graphite furnace and the least common Smith-Hieftje setup.

8. Technological innovations

8.1 Tungsten filament

Studies about the use of tungsten filament, tantalus and other metals date back from 1960. However, Berndt & Schaldach (1988) presented and developed an open system to work with tungsten filaments. Results obtained as reference solutions demonstrated the power of the tungsten filament as an atomizer for Ba, Cd, Co, Cr, Eu, Mn, Pb, Sn and V determination.

The scientific community, therefore, got more interested in this kind of study where tungsten filaments were used as atomizers with different spectrometry techniques. Nowadays, mechanisms to introduce samples in ICP OES and ICP-MS based on electrothermal heat used in graphite and filaments furnace are commercialized. These mechanisms have permitted, besides the liquid sample, solid sampling and slurries samples. However, the literature shows little work related to the use of tapes, tubes or tungsten filaments that can also help the vaporization for sample introduction in different spectrometric techniques.

8.2 Flame tubular furnace and thermal aerosol

The use of thermal aerosol in the sample introduction of the tube atomizer (TS-FFAAS, or thermo spray flame furnace atomic absorption spectrometry) in flame (air-acetylene) was initially proposed by Gaspar & Berndt (2000). The system consists of an auto sampler, a peristaltic pump, a ceramic capilar where the sample is introduced into a flame-heated nickel tube (Figure 10).

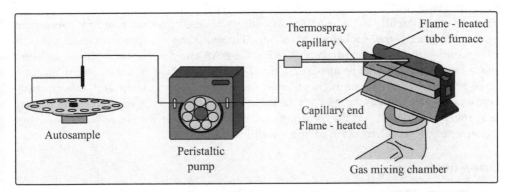

Fig. 10. Representation of a sample introduction system (From: Gáspar & Berndt, 2000).

The complete introduction of the sample in the atomizer tube as well as the increase of the residence time of the atomic cloud in the optical path (137ms) promote the improvement of the analytical signal to various volatile metals as: Ag, As, Au, Bi, Cd, Cu, Hg, In, K, Pb, Pd, Rb, Sb, Se, Te, Tl and Zn. Consequently, a considerable gain-sensitivity is obtained when compared to FAAS. The detection gain varies according to the element, being able to oscillate by a factor of three (Cu, of 14 µg L⁻¹ in FAAS for 4.2 µg L⁻¹ in TS-FF-FAAS) or, as observed for Cd, in a factor of 110 (Cd, of 34 µg L⁻¹ in FAAS for 0.31 µg L⁻¹ in TS-FF-FAAS). The aerosol formation is based on Leidenfrost effect, in which the heated surface of the capillary generates an increasing temperature and a steam layer in the capillary wall that helps with the drop and pulverization. This makes that the steam phase of the sample be superior to the liquid phase present in the tube. Pereira-Filho et al., (2002) published the first work involving analytical application of TS-FF-AAS for the determination of Cd, Cu and Pb in biological samples suspensions. Perspective techniques have found direct injection of slurries as a good alternative to contour problems with the sample preparation, being chemical analysis the most difficult ones. Another possibility aims for synergistic effects with multi-elementary measurements, high sensitivity and selectivity of a high resolution continuum source atomic absorption, because there are no reports applying metallic atomizers with this technique.

8.3 High resolution continuous source atomic

Although the first atomic absorption spectrometry developed in the 19th century used a continuous source. Only at the end of the 20th century, it was possible to use atomic absorption spectrometry with continuum source. Improving the background corrections and diffraction patterns allied to studies and development proposed by the group of Harnly and Becker-Ross, the necessary technological maturity resulted in the atomic absorption spectrometry of high resolution and continuum source (HR-CS-AAS). This equipment is composed by Xenon arc lamp, ellipsoidal mirror, atomizer (flame or graphite furnace); entrance slit, paraboloid mirrors, prism, adjusted intermediate slit, echelle and CCD detector. This equipment contains a small Xenon arc lamp operating at 15 bar high pressure in a hot spot mode with a continuous emission monitoring system between 190 to 850 nm allowing a multi-elementary analytical determination.

Relating to measurements and background correction, the high resolution of the monochromator and CCD detector permit to identify the origin of atoms or molecular and eliminate them with an adequate procedure. Besides that, the atomic absorption measurement and the background are strictly simultaneous, permitting to visualize background signals. Some advantages of HR-CS-AAS and the use on a unique radiation source for all elements can be mentioned, multi-elementary determination, improvement of signal-noise-ratio due to the high emission intensity of Xenon lamp and, consequently, increase of the calibration range with variation until 5 orders of magnitude, simultaneous background correction in the proximity of the analytical line and better analytical performance of trace elements in complex samples.

9. New tendencies

In atomic absorption spectrometry or AAS, both with the classical flame AAS and with furnace AAS innovation took place. Remarkable efforts, however, were made to use all types of methods allowing volatile species generation with metals. Flame atomic absorption spectrometry (FAAS) is a mature analytical method, which is present in almost any analytical laboratory as a working horse for elemental determinations of metals. Innovation, however, is still going on with respect to the introduction of the sample into the atomizer and the increase of the analyte sampling efficiencies and residence times in the atomizer. By the so-called furnace in flame approach for volatile elements or volatile species forming elements especially, the sampling efficiency can be considerably increased as compared to conventional pneumatic nebulization with the improvement of limit of detection for lots of elements. A further technique for sample introduction which remains to attract the interest is thermospray. The use of cloud point extraction for isolating and pre-enriching heavy metals from complex samples is another strategy for sample preparation and interesting for environmental sample application (Bings et al., 2010).

Addition to the elements that form volatile hydrides, also, a number of metals could be employed for analytical purpose. During the last years much research was done to widen the circle of metals which could be determined in this way and to know more about the volatile species generation mechanisms. Room temperature ionic liquids further were found to enhance the chemical vapor generation of transitions elements following reduction in acidified aqueous solution with KBH_4.

In relation to sample introduction by electrothermal systems the background correction further remains a topic of methodological development in GFAAS and SIMAAS (simultaneous atomic absorption spectrometry with electrothermal atomization). In the case of diode laser electrothermal atomization, both Zeeman and wavelength modulated atomic absorption could be shown to be valuable, and the detection limits obtained were shown to be at the same level as in coherent forward scattering in the case of crossed polarizers . With diode lasers in GFAAS, also, isotopic dilution could be applied for calibration, as Doppler-free measurements are possible. Apart from graphite furnace, also, tungsten furnaces remain a topic of investigation. Though they do not have the advantages of graphite furnaces in the sense of a reduction of the analytes from the oxides to the elemental form nor do they have the precision as a result of a good sample uptake, they have the advantage of being cheaper and without risks for stable carbide formation.

Furnace atomic absorption spectrometry retained its original form stemming from L'vov and Massmann's work, however, in every part of the system, considerable innovation occurred. In graphite furnace atomic absorption spectrometry (GFAAS), special attention has been paid to the thermochemical processes responsible for the atom cloud formation, as they are of crucial importance for the signal magnitude and form and, accordingly, for the accuracy and precision achievable. The stabilized temperature platform principle by a number of groups has been shown to be very powerful to eliminate a number of volatilization interferences in GFAAS. Similarly, lots of work have been published with the surface of graphite tube modified, mainly covered with refractory carbides that, besides improving atomization conditions, increases considerable the life cycle of the tube. Works have been published for the formation of refractory carbides and tubes and graphite platforms by using different methodologies and proving the efficiency through physical analyses (Izário Filho et al., 2001a, 2001b, 2001c). Direct solids sampling can be realized in flame AAS but innovation especially occurs in the case of electrothermal atomization. The less manipulation of sample is the great advantage of solid sampling when compared with sample preparation with dry or wet digestion. The use of the solid sample is associated to determinations of semi-metallic and non-metallic elements into the atomic absorption system, the HR-CS AAS, mainly for determination of phosphorous and sulphur in samples of petroleum, fuel and other derivatives. Welz and Curtius are exponents who developed applications of HR-CS AAS for elementary analysis and molecular analysis.

On the other hand, special problem in direct solids analysis is the calibration. In direct solids GFAAS, it has been proposed to use spiked filter papers for the determination of transitions elements in vegetable samples. For a sampling of powders, the use of slurries is very convenient, e.g., in the direct determination of elements in gel forming konjac samples by enzymatic hydrolysis assisted slurry sampling GFAAS. The direct sampling of powder samples and the injection of slurries under the use of a modifier mixture of Pd and Mg salts, where the solid sampling technique was found to be the most sensitive and easy technique.

10. Technical applications and research

As discussed in this text, Atomic Absorption Spectrometry have been used for the quantitative and qualitative determination of chemical elements on samples such as aqueous solutions, waters, sea-waters, metals and alloys, glass, drugs, food, environmental samples, industrial wastes, biological samples among others. Studies have been conducted using variations of the AAS, as flame (FAAS), hydride generation (HG-AAS) and graphite-furnace or electro-thermal (GFAAS or ETAAS) (Vogel et al., 2002). Quite whole work proposed some differential aspects related to the methodology or to the equipment. In some cases, association with other procedures or techniques, results in a hybrid technique, as occurs when AAS is used together with continuous flow-injection (FIA) or high-performance liquid chromatography (HPLC). Other technical advances are proposed too, as the use of integrated contact furnace (ICF), tungsten coil, or the association of TS-FF-AAS, acronym of thermal spray flame furnace AAS.

To estimate the importance of a subject to the scientific world is a very complex matter. Many criteria may be used, each one with vantages and disadvantages. In order to only illustrate the importance and the evolution of the atomic absorption spectrometry, we did a rapid search in the Science Direct <http://www.sciencedirect.com>, considering "atomic absorption spectrometry" as key-word in TITLE-ABSTR-KEY. Two periods were considered: the whole period (about half-century, since the first article was published) and the most recent one (from 2010 until present data, july 2011). Our rapid search returned 5,637 scientific papers, considering the whole period. A significant fraction of the works (463 articles, or 8.2%) concentrates in the most recent period. Considering graphite furnace atomic absorption 1,953 articles (34,6%) were found for the whole period, and 95 articles (4,9%) was found from 2010 until the present data.

11. References

Ajtony, Z.; Bencs, L.B.; Haraszi, R.; Szigeti, J. & Szoboszlai, N. (2007). Study on the simultaneous determination of some essential and toxic trace elements in honey by multi-element graphite furnace atomic absorption spectrometry. *Talanta*, Vol.71, No. 2 , (February 2007), pp. 683-690.

Berndt, H. & Schaldach, G. (1988). Simple low-cost tungstein-coil for electrothermal atomic-absorption spectrometry. *Journal of Analytical Atomic Spectrometry*, Vol.3, No.5, (August 1988), pp.709-712.

Bings, N.H.; Bogaerts, A. & Broekaert, J.A.C. (2010). Atomic Spectroscopy: A Review. *Analytical Chemistry*, Vol.82, No.12, (June 2010), pp.4653-4681.

Binstock, D.A. (2000). A microwave digestion method for total decomposition of lead and ether metals in paint, soil and dust. *Analytical Letters*, Vol.33, No.15, pp.3397-3406.

Costa. L.M.; Silva, F.V., Gouveia, S.T; Nogueira, A.R. & Nóbrega, J.A. (2001). Focused microwave-Assisted Acid Digestion of oils: na Evaluation of the residual carbon content. *Spectrochimica Acta part B – Atomic Spectroscopy*, Vol.56, No.10, pp.1981-1985.

Feng, S.; Huang, Z. & Li, P. (2011). Solid phase extraction and flame atomic absorption spectrometry for the determination of cobalt. *Asian Journal of Chemistry*, Vol.23, No.4, (April 2011), pp.1573-1576.

Figueiredo, B.R.; Jacinto, A.O.; Reis, B.F.; Zagatto, E.A.G.; Krug, F.J.; Giné, M.F.R.; Araújo, M.C.U., Pereira, N.M. & Bruns, R.E. (1985). *Análise química de rochas por ICP-OES; eliminação automática de interferências*. Editora da Unicamp; Campinas.

Gáspar, A. & Berndt, H. (2000). Thermospray flame furnace atomic absorption spectrometry (TS-FFAAS): A simple method for trace element determination with microsamples in the µg/l concentration range. *Spectrochimica Acta B*, Vol.55, pp.587-597.

Hoenig, M. & Kersabiec, A.M. (1996). Sample preparation steps for analysis by atomic spectroscopy methods: Present status. *Spectrochemistry Acta*, Vol.51-B, pp.1297-1307.

Izário Filho, H.J.; Vernilli Junior, F.; Pinto, D.V.B.S.; Baccan, N. & Sartori, A.F. (2001a). Tratamento para obtenção de TaC em superfície de grafite. Parte I: Imersão em Solução Aquosa de TaF_7^{2-}. *Cerâmica*, Vol.47, n.303, pp.144-148.

Izário Filho, H.J.; Vernilli Junior, F.; Pinto, D.V.B.S.; Baccan, N. & Sartori, A.F. (2001b). Tratamento para obtenção de TaC em superfície de grafite. Parte II: eletrodeposição em meio de fluoretos fundidos. *Cerâmica*, Vol.47, No.303, pp.149-152.

Izário Filho, H.J.; Vernilli Junior, F.; Pinto, D.V.B.S.; Baccan, N.; Sartori, A.F. (2001c). Tratamento para obtenção de TaC em superfície de grafite. Parte III: Imersão em Suspensão Aquosa de Ta e Ta_2O_5. *Cerâmica*, Vol.47, No.303, pp.153-157.

Lodders, K. (2003). Solar system abundances and condensation temperatures of the elements. *The Astrophysical Journal*, Vol.591, pp. 1220-1247.

Oliveira, E. (2003). Sample preparation for atomic spectroscopy: Evolution and future trends. *Journal of Brazilian Chemistry Society*, Vol.14, No.2, pp.174-182.

Pereira-Filho, E.R.; Berndt, H. & Arruda, M.A.Z. (2002). Simultaneous sample digestion and determination of Cd, Cu and Pb in biological samples using thermospray flame furnace atomic absorption spectrometry (TS-FFAAS) with slurry sample introduction. *Journal of Analitical Atomic Spectrometry*, Vol.17, n.10, pp.1308-1315.

Pouzar, M.; Cernohorskyt. & Krejcova, A. (2001). Determination of metals in lubricating by X-ray fluorescence. *Spectrometry*, Vol.54, No.5, pp.829-835.

Skoog, D.A.; West, D.M.; Holler, F.J. & Crouch, S.R. (2006). *Fundamentos de química analítica*, 8th Edition, Thomson, São Paulo.

Vogel, A.R.; Mendham, J.; Denney, R.C.; Barnes, J.D. & Thomas, M. (2002). *Análise química quantitativa*, 6th Edition, Editora LTC. 462p.

Wasilewsk, A.M.; Goessler, W.; Zischka, M.; Maichin, B.; Wasilewska, M.; Goessler, W.; Zischka, M.; Maichin, B. & Knapp, G.G. (2002). Efficiency of Oxidation in Wet digestion procedures and influence from the residual organic carbon content on selected techniques for determination of trace elements. *Journal of Analytical atomic Spectrometry*, Vol.17, No.9, pp.1121-1125.

Watson, C.A. (1994). *Official and standardized methods of analysis*. 3rd Edition. The Royal Society of Chemistry, p 461, 749.

Welz, B.; Mores, S.; Carasek, E.; Vale, M.; Okruss, M.; Becker-Ross, H. (2010). High-resolution continuum source atomic and molecular: A review. *Applied Spectroscopy Reviews*, Vol.45, No.5, pp.327-354.

4

Elemental Profiling: Its Role and Regulations

Ajai Prakash Gupta[1] and Suphla Gupta[2]

[1]Patent Cell Division
[2]Plant Biotechnology Department, Indian Institute of Integrative Medicine,
Jammu-180001, Jammu & Kashmir
India

1. Introduction

The commercial significance of medicinal plants was known since millennia but its popularity has grown remarkably in terms of herbal drugs, herbal cosmetics and nutraceuticals during the last few decades. The primary reason being increased interest of developed countries in herbal medicines for safer and natural health care system. This has intensified research on medicinal plants of developing and third world countries for safer and effective drugs for chronic diseases. In fact, more than 60% of the new anti-cancer drugs approved since 1983 were derived from plants. Hence, countries in Asia, Africa and Latin America see greater scope in earning valuable foreign exchange through export of their plant wealth to the western countries. It has been observed, since the last two decades, the interest in the developed countries for complementary and alternative medicines has increased by 60%. In the US, consumer use of herbal products rose to about 50% in 2004 from 5% in 1991. According to the WHO estimate, the world market for herbal medicines and herbal products is worth US$ 62 billion and would hit US$ 5 trillion by 2050. The market is growing @7% per annum. US is the major market for essential oils and herbal tea. Leading markets for herbal products in Europe are Germany, France, UK and Italy with Germany having the largest herbal extraction industry in Europe.

Since many of the traditional herbs do not conform to the perceptions and norms of the western countries as they correspond to a different system (allopathy) and concept, the mismatch may create problem. The latest impact of these factors is the ban imposed by the Canadian government on 'unapproved' Ayurvedic drugs from India, on the plea of hazardous heavy metal concentration in them. The agency banned products manufactured by herbal giants like Dabur, Zandu, Himalaya and Hamdard, although some of these products (like Dabur's 'Shilajit' which is actually a natural rock extract) are claimed by their manufacturers to be free from metals. The manufacturers are of the view that they would take measures like policy advocacy on this issue. However, the ban is not applicable for those Ayurvedic products which have been authorized for sale in Canada. To prevent recurrence of similar types of incidents, the WHO has issued certain guidelines to promote traditional medicines in its traditional medicine strategy (2004-05). The organization has advocated for regulations and other arrangements (like public awareness program) so as to ensure safe utilization of the traditional medicines. If the advice is implemented in a coordinated manner so as to avoid any kind of mismatch, then the objective of providing safer drugs to the people may be fulfilled.

2. Elements in nature

According to the position of metals in the Periodic Table, the metals are named alkali metals, alkaline earth metals, transition metals, and rare earth metals. Four elements (nitrogen, carbon, hydrogen and oxygen) account for 96% of living matter. About 50 of the known elements occur in measurable concentrations in the living systems. In humans and other mammals, 23 elements have known physiological activities (macro nutrients and micro-nutrients). The macronutrients are sodium, calcium, magnesium, potassium, chlorine, etc., which are required in larger quantities by living organism while microelements are 11 in numbers and are   classified as **"trace elements"** because of their essentiality at very limited quantity in humans (less than 100 mg/day). Out of these 11 trace elements, eight are in the period IV of the Periodic table (manganese (Mn), iron (Fe), cobalt (Co), copper (Cu), vanadium (V), chromium (Cr), zinc (Zn) and molybdenum (Mo) and three are non-metals selenium (Se), fluorine (F) and iodine (I). Transition metals that are trace elements of significance for human physiology are, Cobalt (Co), molybdenum (Mo), chromium (Cr) and vanadium (V). In biological systems, trace elements are mostly present as metalloproteins (bound to proteins), or to smaller molecules, such as phosphates, phytates, polyphenols and other chelating compounds. Most of the metals in metalloproteins are part of enzymatic systems and have structural functions or use the protein to be transported to their target site in the organism (Mokdad et al., 2004). Research has indicated association of cancer, diabetes and cardiovascular diseases with diet which has prompted increased consumption of fiber, fatty acids, phytochemicals, and trace elements (Willett, 2002). The role a metal plays, depends on its chemical structure, as well as on the molecule that is chelating the metal (Halliwell & Gutteridge, 1999). For example, Zn, as with other group XII elements, has no unpaired electrons when in the state Zn^{2+}, preventing its participation in redox reactions but Zn has been recognized to act as an antioxidant by replacing metals that are active in catalyzing free radical reactions, such as Fe (Oteiza et al., 2004; Zago & Oteiza, 2001). In enzymes, the metals participate in catalytic processes in any of the following ways:

1. Constituents of enzyme active sites.
2. Stabilizers of enzyme tertiary or quaternary structure.
3. Associates in forming weak bonding complexes with the substrate.
4. Stabilizing charged transition states.

Based on the increased knowledge of the biological mechanisms ruling life we have made a good progress in increasing the life expectancy. However, this has lead to increased incidence of chronic and degenerative diseases, one of the reasons of which could be increasing amount of toxic substances in our body. To deal with this essentiality/toxicity duality, biological systems have developed the ability to recognize a metal, and deliver it to the target without allowing the metal to participate in toxic reactions (Luk et al., 2003). Proteins are primarily responsible for such recognition and transport of these elements thereby making them safe for body. However increase in the intake of certain nutrients as therapeutics or through food may lead to high concentrations of these elements resulting toxicity in the body.

Trace elements are essential components of biological structures, but at the same time they can be toxic   beyond the concentration needed for their biological functions. The toxicity can be extended to other non-essential elements of very similar atomic characteristics that can mimic the reactivity of a trace element.

The presence of trace elements in foods is often determined by the availability of metals in the soil. Thus, within a geographical region with soils deprived/excess of trace elements, its population is at a risk thereby resulting into trace elements deficiency/toxicity. Unfortunately, in recent years the avalanche of uncontrolled supplementation with trace elements has put some trace elements on the border of toxicity in several populations. Thus, it is a crucial priority to define the requirements for trace elements, based on essentiality and health promotion, and the limits for toxicity. Then it becomes necessary either to supplement the basic food by adding the appropriate trace elements (milk, flour, etc.) or counteract/dilute the element in excess (de Romana et al., 2005; Hurrell et al., 2004). These supplements sometimes becomes necessary in several disease treatments, e.g. anemic conditions in kidney dialysis (Locatelli et al., 2004) and physiological conditions, e.g. extensive blood loss during menstruation (Munro, 2000). There are other factors to consider that can define the requirements for essential elements beyond their presence in foods (Table 1):

Element	Antagonists restricting absorption, utilization or retention	Synergists promoting absorption, utilization or retention
Zinc	Phytate with high calcium intake High iron intake Heterologous milks (infants only) [High zinc status; aging]	Low calcium intake, animal proteins Homologous milk (infants only) [Late pregnancy; lactation] [Low zinc status]
Copper	High iron, high zinc intakes [High copper status] High molybdenum with high sulphur intake	High protein intake [Late pregnancy; lactation?] [Low copper status]
Iodine	Elevated goitrogen intake [Low selenium status]	- -
Selenium	Elevated heavy-metal intake	-
Chromium	Oxalates, high iron intake [High chromium status]	[Low chromium status]
Manganese	High calcium intake (infant formulae) [High manganese status]	- -
Cadmium	High calcium intake	Low iron intake, low calcium intake
Lead	Phytate with high calcium intake	Low iron, low calcium and phosphorus intakes

Table 1. Antagonists restricting and synergists promoting absorption, utilization or retention of trace elements in humans

1.   Interaction among nutrients, e.g. interactions between iron and other metals (Aschner, 2000);

2. The presence of certain compounds in the diet, that can impair metal absorption, e.g. phytates bind Zn, preventing absorption (Greger, 1999; Lestienne et al., 2005);
3. Genetic defects, e.g. Zn absorption is decreased in acrodermatitis enteropathica (Wang et al., 2004);
4. drug–nutrient interactions, e.g. penicillamine used in the treatment of Wilsons disease causes Zn deficiency (Schilsky, 2001).

The important variables that should be considered when the levels of trace elements are increased in the body are the effects genetic and individual differences in the targeted population, life-style, nutra-genetic interactions, and other individual factors that can determine the effects of the nutrient on the disease.

3. Role of trace elements

Metals are non-uniformly distributed in the soil and environment. Several factors like industrialization, traffic density, and indiscriminate use of chemical fertilizers, pesticides and eco-geological conditions play deciding role in their quality and quantity. Out of the thirty five metals associated with us, due to residential or occupational exposure, twenty three of them are toxic metals (Punz & Seighardt, 1993). Metals can be accumulated /absorbed by our body as in exposure to sunrays/X-rays/ beverages/ air/water and food products. Bioaccumulation of these elements beyond the safety limits results in various malfunctions leading to several abnormalities. Based on their biological effects, they can be categorized into two types (i) non essential /toxic metals (Pb, Cd, Hg, As) and essential / beneficial metals which includes microelements like Cu, Zn, Mn and macro elements like P, Ca, K, Mg (Abdul –Wahabet al., 2008). Literature cites several reports advocating tendency of plants to absorb and accumulate heavy metals in their tissues (Bunzl et al., 2001; Yusuf et al., 2002). Heavy metal contamination is especially harmful for infants and pregnant woman as it affects the central nervous system, kidney, gastro-intestinal and reproductive systems and joints (Tong et al., 2000). However, plants harbor not only toxic metals but beneficial metals as well. They play a vital role as structural and functional components of metallo-protein and enzyme in the living cell in low doses. The role of these elements are presented very briefly in table 2.

S No.	Metal	Functions	Dietary sources/ Presence	Potential toxicity	References
1	Iron	Hemoglobin myoglobin, catalase, cytochromes Fe-sulfur enzymes proteins for Fe storage and transport and other Fe-containing or Fe-activated enzymes	Hemoglobin and myoglobin from animals. Cereals, seeds of leguminous plants, fruits, vegetables, and dairy products.	Anemia. Fe poisoning	Fraga & Oteiza, 2002; Zimmermann & Kohrle, 2002
2	Copper	Connective tissue, nerve coverings, and bone. Fe and energy metabolism. Reductant in the enzymes and several oxidases that reduce molecular oxygen	Liver and other organ meats, oysters, nuts, seeds, dark chocolate, and whole grains. Transported in the organism by cerulo plasmin. Soluble forms of copper are absorbed from the intestine (40-60%)	Normocytic, hypochromic anemia, leucopenia neuropenia, and inclusive osteoporosis in children. Liver damage	Underwood, 1977; Mason, 1979; King et al.,1978; Hoadley & Cousins, 1988

S No.	Metal	Functions	Dietary sources/ Presence	Potential toxicity	References
3	Zinc	Supports normal growth and development in pregnancy, childhood, and adolescence. Zn is involved in Zn-fingers and activity of about 100 enzymes.	Red meat and poultry, beans, nuts, seafood (oysters are extremely rich in Zn), whole grains, fortified breakfast cereals, and dairy products. Zn is mainly transported by cerulo plasmin. Small intestine is the site of maximum absorption.	Common in underdeveloped countries leading to malnutrition, affecting the immune system, wound healing, impairing DNA synthesis.	Baer et al., 1984; Hess et al., 1977
4	Selenium	Selenoproteins, glutathione peroxidase, thioredoxins,	Grains, cereals, red meats, nuts and seafood. Very efficiently absorbed by humans.	Weakened immune system, nutritional, biochemical or infectious stresses, gastrointestinal upsets, hair loss, white blotchy nails, garlic breath odor, fatigue.	Combs et al., 1986; Levander et al., 1987
6	Molyb-denum	Electron transfer agent in enzymes such as oxidase and sulphite reductase xanthine dehydrogenase/oxidase, aldehyde oxidase and sulfite oxidase share a common cofactor, molybdo -pterin, a substituted pterin to which molybdenum is bound by two sulfur atoms.	Conversion of tissue purines to uric acid.	Xanthinuria, low xanthine dehydrogenase activities High intake leads to molybdenosis spontaneous fractures, and mandibular exostoses osteogenesis.	Rajagopalan, 1984; Ostrom et al., 1961; WHO, 1973
7	Chro-mium	Potentiates insulin action and thus influences carbohydrate, lipid and protein metabolism. Affects the ability of the insulin receptor to interact with insulin.	Processed meats, whole grain products, pulses and spices are the best sources of chromium blood plasma.	Include impaired growth, elevated serum cholesterol and triglycerides, increased incidence of aortic plaques, cornea lesions and decreased fertility and sperm count.	Anderson, 1988; Okada el al., 1981; Tuman & Doisy, 1977; Offenbacher et al., 1986
8	Manga-nese	Activator and a constituent of several enzymes	Unrefined cereals, nuts, leafy vegetables and tea will be high in manganese. Manganese absorption is independent both of body manganese status and of dietary manganese content.	Impaired growth, skeletal abnormalities, disturbed reproductive function, ataxia of the newborn, defects in lipid and carbohydrate metabolism, impaired iron metabolism and altered brain function	Hurley, 1987; Weigand et al., 1986

S No.	Metal	Functions	Dietary sources/ Presence	Potential toxicity	References
9	Nickel	Typical nickel-containing enzymes found in plants and microorganisms, namely urease, hydrogenase, methyl coenzyme M reductase and carbon-monoxide dehydrogenase	foodstuffs and simple substances, including milk, coffee, tea, orange juice, ascorbic acid and ethylenediaminetetra- acetate depress this high absorption	No report for deficiency of nickel	Thauer, 1985; Hausinger, 1987; Walsh & Orme- Johnson, 1987; Nielsen, 1982, 1984
10.	Boron	Steroid hormone metabolism	Fruits, leafy vegetables, nuts and legumes are rich sources. Wine, cider and beer are also high in boron. Boron is distributed throughout the tissues, organs and bones of animals and humans.	Induce secondary hyperpara- thyroidism affects steroid hormone metabolism in humans and animals.	Nielsen et al., 1982, Hunt, 1981, Nielsen , 1988
11.	Vana- dium	Regulation of Na+/K+- exchanging ATPase, phosphoryl-transfer enzymes, adenylate cyclase and protein kinases. enzyme cofactor, and in hormone, glucose, lipid, bone and tooth metabolism	Whole grains, seafood, meats and dairy products. Spinach, parsley, mushrooms and oysters.	no significant report of deficiency of vanadium is available	Byrne & Kosta 1978; Nielsen, 1982; Bennet, 1984
12.	Cadmium	Cadmium metallothionein	Rock, soil, mining	kidney and possibly the skeleton.	Punz & Seighardt, 1993
13.	Lead	Pesticides, Industrial waste	Pesticides, geochemical mineralization and industrial waste. --	nervous system of infants and children is particularly sensitive to lead toxicity.	National research council, 1989
14.	Arsenic	Soil, food and water	Soil, food and water --	increased incidence of keratinization and pigmentation of the skin, together with an increased risk of skin cancer.	Kovalskij, 1977; National Academy of Sciences, 1980
15.	Mercury	Pesticides, food, water and soil	Fishes, pesticides, food, water and soil --	Kidneys central nervous system brain	Bennet, 1984; WHO, 1989, 1976, 1990

Table 2. Functions, Dietary source, Biochemistry and Potential toxicity of some of the metals.

In general, all medicinal plants/herbs/spices/food/water/soil and its products (for human and animals use) must meet regulatory guidelines for quality, safety and efficacy. Monitoring of these metals, using advanced techniques, in the plant is important for protecting public against the hazards of metal toxicity and also in creating awareness towards its nutritional qualities.

4. Elemental profiling

Elemental profiling specially of trace elements can be divided into three subgroups:
1. Easy to determine routinely by several techniques (e.g. iron and zinc);
2. Not always easy to assay, particularly at low concentrations (e.g. arsenic, selenium and tin); and
3. Expert handling (e.g. cadmium, chromium, lead, manganese, mercury, molybdenum and nickel) which require a high level of analytical expertise because of the low concentrations present, detection limit problems, matrix interferences, incomplete recoveries and related methodological difficulties. At low concentrations, the analysis of dietary material presents considerable difficulties, depending on whether the matrix is simple (e.g. drinking-water and beverages) or complex (dairy products). Meat and a few other food products contain some trace elements at very high concentrations and are generally easy to analyse. The implications of these various conditions are important when single foods are analysed, something that may present an array of problems. For example, moisture content of foods varies widely, ranging from 94% in leafy vegetables to 60% in meat, 20% in grains and cereals, and up to 10% in oils and fats. Fat content of foods can cause difficulty, e.g. a high-fat product such as cheese will present a real challenge, if dissolution steps are involved. (Oxidation of samples rich in fats and oils with perchloric acid should be avoided because of explosion risks). The concentrations of several trace elements vary considerably even in foods belonging to similar groups belonging to different groups. The problem of trace element analysis can be overcome by freeze-drying of mixed diets as this may lead to six fold enrichment of the component (Benramdane, 1999; Aceto, 2002).

5. Analytical techniques

Analytical techniques such as atomic absorption spectrophotometry (AAS) (flame and flameless), atomic emission spectroscopy (direct-current and inductively coupled plasma), chemical and electro-analytical methods, gas and liquid chromatography, mass spectrometry (in different modes) nuclear-activation techniques and X-ray fluorescence offer sufficiently low detection limits to make them suitable for investigating a variety of biomatrices.

Low detection limits alone are not sufficient to answer all the questions as analytical data on trace elements are mostly regarded with skepticism. Ignorance of various interferences, e.g. matrix-related problems, flaws in sample and standard preparation and inadequate calibration procedures all contribute to this regrettable situation. The analyst is therefore by far the most important component of any analytical system (Ma, 2004; Hiefje, 2000).

5.1 Choice of type of assay

The analyst is faced with the choice between multielement and single-element assays, which is affected by a number of factors. Thus, even though sometimes only partly quantitative, multielement assays are useful in obtaining simultaneous elemental composition profiles of a given specimen. For example, the non-destructive procedures offer the possibility of generating data simultaneously (including repeated determinations on the same test portion) for several elements for purposes of comparison. They also offer the possibility of internal quality control so that unusual situations involving any specific element can be evaluated. Moreover, in a carefully designed study, multielement assays can provide very useful information at relatively low cost. However, some elements must be determined alone because of serious analytical problems. Clinical, environmental and nutritional laboratories dealing with specific elements frequently need single-element assays. In a laboratory performing a wide range of analyses, therefore, a combination of both single- and multi element capability may be essential for effective functioning.

5.2 Choice of analytical technique

The choice of an analytical technique depends on a number of factors, including:
1.   susceptibility to matrix effects;
2.   range of elements covered;
3.   detection limits; and
4.   suitability for the matrix of interest.

The susceptibility of an analytical technique to matrix effects depends on the sample composition. With some matrices, these effects are of major importance, but others can be avoided by a modification of the technique. The usefulness of an analytical method for trace-element analysis, also depends on the range of elements covered and the order of magnitude of its detection limits for the elements at the top and bottom of its sensitivity range. Detection limits will not be the same for all elements, so that simultaneous multielement determination will require compromises in experimental conditions that will affect the accuracy and precision of at least a few elements. Even when there is a method of choice for the analysis of a particular element, its performance will depend on the concentration of the element in question and that of others in the matrix (Toelg, 1988). Concentration ranges also vary widely between different types of biomaterials and foods (Kumpulainen, 1980). These changes in relationships between elements may necessitate modifications to the technique for specific applications in order to maintain optimum performance and prevent any decline in detection limits.

The most important criterion of the suitability of a method, however, is whether it is appropriate for the matrix of interest. Using a set of four representative biological matrices, namely bovine liver, porcine muscle, Bowen's kale and human serum, an advisory group designated by the International Atomic Energy Agency evaluated the performance of different analytical techniques (Cornelis, 1980). For elements such as copper, iron and zinc, several methods were suitable. Thus, for zinc, many methods can generate results with a 1% CV. On the other hand, for elements such as fluorine, iodine, tin and vanadium, the choice was limited . The analytical techniques studied nevertheless reached detection limits below the ng/g level for chromium, manganese and vanadium, i.e. the level at which these elements are expected to occur in some specimens (Jones, 1992; Watson, 1998; Bernazzani, 2001).

6. Introduction of atomic absorption spectrometry

The atom is a nucleus surrounded by electrons which travel around the nucleus in discrete orbitals. Every atom has a number of orbital's in which it is possible for electrons to travel. Each of these electron orbital's has an energy level associated with it. In general, the further away from the nucleus an orbital, the higher its energy level. When the electrons of an atom are in the orbital's closest to the nucleus and lowest in energy, the atom is in its most preferred and stable state, known as its *ground state*. When energy is added to the atom as the result of absorption of electromagnetic radiation or a collision with another particle (electron, atom, ion, or molecule), one or more of several possible phenomena take place. The two most probable events are for the energy to be used to increase the kinetic energy of the atom (*i.e.*, increase the velocity of the atom) or for the atom to absorb the energy and become excited. This later process is known as *excitation*.

When an atom becomes excited, an electron from that atom is promoted from its ground state orbital into an orbital farther from the nucleus and with a higher energy level. Such an atom is said to be in an *excited state*. An atom is less stable in its excited state and will thus decay back to a less excited state by losing energy through a collision with another particle or by emission of a "particle" of electromagnetic radiation, known as a *photon*. As a result of this energy loss, the electron returns to an orbital closer to the nucleus.

If the energy absorbed by an atom is high enough, an electron may be completely dissociated from the atom, leaving an ion with a net positive charge. The energy required for this process, known as ionization, is called the ionization potential and is different for each element. Ions also have ground and excited states through which they can absorb and emit energy by the same excitation and decay processes as an atom. The difference in energy between the upper and lower energy levels of a radiative transition defines the wavelength of the radiation that is involved in that transition (Frank, 1998; Skoog, Holler & Nieman, 1998). The phenomenon of atomic absorption observed in 1802 with the discovery of the Fraunhofer lines in the sun's spectrum. It took more than half century to utilize observed lines for quantitative chemical analysis. Atomic absorption analysis involves measuring the absorption of light by vaporized ground state atoms and relating the absorption to concentration governed by Beer's law.

7. Atomic Absorption Spectroscopy (AAS)

The basic aim of analytical atomic absorption spectroscopy is to identify elements and quantify their concentrations in various media. The procedure consists of three general steps: atom formation, excitation, and emission. For UV and visible spectroscopy, the input energy must be sufficient to raise an electron from the ground state to the excited state. Once the electron is in the excited state, the atom emits light, which is characteristic of that particular element. Before excitation, an element that is bound in a specific matrix must be separated from that matrix so that its atomic emission spectra are free from interferences.

The elements present in a sample are converted to gas phase atoms in the ground state. The UV-Vis absorption of these gas phase atoms are then measured by irradiation of light at a highly specific wavelength causing transition of some of the gas phase atoms to a higher energy level. The extent to which light is absorbed is related to the original concentration of ground state atoms. This situation is completely analogous to the Beer-Lambert law in conventional liquid UV-Vis absorption spectrophotometry. Conversion of the sample from its native state to the atomic state can be achieved using a flame (flame-AAS) or an electric furnace (electro-thermal or graphite furnace AAS). The later will be studied herein. In the furnace, the sample undergoes a number of pretreatment steps prior to analysis.

1. Sample is dried by evaporating the solvent (in this case the water).
2. The organic matrix is decomposed by heating of the sample ≥1000°C. (taking care not to lose any of the analyte through evaporation processes).
3. Furnace is rapidly heated to temperatures around 2400°C to produce vaporized neutral atoms.

This method provides both sensitivity and selectivity since other elements in the sample will not generally absorb the chosen wavelength and thus, will not interfere with the measurement. However, molecular species may also be formed during the atomization step. Which can alter the spectral characteristics of the analyte metal or can cause spectral interference at the wavelength being monitored. To reduce background interference, the wavelength of interest is isolated by a mono-chromator placed between the sample and the detector. Additional techniques such as D2 or Zeeman background correction may also be used for complex matrices such as beer.

The most common instrument components consist of a hollow cathode lamp source, a pneumatic nebulizer for an atomizer, a conventional grating mono-chromator and photomultiplier tube detector. The hollow cathode lamp is made of a glass envelope with a quartz window filled with an inert gas at slightly above atmospheric pressure. The cathode is made of the pure metal of interest. The pneumatic nebulizer aspirates and nebulizes the liquid sample solution when the sample is sucked through a capillary tube. The grating mono-chromator eliminates much of the background light from the flame and the photomultiplier tube detector detects that light from the hollow cathode lamp which passes through the flame.

Atomic absorption analyses are most commonly and routinely performed on solutions. Therefore a sample must be converted to liquid form prior to analysis using a microwave to digest the sample, leaving a solution that can then be analyzed. However this method has the following limitations (Ingle & Crouch, 1988; Lajunen, 1992; Frank, 1997; Skoog et al., 1998).

7.1 Limitations
1.   Elemental range is limited to metals and metalloids.
2.   Sample preparation is tedious and time consuming.
3.   The sample is destroyed by the analysis.
4.   Only one element at a time can be measured.

8. Techniques based on atomic spectrometry

In *atomic absorption spectrometry* (AAS), light of a wavelength characteristic of the element of interest is shone through this atomic vapor. Some of this light is then absorbed by the atoms of that element. The amount of light that is absorbed by these atoms is then measured and used to determine the concentration of that element in the sample.

In *optical emission spectrometry* (OES), the sample is subjected to temperatures high enough to cause not only dissociation into atoms but to cause significant amounts of collisional excitation (and ionization) of the sample atoms to take place. Once the atoms or ions are in their excited states, they can decay to lower states through thermal or radiative (emission) energy transitions. In OES, the intensity of the light emitted at specific wavelengths is measured and used to determine the concentrations of the elements of interest.

One of the most important advantages of OES results from the excitation properties of the high temperature sources used in OES. These thermal excitation sources can populate a large number of different energy levels for several different elements at the same time. All of the excited atoms and ions can then emit their characteristic radiation at nearly the same time. This results in the flexibility to choose from several different emission wavelengths for an element and the ability to measure emission from several different elements concurrently. However, a disadvantage associated with this feature is that as the number of emission wavelengths increases, the probability also increases for interferences that may arise from emission lines that are too close in wavelength to be measured separately.

In *atomic fluorescence spectrometry* (AFS), a light source, such as that used for AAS, is used to excite atoms only of the element of interest through radiative absorption transitions. When these selectively excited atoms decay through radiative transitions to lower levels, their emission is measured to determine concentration, much the same as in OES. The selective excitation of the AFS technique can lead to fewer spectral interferences than in OES. However, it is difficult to detect a large number of elements in a single run using AFS, as the number of spectral excitation sources and detectors that can be used at one time is limited by the instrument (Smith et al., 1995; Bernazzani & Paquin, 2001).

Another technique, called *atomic mass spectrometry*, is related to three atomic spectroscopy techniques described above. Instead of measuring the absorption, emission or fluorescence of radiation from a high temperature source, such as a flame or plasma. Mass spectrometry measures the number of singly charged ions from the elemental species within a sample. Similar to the function of a monochromator in emission/absorption spectrometry that separates light according to wavelength, a quadrupole mass spectrometer separates the ions of various elements according to their mass-to-charge ratio in atomic mass spectrometry (Jones, 1992).

8.1 Atomic Emission Spectroscopy (AES)

Atomic emission spectroscopy (AES) is one of the most important techniques of elemental analysis. One of its advantages over atomic absorption is the capability for simultaneous multi-element analysis. It can be used for the analysis of major components of the sample as well as for trace analysis, because calibration curves are linear over several orders of magnitude. As a result of these advantages AES technique are popular in analytical laboratories. Over the past decade, a new generation of spectrometer configurations based on charge coupled device (CCD) detectors has appeared (Hiefje, 2000; Bernazzani & Paquin, 2001) Some of them are simple and low-cost. The perfect atomic emission source would have the following characteristics:

1. Easy to operate.
2. Inexpensive to purchase and maintain.
3. A source that can handle a range of solvents, both organic and inorganic in nature.
4. A source that is adjustable to handle solids, slurries, liquids, or gases.
5. Complete removal of the sample from its original matrix thus minimum interferences.
6. Complete atomization but minimum ionization of all elements to be analyzed.
7. A controllable energy source for excitation, which allows the proper energy needed to excite all elements without appreciable ionization.
8. An inert chemical environment, which prohibits the formation of undesirable molecular species (e.g. oxides, carbides, etc.) that may affect the accuracy of the measurement.
9. No background radiation from the source (unwanted atomic or molecular emission that could interfere with the analytical wavelengths).

Every element has its own characteristic set of energy levels and has unique set of absorption and emission wavelengths. This property of the element makes atomic spectrometry useful for element-specific analytical techniques. The ultraviolet (UV)/visible region (160 - 800 nm) of the electromagnetic spectrum is the region used for analytical atomic spectrometry. This is also the region of the electromagnetic spectrum. The main reasons for the popularity of analytical techniques that use the UV/visible region are that these techniques are accurate, precise, flexible and relatively inexpensive compared to techniques which use other regions, such as gamma ray spectrometry and X-ray spectrometry (Watson et al., 1998; DeGraff et al., 2002).

8.2 Inductively Coupled Plasma-Mass Spectrometry (ICP-MS)
One of the largest volume uses for ICP-MS is in the medical and forensic field, specifically, toxicology. A physician may order a metal assay for a number of reasons, such as suspicion of heavy metal poisoning, metabolic concerns, and even hepatological issues. Depending on the specific parameters unique to each patient's diagnostic plan, samples collected for analysis can range from whole blood, urine, plasma, serum, to even packed red blood cells. Another primary use for this instrument lies in the environmental field. Such applications include water testing for municipalities or private individuals all the way to soil, water and other material analysis for industrial purposes. This technique has been utilized in food, herbs, herbal drug analysis for safety and efficacy. This technique is also widely used the field of radiometric dating, in which it is used to analyze relative abundance of different isotopes. ICP-MS is more suitable for this application than the previously used Thermal Ionization Mass Spectrometry, as species with high ionization energy such as Osmium (Os) and Tungsten (Hf-W) can be easily ionized(Vladimir, 2007, Eliot, 2007).

8.3 Inductively Coupled Plasma-Atomic Emission Spectroscopy (ICP-AES)
An inductively coupled plasma spectrometer is a tool for trace detection of metals in solution, in which a liquid sample is injected into argon gas plasma contained by a strong magnetic field. The elements in the sample become excited and the electrons emit energy at a characteristic wavelength as they return to ground state. The emitted light is then measured by optical spectrometry. This method, known as inductively coupled plasma atomic emission spectrometry (ICP-AES) or inductively coupled optical emission spectrometry (ICP-OES). ICP Spectrometers can be used for the analysis of environmental samples, contaminants in food or water, metalloproteins in biological samples, and similar studies. Most ICP-AES instruments are designed to detect a single wavelength at a time (mono-chromator). Since an element can emit at multiple wavelengths, it is sometimes desirable to detect more than one wavelength at a time. This can be done by sequential scanning or by using a spectrometer that is designed to capture emissions of several wavelengths simultaneously (poly-chromator). Detection limits typically range from parts per million (ppm) to parts per billion (ppb), although depending on the element and instrument, it can sometimes achieve even less than ppb detection.

8.3.1 Interferences
Any chemical or physical process that adversely affects the measurement of the radiation of interest can be classified as interference. Interferences in ICP-AES may start in the sample preparation stage and extend to the plasma operating conditions(Montaser & Golightly 1988).

8.3.2 Applications
Pharmaceutical industries (metals in wine (Aceto, 2002), arsenic in food (Benramdane, 1999), and trace elements bound to proteins Biological samples(Ma, 2004). Precious metal

estimation at low level, Heavy metal estimation at sub ppm level Rock, Soil, Fly ash (Complete analysis), Environmental sample analysis (Water, Air, Soil, sediments, etc.) and Polymer industries.

In plasma mass spectroscopy (MS), the inductively coupled argon plasma (ICP) is once again used as an excitation source for the elements of interest. However in contrast to OES, the plasma in ICP-MS is used to generate ions that are then introduced to the mass analyzer. These ions are then separated and collected according to their mass to charge ratios. The constituents of an unknown sample can then be identified and measured. ICP-MS offers extremely high sensitivity to a wide range of elements.

8.4 Inductively Coupled Plasma Optical Emission Spectroscopy (ICP-OES Analysis)

ICP technology has evolved significantly since its inception in 1960's, when the first ICP prototypes emerged to provide better analytical sensitivity and capacity than Atomic Absorption Spectrometry. During the course of time this technique showed significant improvements in technology, capability and price. Nowadays, ICP-OES is commonly found in contract analytical laboratories and academic facilities where routine analysis needing high sensitivity throughput.

Plasma is an ionized gas that in addition to atoms also contains electrons and ions. After ignition with a Tesla spark the energy transfer is via the high frequency field in the coil that is surrounding the plasma. Free electrons are accelerated and heat the plasma by collision with argon atoms. We distinguish between ionization, electron and excitation temperature, which are different at different locations of the plasma. The sample aerosol is introduced through the center of the plasma flow without affecting its stability and equilibrium.

In the plasma the atoms and particularly the ions are excited to emission. After spectral dispersion of the emitted radiation in a powerful optical system the element-specific wavelengths are used for identification and quantification. Samples are introduced into the plasma in a process that desolvates, ionises, and excites them. The constituent elements can be identified by their characteristic emission lines, and quantified by the intensity of the same lines. The method has the following advantages (Bafley, 1989):

- High sample throughput enabling the efficient analysis of large batches
- Simultaneous determination of multiple elements in each sample
- Complementary analysis to techniques like XRF
- Large dynamic linear range
- Low chemical and matrix interference effects.

9. High-Performance Inductively-Coupled Plasma Optical Emission Spectrometry (HP-ICP-OES) through exact matching

Using high-performance inductively coupled plasma optical emission spectrometry (HP-ICP-OES), relative expanded uncertainties on the order of 0.2 % can be routinely achieved. Nevertheless, analysis results can be improved by implementing "exact matching" with the HP-ICP-OES protocol. This should be very careful matching of the analyte mass fractions, internal standard element mass fractions, and solution matrices of the calibration solutions to the samples. The analytical benefits of this approach are being systematically investigated. Results show that the primary benefit is mitigation of the deleterious effects of nonlinearities of the ICP-OES instrument responses to the analyte and/or internal standard element mass fractions.

10. Regulatory and dietary recommendations

The side effects against heavy metals have built up over time. Close examination by toxicologists studying cases of poisoning from heavy metals has revealed that only in cases of high exposures, visible clinical symptoms are likely. At lower but still unacceptable 'levels of exposure, as in consumption of certain foods, effects may be restricted at physiological or biochemical level only, (Hutton, 1987). Hamilton (1988) noted that the first legislation to control the adulteration of food or drink occurred in Britain in 1860, in response to long use of heavy metal salts as coloring matter in food. Regulatory agencies in most countries now seek to protect public health by exercising control limits over the chemical composition of specific food types. The process typically involves setting appropriate standards for potentially toxic chemicals in foods which by law should not be exceeded. These or similar agencies were setup to overview random and/or periodic chemical testing of appropriate samples to ensure compliance of the law. For trace elements, given the absence of metabolization of the metal or nonmetal, it is possible to establish clear separations among essentially, health benefits and toxicity. Many countries and regions have defined the requirements and limits of supplements for trace elements. Walker (1988) has summarized international food standards for cadmium applicable in 1986. 19 countries had set regulatory limits for cadmium in foods, but with some exceptions. Australia, Denmark, the Netherlands and Hungry had set limits for cadmium in particular foods.

11. Initiation for scientific and regulatory bodies

The report of the WHO Expert Committee on Trace Elements in Human Nutrition that met in 1973 ended with six important general recommendations for future national and international activities in trace-element nutrition. They are briefly given as under:
1.   Need to obtain reliable information on the trace-element content of foods, especially milk
2.   To monitor contents in relation to future changes in agricultural and industrial practices.
3.   Trace-element requirements should be taken into account in food standards and especially in those for formulated foods designed for infants and young children.
4.   Need for international centers for the study of trace elements in humans and for international analytical reference laboratories.
5.   Further review of new findings in trace-element nutrition in order to update the recommended levels of intake.
The two dietary standards -the 1990 version of the RNIs and 1989 RDAs, did not differ much in the described derivations of the recommended intakes but differences remain about how intended uses are described, resulting in some confusion for the users of both reports. The answer was sought in the joint U.S. and Canadian development of the new Dietary Reference Intakes/Recommended dietary Allowances (DRIs/RDAs). The Food and Nutrition Board of the United States National Academy of Sciences has taken responsibility for establishing guidelines on what quantities of the various nutrients should be eaten by human males and females at various ages. These were called RDAs (for Recommended Dietary Allowances, and often referred to as Recommended Daily Allowances). They provide the data on which food labels are based. To avoid any further confusions, the terms used in the guidelines were explained explicitly.

12. Needs of individuals requirement

This is the lowest continuing level of nutrient intake that, at a specified efficiency of utilization, will maintain the defined level of nutrient in the individual.

12.1 Basal requirement
This refers to the intake needed to prevent pathologically relevant and clinically detectable signs of impaired function attributable to inadequacy of the nutrient.

12.2 Normative requirement
This refers to the level of intake that serves to maintain a level of tissue storage or other reserve that is judged by the expert consultation to be desirable. The essential difference between the basal requirement and the normative requirement is that the latter usually facilitates the maintenance of a desirable level of tissue stores. For most trace elements, metabolic and tissue-composition studies indicate the existence of discrete stores which, by undergoing depletion at times of reduced intake or high demand, can provide protection for a certain period against the development of pathological responses to trace-element deficiency. Since higher levels of intake are needed to maintain these reserves the normative requirement is necessarily higher than the basal requirement.

12.3 Daily intake
It is defined as the individual's average intake persisting over moderate periods of time without necessarily being present in those amounts each day. Individuals differ in their requirements, even though they may have the same general characteristic (e.g. age, sex, physiological state and body size). One may therefore speak of the *average requirement* of a group of individuals (e.g. young adult men) or of the level that marks a point in the upper tail of the requirement distribution, the level previously identified as the recommended or safe level of intake. Except where specifically indicated, the estimates refer to the maintenance of a defined level of nutritional status in individuals already in that state. They refer to healthy individuals, and the estimated requirements may be altered by disease or other conditions.

12.4 A nutrient intake value
It is estimated to meet the requirement of half of the healthy individuals in a life stage and gender group.

12.5 Recommended Dietary Allowance (RDA)
The dietary intake level that is sufficient to meet the nutrient requirements of nearly all healthy individuals in a life stage and gender group.

12.6 Adequate Intake (AI)
A recommended intake value based on observed or experimentally determined approximations or estimates of nutrient intake by a group (or groups) of healthy people that are assumed to be adequate (used when an RDA cannot be determined).

12.7 Tolerable upper intake level (UL)
The highest level of nutrient intake that is likely to pose no risk of adverse health effects for almost all individuals in the general population. As intakes increase above the UL, the risk of adverse effects increases.

13. UK Recommended Daily Allowance (RDA) for minerals

The Recommended Daily Allowance (RDA) of foods and supplements has traditionally been set by government health bodies in the UK, Europe and USA. In the UK the department of health gave the RDA for vitamins A, C, D, three of the B vitamins, and three minerals in 1979. However, in 1993 the European Union (EU) issued a directive on food labelling for its members, which included RDAs for twelve vitamins and six minerals. As a result, UK food labels are being revised to include the EU recommendations. In addition, the government of the UK issued a report in 1991 suggesting new guidelines for daily requirements of vitamins and minerals, which would replace the RDAs. These are called, collectively, Dietary Reference Values (DRVs), and consist of these terms:

Estimated Average Requirement (EAR), which should meet the requirements of half of the population.

Reference Nutrient Intake (KNI), which replaces the former RDA, is meant to represent the nutrient requirements of some 97 percent of the population. The amount recommended is higher than most people actually need. The EAR meets the requirements of half of the population, of the remaining half some need more and some need less than the EAR.

Lower Reference Nutrient Intake (LRNI), which is the nutritional requirement for those whose needs are low. Most people will need more than this amount in order to maintain their health. Anyone who is getting less than the LRNI might be in danger of nutritional deficiency.

The following list describes those vitamins and minerals that should be present in the daily diet. The descriptions include the RDAs for the EU and USA. The amounts given are those that should be taken to prevent a mineral deficiency, not those needed to improve health or prevent non-deficiency diseases:

Recommended Daily Allowances / Dietary Reference Intake (Table 3-6)

In the Recommended Dietary Allowance charts below, amounts marked with a * indicate AI (Adequate Intake). Figures taken from the Dietary Reference Intakes (DRI, 1997, 98, 2000, 2001, 2003 and 2004).

Minerals	0-6 months	7-12months	1-3 years	4-8 years
Calcium	210* mg	270* mg	500* mg	800* mg
Chromium	0.2* µg	5.5* µg	11* µg	15* µg
Copper	200* µg	220* µg	340 µg	440 µg
Fluoride	0.01* mg	0.5* mg	0.7* mg	1* mg
Iodine	110* µg	130* µg	90 µg	90 µg
Iron	0.27* mg	11 mg	7 mg	10 mg
Magnesium	30* mg	75* mg	80 mg	130 mg
Manganese	0.003* mg	0.6* mg	1.2* mg	1.5* mg
Molybdenum	2* µg	3* µg	17 µg	22 µg
Phosphorus	100* mg	275* mg	460 mg	500 mg
Selenium	15* µg	20* µg	20 µg	30 µg
Zinc	2* mg	3 mg	3 mg	5 mg
Potassium	0.4* g	0.7* g	3.0* g	3.8* g
Sodium	0.12* g	0.37* g	1.0* g	1.2* g
Chloride	0.18* g	0.57* g	1.5* g	1.9* g

1µg = 1mcg = 1microgram = 1/1,000,000 of a gram;1mg = 1milligram =1/1,000 of a gram;1g =1gram

Table 3. Recommended Daily Allowances (RDA) Chart for Infants & Children

Minerals	Male 9-13 Yrs	Male 14-18 Yrs	Female 9-13 Yrs	Female 14-18 Yrs
Calcium	1300* mg	1300* mg	1300* mg	1300* mg
Chromium	25* µg	35* µg	21* µg	24* µg
Copper	700 µg	890 µg	700 µg	890 µg
Fluoride	2* mg	3* mg	2* mg	3* mg
Iodine	120 µg	150 µg	120 µg	150 µg
Iron	8 mg	11 mg	8 mg	15 mg
Magnesium	240 mg	410 mg	240 mg	360 mg
Manganese	1.9* mg	2.2* mg	1.6* mg	1.6* mg
Molybdenum	34 µg	43 µg	34 µg	43 µg
Phosphorus	1250 mg	1250 mg	1250 mg	1250 mg
Selenium	40 µg	55 µg	40 µg	55 µg
Zinc	8 mg	11 mg	8 mg	9 mg
Potassium	4.5* g	4.7* g	4.5* g	4.7* g
Sodium	1.5* g	1.5* g	1.5* g	1.5* g
Chloride	2.3* g	2.3* g	2.3* g	2.3* g

Table 4. Recommended Daily Allowances for Older Children (9 to 18 Years)

Minerals	Male 19-50 Yrs	Male >50 Yrs	Female 19-50 Yrs	Female >50 Yrs
Calcium	1000* mg	1200* mg	1000* mg	1200* mg
Chromium	35* µg	30* µg	25* µg	20* µg
Copper	900 µg	900 µg	900 µg	900 µg
Fluoride	4* mg	4* mg	3* mg	3* mg
Iodine	150 µg	150 µg	150 µg	150 µg
Iron	8 mg	8 mg	18 mg	8 mg
Magnesium #1	400/420 mg	420 mg	310/320 mg	320 mg
Manganese	2.3* mg	2.3* mg	1.8* mg	1.8* mg
Molybdenum	45 µg	45 µg	45 µg	45 µg
Phosphorus	700 mg	700 mg	700 mg	700 mg
Selenium	55 µg	55 µg	55 µg	55 µg
Zinc	11 mg	11 mg	8 mg	8 mg
Potassium	4.7* g	4.7* g	4.7* g	4.7* g
Sodium #2	1.5* g	1.3* g	1.5* g	1.3* g
Chloride #2	2.3* g	2.0* g	2.3* g	2.0* g

As per the Food and Nutrition Board (FNB) Recommendations : Men from 31 to 50 need slightly more magnesium (420 mg) than those from 19 to 30 years old (400 mg). Women from 31 to 50 also need slightly more magnesium (320 mg) than those from 19 to 30 years old (310 mg). Adults over 70 years need slightly different levels of sodium (1.2 g) and chloride (1.8 g). Pregnant women from 31 to 50 need slightly more magnesium (360 mg) than those between 19 to 30 years old (350 mg). Women from 31 to 50 who are breastfeeding also require slightly more magnesium (320 mg) than those between 19 to 30 years old (310 mg).

Table 5. Recommended Daily Allowances for Adults (19 Years and Up)

Minerals	Pregnancy 14-18 Yrs	Pregnancy 19-50 Yrs	Lactation 14-18 Yrs	Lactation 19-50 Yrs
Calcium	1300* mg	1000* mg	1300* mg	1000* mg
Chromium	29* µg	30* µg	44* µg	45* µg
Copper	1000 µg	1000 µg	1300 µg	1300 µg
Fluoride	3* mg	3* mg	3* mg	3* mg
Iodine	220 µg	220 µg	290 µg	290 µg
Iron	27 mg	27 mg	10 mg	9 mg
Magnesium #3	400 mg	350/360 mg	360 mg	310/320 mg
Manganese	2.0* mg	2.0* mg	2.6* mg	2.6* mg
Molybdenum	50 µg	50 µg	50 µg	50 µg
Phosphorus	1250 mg	700 mg	1250 mg	700 mg
Selenium	60 µg	60 µg	70 µg	70 µg
Zinc	12 mg	11 mg	13 mg	12 mg
Potassium	4.7* g	4.7* g	5.1* g	5.1* g
Sodium	1.5* g	1.5* g	1.5* g	1.5* g
Chloride	2.3* g	2.3* g	2.3* g	2.3* g

Table 6. Recommended Daily Allowances for Pregnancy / Lactating Mothers

14. Trace minerals in human milk and in drinking water guidelines (Table 7 and 8)

In considering the consumption of drinking water by vulnerable populations, a figure of 0.75 liters per day has been used for a 5kg child and a figure of one liter per day for a 10kg

Mineral	Mature human milk (Lawrence & Lawrence, 1999)	Drinking water guidelines (WHO, 1996)	EC, SCF, 2003 (Recommended energy content: 60-70kcal/dl; based on 65kcal/dl, cow's-milk-protein based formula)
	mg/L	mg/L	mg/L
Calcium	280	--	325-910 (Ca: P=1-2)
Iron	0.40	0.3 b	1.95-8.45
Zinc	1.2	3.0 b	3.25-9.75
Copper	0.25	1.0 b; 2.0 * (P)	0.228-0.65
Selenium	20	10*	20-59
Fluoride	0.016	1.5 (P)	≤ 0.65
Magnesium	30	--	33-98
Sodium	180	200*	130-390
Sulphate	140 (sulphur)	250*	--
Chloride	420 (chlorine)	250 *	325-1040
Manganese (µg/L)	6	100 b; 500 * (P)	6.5-650
Molybdenum (µg/L)	2µg/dc	70	--

* Health-based guideline value, (P): provisional; b Parameters in drinking water that may give rise to complaints from consumers, c FNB

Table 7. Trace minerals in human milk and in drinking water guidelines

Inland surface water		Drinking water			WHO (2006)	
BIS/CPCB	WHO	BIS	CPCB	WHO(1993)	Normal	Health based
Cd	2.0	0.1	2.0	2.0	0.003-0.005	<1 µg/l	0.003
Cu	3.0	0.05-1.5	3.0	3.0	2.0	-	2.0
Fe	3.0	0.1-1.0	-	3.0	0.2	0.5-5.0	-
Mn	2.0	0.05-0.5	0.1	0.1	0.5-0.05	-	0.4
Ni	2.0	-	-	2.0	0.02	0.02	0.07
Pb	0.1	0.1	2.0	3.0	0.01	-	0.01

Table 8. Recent standards for heavy metals (mg/l) in Inland surface and drinking water (Bharti 2007)

child. Although these figures may be applicable for standard calculations, the range of quantitative water intake observed in populations at that age might be considerable according to the Food and Nutrition Board of the Institute of Medicine.

15. References

Abdul-Wahab O. El-Rjoob; Adnan M. Massadeh & Mohammad N. Omari (2008). Evaluation of Pb, Cu, Zn, Cd, Ni and Fe levels in Rosmarinus officinalis labaiatae (Rosemary) medicinal plant and soils in selected zones in Jordan, *Environmental Monitoring and Assessment* Vol. 140 (No. 1-3): 61-68.

Aceto, M., Abollino, O., Bruzzoniti, M.C., Mentasti, E., Sarzanini, C. & Malandrino, M. (2002). Determination of metals in wine with atomic spectroscopy (flame-AAS, GF-AAS and ICP-AES); a review. *Food additives and contaminants.* Vol.19 (No. 2): 126–33.

Anderson, R.A. (1988). *Chromium.* Ed. Smith, K., Trace minerals in foods. New York, Marcel Dekker, 231-247.

Arsenic, (1980). *Mineral tolerances of domestic animals.* Washington, DC, National Academy of Sciences, 40-53.

Aschner, M. (2000). Manganese: brain transport and emerging research needs. *Environment Health Perspect.* 108 (Suppl. 3), 429–432.

Baer, M.J. & King, J.C. (1984).Tissue zinc levels and zinc excretion during experimental zinc depletion in young men, *American journal of clinical nutrition* Vol.39: 556-570.

Bennet EG. *Modelling exposure routes of trace metals from sources to man. In:* Nriagu JO ed. *Changing metal cycles and human health.* Berlin, Springer-Verlag,1984: 345-356.

Benramdane, L., Bressolle, F. & Vallon, J.J. (1999). Arsenic speciation in humans and food products: a review, *Journal of chromatographic science* Vol.37(No.9): 330–44.

Bernazzani, P. & Paquin, F. (2001). Modular spectrometers in the undergraduate chemistry laboratory, *J. Chem. Educ.,* Vol.78: 796–798.

Bharti, P. K. (2007). *Current Science* VOL. 93(No. 9), 10-18.

Bunzl, K., Trautmannsheimer, M., Schramel, P. & W. Reifenhäuser. (2001). Availability of Arsenic, Copper, Lead, Thallium, and Zinc to Various Vegetables Grown in Slag-Contaminated Soils, *Journal of Environmental Quality* Vol.30: 934-939.

Byrne, A.R. & Kosta, L. (1978). Vanadium in foods and in human body fluids and tissues, *Science of the total environment* Vol. 10: 17-30.

Combs, G.F. Jr & Combs, S.B. (1986). *The role of selenium in nutrition*. Orlando, FL, Academic Press.

Cornelis, R. & Versieck, J. (1980). *Critical evaluation of the literature values of 18 trace elements in human serum or plasma*, Ed. Bratter, P. & Schramel, P. Trace element analytical chemistry in medicine and biology. Berlin, de Gruyter, 587-600.

de Romana, D.L., Salazar, M., Hambidge, K.M., Penny, M.E., Peerson, J.M., Krebs, N.F.& Brown, K.H. (2005). Longitudinal measurements of zinc absorption in Peruvian children consuming wheat products fortified with iron only or iron and 1 of 2 amounts of zinc, *Am. J. Clin. Nutr.* Vol.81: 637–647.

DeGraff, B.A., Hennip, M., Jones, J.M., Salter, C. & Schaertel, S.A. (2002). An inexpensive laser Raman spectrometer based on CCD detection, *Chem. Educator* Vol.7: 15–18.

Elliott, S.; Sturman, B.; Anderson, S.; Brouwers, E. & Beijnen, J. (2007). ICP-MS: When Sensitivity Does Matter, *Spectroscopy Magazine* April 1, 2007.

FNB, (2001).Food and Nutrition Board, Institute of Medicine. *Dietary Reference Intakes: Vitamin A, Vitamin K, Arsenic, Boron, Chromium, Copper, Iodine, Iron, Manganese, Molybdenum, Nickel, Silicon, Vanadium and Zinc*. Washington DC: National Academy Press.

Food Nutrition Board (2003), Institute of Medicine. *Dietary reference intakes for water,potassium, sodium, chloride, and sulphate*. Washington, D.C.: The National Academies Press.

Fraga, C.G. & Oteiza, P.I. (2002). Iron toxicity and antioxidant nutrients, *Toxicology* Vol.180: 23–32.

Greger, J.L. (1999). Nondigestible carbohydrates and mineral bioavailability, *J. Nutr.* Vol.129: 1434S–1435S.

Halliwell, B. & Gutteridge, J.M.C. (1999). *Free Radicals in Biology and Medicine*. Oxford university press, Oxford.

Frank S. *Handbook of Instrumental Techniques for AnalyticalChemistry*(1997)., Ed. X-Ray Fluorescence Spectrometry", G.J. Havrilla. Prentice Hall.

Hausinger, R.P. (1987). Nickel utilization by microorganisms, *Microbiological reviews* Vol.51: 22-42.

Hess, F.M., King, J.C. & Margen S. (1977). Zinc excretion in young women on low zinc intakes and oral contraceptive agents, *Journal of nutrition* Vol.107: 1610-1620.

Hiefje, G.M. (2000). Atomic emission spectroscopy — it lasts and lasts and lasts, *J. Chem. Educ.* Vol.77: 577–583.

Hoadley, J.E. & Cousins, R.J. (1988). *Regulatory mechanisms for intestinal transport of zinc and copper*. Ed. Prasad AS, Essential and toxic trace elements in human health and disease, New York, Alan R. Liss pp141-155.

Hunt, C.D. & Nielsen, F.H. (1981). *Interaction between boron and cholecalciferol in the chick*. Eds. Howell, J. McC.; Gawthorne, J.M. & White, C.L., Trace element metabolism in man and animals. Canberra, Australian Academy of Science, 597-600.

Hunt, C.D. (1988). Boron homeostasis in the cholecalciferol-deficient chick, *Proceedings of the North Dakota Academy of Sciences* Vol.42: 60-89.

Hurley, L.S. & Keen, C.L. (1987). *Manganese*. Ed. Mertz W., *Trace elements in human and animal nutrition*, 5th ed., Vol 1. San Diego, Academic Press, 185-223.

Hurrell, R.F., Lynch, S., Bothwell, T., Cori, H., Glahn, R., Hertrampf, E., Kratky, Z., Miller, D., Rodenstein, M., Streekstra, H., Teucher, B., Turner, E., Yeung, C.K. & Zimmermann, M.B. (2004). Enhancing the absorption of fortification iron, A sustain task force report, *Int. J. Vitam. Nutr. Res*. Vol.74: 387–401.

Hutton, H. (1987). *Human health concerns of lead, mercury, cadmium and Arsenic, Lead, Mercury, Cadmium and Arsenic in the Environment*, Ed. Hutchinson, T.C., John Wiley & Sons, Brisbane.

Institute of Medicine. (1997). *Dietary Reference Intakes for Calcium, Phosphorous, Magnesium, Vitamin D, and Fluoride*. Washington DC: National Academy Press.

Institute of Medicine. (1998). *Dietary Reference Intakes for Thiamin, Riboflavin, Niacin, Vitamin B6, Folate, Vitamin B12, Pantothenic Acid, Biotin, and Choline*. Washington, DC: National Academy Press.

Institute of Medicine. (2000). *Dietary Reference Intakes for Vitamin C, Vitamin E, Selenium, and Carotenoids*. Washington, DC: National Academy Press.

Institute of Medicine. (2001). *Dietary Reference Intakes for Vitamin A, Vitamin K, Arsenic, Boron, Chromium, Copper, Iodine, Iron, Manganese, Molybdenum, Nickel, Silicon, Vanadium, and Zinc*. Washington, DC: National Academy Press.

Institute of Medicine. (2004). *Dietary Reference Intakes for Water, Potassium, Sodium, Chloride, and Sulfate*. Washington, DC: National Academy Press.

Jones, B.T. (1992). Instrumental analysis courses. The choice and use of instrumentation. *J. Chem. Educ*. Vol. 69: A268–A269.

Jones, B.T. (1992). Instrumental analysis courses. The choice and use of instrumentation, *J. Chem. Educ*. Vol.69: A268–A269.

King, J.C., Reynolds, W.L. & Margen, S. (1978). Absorption of stable isotopes of iron, copper and zinc during oral contraceptive use, *American journal of clinical nutrition* Vol.31: 1198-1203.

Kovalskij, V.V.M. (1977).*Geochemische Okologie, Biogeochemie. [Geochemical ecology,biogeochemistry.]* Berlin, Deutscher Landwirtschaftsverlag.

Kumpulainen, J. (1980). *Chromium*. Ed. McKenzie, H.A. & Smythe, L.E., Quantitative trace analysis of biological materials. Amsterdam, Elsevier, 451-462.

Lawrence, R.A. & Lawrence, R.M. (1999). *Breastfeeding. A guide for the medical profession*. St.Louis, Missouri: Mosby Inc..

Lestienne, I., Besancon, P., Caporiccio, B., Lullien-Pellerin, V. & Treche, S. (2005). Iron and zinc in vitro availability in pearl millet flours (*Pennisetum glaucum*) with varying phytate, tannin, and fiber contents, *J. Agric. Food Chem*. Vol.53: 3240–3247.

Levander, O.A. (1987). A global view of human selenium nutrition, *Annual review of nutrition* Vol. 7: 227-250.

Luk, E.; Jensen, L.T. & Culotta, V.C. (2003). The many highways for intracellular trafficking of metals, *J. Biol. Inorg. Chem*. Vol.8: 803–809.

Ma R, McLeod C.W.; Tomlinson, K. & Poole R.K. (2004). Speciation of protein-bound trace elements by gel electrophoresis and atomic spectrometry. *Electrophoresis* Vol. 25(No.15): 2469–77.

Mason, K.E. (1979). A conspectus of research on copper metabolism and requirements of man, *Journal of nutrition* Vol.109: 1979-2006.

Mercury. Geneva, World Health Organization, 1976: 131 (Environmental Health Criteria 1).

Mokdad, A.H.; Marks, J.S.; Stroup, D.F. & Gerberding, J.L. (2004). *Actual causes of death in the United States, 2000.* J. Am. Med. Assoc. Vol.291: 1238-1245.

Montaser, A. & Golightly, D. W. (1988). *Inductively Coupled Plasmas in Analytical Atomic Spectrometry,* VCH, New York.

National Research Council, (1989). *Diet and health, implications for reducing chromic disease risk.* Washington, DC, National Academy of Sciences.

National Research Council. *Recommended dietary allowances,* 10th ed. Washington, DC, National Academy of Sciences, 1989.

Nielsen, F.H. (1982). *Possible future implications of nickel, arsenic, silicon, vanadium, and other ultratrace elements in human nutrition,* Ed. Prasad AS, Clinical, biochemical, and nutritional aspects of trace elements. New York, Alan R. Liss, 379-404.

Nielsen, F.H. (1984). Nickel. Ed. Frieden, E., *Biochemistry of the essential ultratrace elements,* New York, Plenum Press, 293-308.

Okada, S., Ohba, H. & Taniyama, M. (1981). Alterations in ribonucleic acid synthesis by chromium(III), *Journal of inorganic biochemistry* Vol.15: 223-231.

Ostrom, C.A., Van Reen, R. & Miller, C.W. (1961). Changes in the connective tissue of rats fed toxic diets containing molybdenum salts, *Journal of dental research* Vol.40: 520-527.

Oteiza, P.I.; Mackenzie, G.G.& Verstraeten, S.V. (2004). Metals in neurodegeneration: involvement of oxidants and oxidant-sensitive transcription factors, *Mol. Aspects Med.* Vol.25: 103-115. *Principles of Instrumental Analysis* 5th Edition by Skoog/Holler/Nieman, Saunders College Publishing 1998.

Punz, W.F. & H. Sieghardt. (1993). The response of roots of herbaceous plant species to heavy metals. *Environment and Experimental Botany* Vol.33(No.1): 85-98.

Rajagopalan, K.V. (1984). *Molybdenum. Biochemistry of the essential ultratrace elements.* Ed. Frieden E. New York, Plenum Press, 147-174.

Schilsky, M.L. (2001). Treatment of Wilson's disease: what are the relative roles of penicillamine, trientine and zinc supplementation?, *Curr. Gastroenterol. Rep.* Vol.3: 54-59.

Scientific Committee on Food (2003) *Report of the Scientific Committee on Food on the Revision of Essential Requirements of Infant Formulae and Follow-on Formulae.* SCF/CS/NUT/IF/65.

Smith, G.D.; Sanford, C.L. & Jones, B.T. (1995).Continuous liquid-sample introduction for Bunsen burner atomic emission spectrometry, *J. Chem. Educ.* Vol.72: 438-439. *Spectrochemical Analysis by Atomic Absorption and Emission,* L.H.J. Lajunen, Royal Society of Chemistry, 1992. *Spectrochemical Analysis* Ingle and Crouch, Prentice Hall, 1988.

Thauer, R.K. (1985). Nickel enzyme in Stoffwechsel von methanogenen Bakterien. [Nickel enzymes in the metabolism of methanogenic bacteria.], *Biological chemistry* Vol.366: 103-112.

Toelg G. (1988). *Where is analysis of trace elements in biotic matrices going to? In:* Bratter P, Schramel P, eds. Trace element analytical chemistry in medicine and biology. Berlin, de Gruyter,: 1-24.

Tong, S. von Schirnding, Y.E. & Prapamontol, T. (2000). Environmental lead exposure: a public health problem of global dimensions, *Bull World Health Organ* Vol.78(No.9):1068–1077.

Trace elements in human nutrition. Report of a WHO Expert Committee. Geneva, World Health Organization, 1973 (WHO Technical Report Series, No. 532).

Tuman, R.W. & Doisy, R.J. (1977). Metabolic effects of the glucose tolerance factor (GTF) in normal and genetically diabetic mice, *Diabetes* Vol.26: 820-826.

Underwood, E.J. (1977). *Trace elements in human and animal nutrition*, 4th ed. New York, Academic Press, pp 56-108.

Uthus, E.O. & Nielsen, F.H. (1989).*The effect of vanadium, iodine, and their interaction on thyroid status indices*, Ed. Anke, M., Proceedings of the Sixth International Trace Element Symposium, Vol. l. Jena, Friedrich-Schiller-Universitat, 44-49.

Vladimir, N., Epov, R., Douglas, E., Jian Zheng, O. F. X., Donard, C. & Yamada M. (2007). Rapid fingerprinting of ^{239}Pu and ^{240}Pu in environmental samples with high U levels using on-line ion chromatography coupled with high-sensitivity quadrupole ICP-MS detection, *J. Anal. At. Spectrom.* Vol.22 (No.9): 1131–1137.

Walker, J.M. (1988). *Regulation by other countries of cadmium in foods and the human environment*, Proceedings No 2, Cadmium Accumulations in Australtin Agriculture: National Symposium, Canberra, 1-2 March 1988, Australian Government PublishingService, Canberra, 176-185.

Walsh, C.T. & Orme-Johnson, W.H. (1987). Nickel enzymes, *Biochemistry* Vol.26: 4901-4906.

Wang, F., Kim, B.E., Dufner-Beattie, J., Petris, M.J., Andrews, G. & Eide, D.J. (2004). Acrodermatitis enteropathica mutations affect transport activity, localization and zinc-responsive trafficking of the mouse zip4 zinc transporter, *Hum. Mol. Genet.* Vol.13, 563–571.

Watson, K.A.; Levine, K.E. & Jones, B.T. (1988). A simple, low cost, multielementatomic-absorption spectrometer with a tungsten coil atomizer, *Spectrochim.Acta Part B* Vol.53B: 1507–1511.

Watson, K.A.; Levine, K.E. & Jones, B.T. (1998). A simple, low cost, multielement atomic-absorption spectrometer with a tungsten coil atomizer, *Spectrochim.Acta Part B*, Vol. 53B: 1507–1511.

Weigand, E., Kirchgessner, M. & Helbig, U. (1986). True absorption and endogenous fecal excretion of manganese in relation to its dietary supply in growing rats, *Biological trace element research* Vol.10: 265-279.

WHO and World Health Organization (1996). *Guidelines for drinking-water quality. Health criteria and other supporting information*. Geneva.

Willett, W.C. (2002). Balancing life-style and genomics research for disease prevention, *Science* Vol.296: 695–698.

Yusuf, A.A., Arowolo, T.O.A. & Bamgbose, O. (2002). Cadmium, copper and nickel levels in vegetables from industrial and residential areas of Lagos City, Nigeria, *Global Journal of Environmental Science* Vol.1(No.1): 1-6.

Zago, M.P. & Oteiza, P.I. (2001). The antioxidant properties of zinc: interactions with iron and antioxidants, *Free Radic. Biol. Med.* Vol.31: 266–274.

Zimmermann, M.B. & Kohrle J. (2002): The impact of iron and selenium deficiencies on iodine and thyroid metabolism: biochemistry and relevance to public health, *Thyroid* Vol.12: 867–878.

5

Flame Spectrometry in Analysis of Refractory Oxide Single Crystals

T.V. Sheina and K.N. Belikov
State Scientific Institution "Institute for Single Crystals"
of National Academy of Sciences
Ukraine

1. Introduction

This chapter is devoted to the use of flame atomic-emission and atomic-absorption spectrometry techniques to determine the stoichiometric composition, as well as studying the regularities of distribution of impurities and dopants in single crystals based on refractory oxides used as active media for solid-state lasers (leucosapphire, ruby, yttrium-aluminium garnet), laser light modulators (strontium titanate), scintillation detectors (cadmium tungstate), solid electrolytes (β-alumina), medical implants (leucosapphire).

Growth of high quality single crystals requires not only adherence to specifications and the growing parameters but reliable methods of analytical control of their composition. Last years along with so-called F-centre forming elements, the increasing attention is given to alkali and alkali-earth impurities which also may cause worsening of quality of single crystals (Dobrovinskaya et al.,2007; Nagornaya et al., 2005; Tupitsyna et al., 2009).

Digestion of refractory oxide single crystals is one of the most complicated and important stage of the procedure of analysis. Condensed phosphoric acid is recommended as an effective reagent for sample preparation of single crystals of oxide compounds. For the rapid flame spectrometry determination of alkali metals and calcium impurities, procedures of the sample preparation of water extracts of a- and γ-forms of aluminum oxide, yttrium-aluminum garnet and magnesium-aluminates used as raw materials for single crystals growing are proposed. It is also shown the effectiveness of ultrasonic sample preparation technique followed by the direct determination of Co and Ni in suspensions of a-Al_2O_3.

2. Phosphoric acid digestion of refractory oxide single crystals followed by flame spectrometry analysis

The most commonly used methods of digestion of sparingly soluble substances have a number of shortcomings. Alkali fusion techniques result in high blank values for alkali-earth elements and totally unsuitable for alkali elements determination. Heating of single crystals samples with different acid mixtures is time consuming and not always effective procedure even when autoclave digestion method is used (Foner,1984; Haines, 1988; Homeier et al., 1988; Krasil'shchik et al., 1986; Morikava, 1987; Otruba, 1990). It is shown, that for the acid digestion of minerals and oxides condensed phosphoric acid can be successfully applied (Bock, 1979; Hannaker, 1984; Mizoguchi, 1978). It can be obtained by dissolving of P_2O_5 in

75 % orthophosphoric acid. Also, upon heating orthophosphoric acid, condensation of the phosphoric groups can be induced by driving off the water formed from condensation (Mizoguchi, 1978; Zolotovitskaya et al., 1984). Sometimes, dehydration is performed at the reduced pressure and temperature 260…300 °C (Corbridge, 1990). Condensation of orthophosphoric acid leads to forming the mix of polyphosphoric acids which are strong complexing agents having high dissolving ability (Zolotovitskaya et al., 1984, 1997; Trachevskii et al. 1996).

In this section the main results on application of condensed phosphoric acid for the digestion of wide range of compounds are presented. Among the investigated materials were aluminium oxide single crystals and fusion mixtures; magnesium aluminium spinel doped with Fe, Ni, Co, V, Ti; yttrium aluminium garnet; gallium scandium gadolinium garnet doped with Nd; cadmium tungstate; strontium titanate and other functional materials.

2.1 Thermally activated acid-base transformations in the system "condensed phosphoric acid – oxide material"

The composition of the polyacids obtained after heating of an orthophosphoric acid, the mechanism of solvent action of polyphosphoric acids as well as the composition of complex compounds formed during digestion of analysed materials in condensed phosphoric acid has been studied by heteronuclear NMR spectroscopy (Trachevskii et al., 1996). The NMR spectra (^{31}P, ^{17}O, ^{27}Al, ^{113}Cd) of liquid and solid samples were recorded on Bruker CXP-200 spectrometer using single and multipulse sequences as well as magic angle spinning techniques.

Heating-up to 400 °C of 75 % orthophosphoric acid leads to driving off the water and to forming linear polyphosphoric acids. Identification of the obtained substances was performed using ^{31}P NMR spectroscopy of water solutions of corresponding salts as well as water solutions derived from melts partly neutralized by ammonia.

In the investigated temperature range (20…400 °C) a set of processes of orthophosphoric acid transformations can be expressed by the following equation:

$$n\ H_3PO_4 \leftrightarrow H_{n+2}P_nO_{3n+1} + (n\text{-}1)\ H_2O$$

where $n = 1…12$.

From the point of view of prediction of reactivity of dissolvent the effect of upfield shift of signals in ^{31}P NMR spectra, corresponding to consecutive thermal generation of polyphosphatic homologues (with $n=2, 3, 4$) (Trachevskii et al., 1996) is important. The analogous trend of change of signals was observed at shift of acid-base equilibriums towards augmentation of protonation of anions:

$$PO_4^{3-} + n\ H^+ \leftrightarrow H_{3\text{-}n}PO_4, \qquad\qquad n = 0…3$$

$$P_2O_4^{4-} + n\ H^+ \leftrightarrow H_{4\text{-}n}P_2O_7, \qquad\qquad n = 0…4$$

$$P_3O_{10}^{5-} + n\ H^+ \leftrightarrow H_{5\text{-}n}P_2O_7, \qquad\qquad n = 0…5$$

With driving off the water, the equilibrium $H_3PO_4 + H_2O \leftrightarrow H_3O^+ + H_2PO_4^-$ is shifting towards formation of molecular form H_3PO_4. At these conditions the strongest base remaining in system is a orthophosphoric acid. The nearest homologues have more strongly pronounced acid function and the following equilibriums take place:

$$H_3PO_4 + H_4P_2O_7 \leftrightarrow H_4PO_4^+ + H_3P_2O_7^-$$

$$H_3PO_4 + H_5P_3O_{10} \leftrightarrow H_4PO_4^+ + H_4P_3O_{10}^-$$

Thus, along with consumption of H_3PO_4 in condensation processes the yield of the molecular di - and triphosphate forms increases. Also, formation of the highest, more acidic homologues leads to augmentation of probability of formation $H_4PO_4^+$.

The obtained data allows us to consider that the solvent action factors are: (1) acid-base transformations of O^{2-} anion in the reaction with $H_4PO_4^+$ cation with formation of the following forms where acid function increases in the order: $O^{2-} < OH^- < H_2O < H_3O^+$, (2) complexing properties of di - and triphosphate anions. A complex formation with other polyphosphates is less probable owing to a smaller yield of these compounds and high viscosity of medium.

The following empirical characteristics describing processes of digestion of different compounds in condensed phosphoric acid were obtained: (a) conditional solubility (mass of dissolved material per 1 cm^3 of H_3PO_4); (b) optimum temperature of dissolution (t_{opt}, °C); (c) average dissolution rate at fixed temperature (V, g min^{-1}); (d) degree of dissolution (a, %). Kinetics of dissolving a-Al_2O_3 and $CdWO_4$ was investigated. A mixture of 0.3 g of a-Al_2O_3 and 10 ml of H_3PO_4 or 0.6 g of $CdWO_4$ and 5 ml of H_3PO_4 was kept at fixed temperature for a definite time. Then, after centrifugation, concentration of Al in solution was determined by flame atomic absorption spectrometry and concentrations of Cd and W were determined by X-ray fluorescence spectrometry (Mirenskaya et al., 1994).

Kinetic curves of dissolving Al_2O_3 and $CdWO_4$ (Fig. 1) are s-shaped where initial and finite parts correspond to low dissolution rate which is limited by evaporation of water and decreasing of content of solid phase in reaction volume, respectively. A middle part of curves is linear and corresponds to the maximum dissolution rate (97 % of $CdWO_4$ and 80-85 % of Al_2O_3 go in solution) when simultaneous processes of acid-base transformations, polycondensation of H_3PO_4 and forming of metal phosphate complexes are take place.

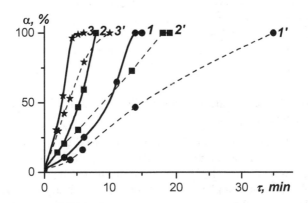

Fig. 1. Kinetics of dissolving Al_2O_3 (1' - 3') and $CdWO_4$ (1 -3) in condensed phosphoric acid: 215 °C (1, 1'), 270 °C (2, 2'), 330 °C (3, 3').

In Fig. 2 dependences of dissolution rate of different fusion mixtures and single crystals on temperature are presented. The mass of samples was 0.05...2.00 g and volume of H_3PO_3 was 10 ml for Al_2O_3 or TiO_2 and 5 ml for WO_3. Dramatic increase in dissolution rate is observed at 210...270 °C for charge and at 250...300 °C for single crystals. Optimum values of temperature for dissolving of fusion mixtures and single crystals were 270 and 300 °C, respectively. It was unreasonable to dissolve such materials at higher temperatures due to formation of sparingly soluble vitreous products. The obtained data were used for optimization of sample preparation procedures of some oxide materials (Table 1).

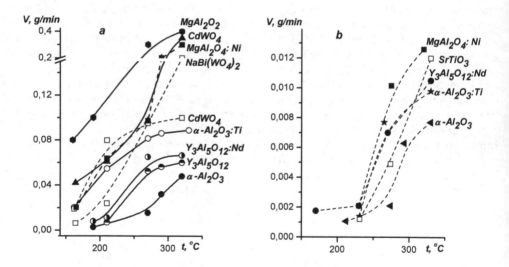

Fig. 2. Dependence of dissolution rate of different fusion mixtures (–) and single crystals (- - -) on temperature

NMR-spectroscopy data (Trachevskii et al., 1996) has allowed us to explain high hydrolytic stability of phosphate decomposition products. Analysis of ^{31}P NMR spectra for the system of K_2WO_4 - H_3PO_4 has shown that after dilution there are significant amounts of diphosphate anions and their complexes with tungsten in solutions. Presence of specified diphosphate complexes as well as solvable orthophosphate complexes is the main reason of stability of diluted solutions. Obviously, similar processes cause the same behavior of other oxide materials studied in this work.

Material	Sample mass, g	Temperature, °C	Digestion time, minutes	Conditional solubility, g cm^{-3}
α-Al$_2$O$_3$ charge	0.3...0.4	270	16	0.05
α-Al$_2$O$_3$ single crystal	0.15	290...300	30	0.016
α-Al$_2$O$_3$:Ti charge	0.3	270	13	0.09
α-Al$_2$O$_3$:Ti single crystal	0.2	300	25	0.02
mAl$_2$O$_3$ nY$_2$O$_3$ charge	0.3...0.4	270	11	0.07
mAl$_2$O$_3$ nY$_2$O$_3$ kNd$_2$O$_3$ charge	0.3...0.5	270	10	0.09
mAl$_2$O$_3$ nY$_2$O$_3$ kNd$_2$O$_3$ single crystal	0.2...0.3	300...320	20	0.05
mAl$_2$O$_3$ nMgO charge	0.3...0.6	270	5	0.20
mAl$_2$O$_3$ nMgO:Ni charge	0.3	270	6	0.20
mAl$_2$O$_3$ nMgO:Ni single crystal	0.1	270	18	0.03
SrTiO$_3$ charge	0.1	330	9	0.010
CdWO$_4$ charge	0.6...0.8	265	9	> 0.40
CdWO$_4$ single crystal	0.6...0.8	270	15	0.40
NaBi(WO$_4$)$_2$ single crystal	0.4...0.8	300...330	8	> 0.40

Table 1. Optimum conditions of digestion of some single crystals and raw materials in condensed phosphoric acid

2.2 Flame spectrophotometric analysis of phosphate solutions of refractory oxides

The combination of a highly effective digestion technique with flame atomic-emission (FAES) and atomic-absorption (FAAS) analysis of the phosphate solutions was used for the determination of Na, K, Ca in α-Al$_2$O$_3$; Ca in SrTiO$_3$; Na, K, Ca, Mg in CdWO$_4$; dopants Li, Na, Mg in β-Al$_2$O$_3$; Nd in yttrium aluminium garnet; Ti, V, Fe, Co, Ni in corundum; Li, Rb, Cs, Sr, Ba in CdWO$_4$; a stoichiometric composition of garnets $(n-x)$Y$_2$O$_3$ xNd$_2$O$_3$ mAl$_2$O$_3$ and MgO kAl$_2$O$_3$ where n can vary from 1 to 3, m and k - from 1 to 5, x - from 0 to 0,2.

Flame spectrometry measurements were performed on a spectrophotometer "Saturn" (Ukraine) with use of acetylene-air, propane-butane-nitrous oxide and acetylene-nitrous oxide gas mixtures. The processes taking place during flame atomization of materials and excitation are well studied (Havezov et al., 1983; Haswell, 1991; Hill, 2005; L'vov et al., 1975; Magyar, 1987; Welz , 1999). However, application of flame spectrometry in analysis of solutions of the complex composition especially containing metal phosphate complexes, demands to study of influence of various factors on absorption and emission signals of analytes. Most essential of them are the main composition of solution and conditions of atomization of material in a flame. The last depends on the chosen gas mixture, a fuel/oxidant ratio, zone of photometric observations. The numerous data available in the literature on this problem are ambiguous.

2.2.1 Effects of redox characteristics of the flame and the height of the burner on analytical signals

The redox characteristics of a flame defined by the fuel/oxidant ratio were estimated visually, basing on the height of the inner cone (l, mm) which was changed over a range from 1 to 5 for acetylene-air, propane-butane-nitrous oxide flames and from 1 to 40 for acetylene-nitrous oxide flame. It was considered that acetylene-air and propane-butane-nitrous oxide flames were oxidative, stoichiometrical or reducing when l was equal to 1...2, 3...4 and 5...6, respectively. The corresponding values for the acetylene-nitrous oxide flame were 1...2, 2...35 and 40...45.

In order to investigate effect of redox characteristics of the flame on analytical signals they were recorded at different distances above the burner orifice while l value was constant. It has been found that analytical signals of all elements of interest depend on redox characteristics of a flame. Most distinctly these dependences are appeared for the metals forming hardly dissociated compounds (alkali-earth and rare-earth elements, Al) even when a high-temperature flame is used (Maitra, 1987; Pupyshev. et al., 1990).

As an example, in Fig. 3 dependences of analytical signals of Al and Y on the height of the inner cone of the flame are presented.

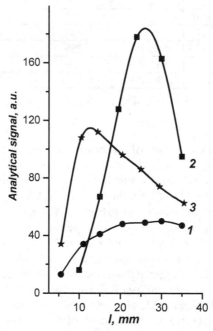

Fig. 3. Dependence of absorption signals of Al (1) and Y (2) and an emission signal of Al (3) in phosphate solutions of $Y_3Al_5O_{12}$ on the inner cone of the flame (l, mm)

It can be seen that at determination of these elements, especially by atomic-emission method, it is necessary to select a fuel/oxidant ratio very carefully. For alkali-earth metals, Mg and metals of subgroup Fe composition of gas mixture is less critical, however also is important enough. So, sensitivity of the determination of alkali-earth metals in the presence of H_3PO_4 in acetylene-air oxidizing flame is ~2 times lower, than in stoichiometrical or reducing flame. In Table 2 optimum operating conditions for the determination a number of elements in phosphate solutions are resulted. It is found that the optimal fuel/oxidant ratio does not depend on the sample matrix. Fig. 4 demonstrates dependences of analytical signals of Mg, Ca, Sr, Ba, Al and Y from the height of the burner for the solutions of different composition. It can be seen that there is a correlation between the maximum of analytical signal and the height of the inner cone. In the presence of a "heavy" matrix the specified maximums shift to a hotter flame zone which is located directly over an inner cone or close to it. In some cases (see curves a, e) maximums are diffuse. Buffer additions of chlorides of alkali metals have no effect on the position of maximum. Also, it is found that analytical signals of alkali metals do not depend considerably on the height of the burner.

Element	Method	Wavelength, nm	Flame[1]	Gas flow rate, L h[-1]		Height of inner cone, mm	Burner height, mm
				Combustible gas	Oxidant		
Li	FAES	670.8	AA (PBNO)	100 (85)	620 (600)	3	6...9
Na		589.0					
K		766.5					
Rb		780.0					
Cs		852.1					
Mg	FAAS	285.2	AA	95	620	2...3	8...9
			PBNO	60	600	2...3	7...8
			ANO	480	580	10	9
Ca	FAES	422.7	ANO	425	580	1...2	5...6
Sr	FAES	460.7	ANO	440	580	2	5...6
Ba	FAES	535.5	ANO	440...450	580	2...4	5...6
Al	FAES	396.1	ANO	490	560	5...10	9
	FAAS	309.2	ANO	530	560	15...20	8
Y	FAAS	410.2	ANO	550	560	25...30	12
Nd	FAES	492.4	ANO	515...530	580	30	10
Ti	FAAS	365.3	ANO	515...530	580	30...35	10
V	FAAS	318.4	ANO	515...530	580	30...35	10
Co	FAAS	240.7	ANO (PBNO)	70 (60)	690 (600)	1...2	6...9
Ni	FAAS	232.0	ANO (PBNO)	70 (60)	690 (600)	1...2	6...9

Table 2. Optimum operating conditions of the atomizer

Generally, change of optimum fuel/oxidant ratio as well as the height of the burner can lead to the considerable decrease in sensitivity and increase uncertainty in FAES and FAAS analyses of phosphate solutions.

---

[1] AA - acetylene-air, PBNO - propane-butane-nitrous oxide, ANO - acetylene-nitrous oxide

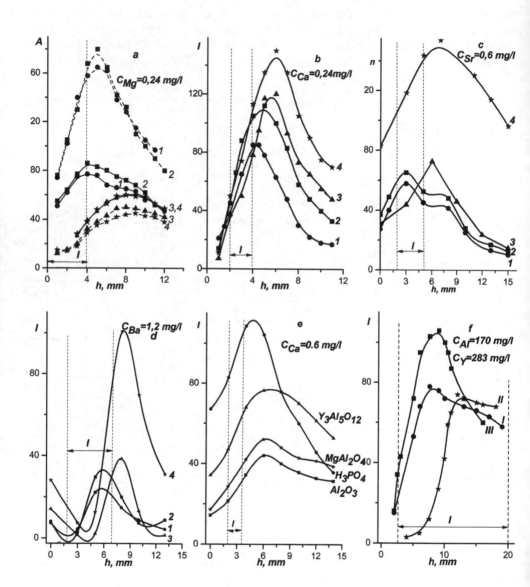

Fig. 4. Dependence of absorption signals (A) of Mg (a), Al^I and Y^{II} (f) and emission signals (I) of Ca (b, e), Sr (c), Ba (d), Al^{III} on the height of the burner (h, mm). Flames: acetylene-air (- · - · -), acetylene – nitrous oxide (---), propane – butane - nitrous oxide (- - -). 1 – model solution of element; 2 – Element + H_3PO_4; 3 - Element + H_3PO_4 + $CdWO_4$; 4 - Element + H_3PO_4 + $CdWO_4$ + NaCl. l – position of inner cone

2.2.2 Solvent effects on the absorption and emission spectra of analysed elements

Effect of phosphate ions on analytical signals of alkali, alkali-earth, transition and other elements was studied in detail in many papers (Magyar, 1987; Smets, 1980). However, these results look rather discordant. This fact can be explained by variety of spectrometers and hardware parameters used as well as by quite different concentrations of phosphate ions and analysed elements.

Therefore, effect of phosphate ions on emission signals of Al, alkali, alkali-earth elements and absorption signals of Mg, Al, Y, Ti, V, Fe, Co, Ni in different flames at the constant concentration of the phosphoric acid (12.4 mas. %) has been investigated. The results presented in Fig. 5 show that effect of a phosphoric acid on analytical signals is various.

In an acetylene-air flame H_3PO_4 shows strong depressing effect on all alkali elements. This effect increases in the order Li <Na, K <Rb <Cs, that is in a good agreement with the literary data (Henrion et al., 1979). In a propane-butane-nitrous oxide flame effect of H_3PO_4 on Li, K signals practically misses. For Rb and Cs the depressing effect is smaller, and for Na is greater, than in an acetylene-air flame.

It is noted that effect of phosphate ions on analytical signals of analysed elements depends essentially on a composition of gas mixture. For example, in an oxidative acetylene-air flames an emission signal of K in the presence of H_3PO_4 decreases for 90 %, whereas in a stoichiometrical flame it decreases for 70 %. It is noted also that a continuous background caused by presence of a phosphoric acid is much lower in a stoichiometrical flame, rather than in oxidative or reducing one. Apparently, character of effect of phosphate ions on analytical signals in various flames is caused not only their temperature, but composition also. It is possible to explain depressing activity of phosphates by formation of hardly vaporable compounds. Such effect is more typical for an acetylene-air flame.

It is found that phosphoric acid in an acetylene-air flame has small depressing effect on an absorption signal of Mg which is incremented with augmentation of the content of this element. In a propane-butane-nitrous oxide flame, effect of phosphates on an absorption signals of traces of Mg is absent, whereas at high concentrations of Mg its analytical signal slightly decreases. Emission signals of Ca, Sr, Ba, Nd and absorption signals of Ti and V in an acetylene-nitrous oxide flame in the presence of a phosphoric acid increase by 20, 22, 13 and 9 %, respectively. Apparently, it is possible to explain these phenomena by the fact that a high temperature of a flame and a reducing medium (red zone) promotes more effective atomisation of substance. Phosphoric acid can interfere with formation of carbides of the Ti and V in a reducing red zone and as it stated in (Magyar, 1987), promotes formation more volatile oxides of metals (for example, V_2O_5 is more flying oxide than other oxides of this element).

The phosphoric acid depresses absorption signals of Fe, Ni, Co (by 20-23 %) both in acetylene-air and propane-butane-nitrous oxide flames, irrespective of their redox characteristics. The received results differ from ones described in (L'vov et al., 1975) in which it is specified that effect of acids on Fe group elements in an oxidative acetylene-air flame misses.

Thus, effect of phosphate ions on analytical signals of analysed elements depends on the physicochemical characteristics of elements, their concentration in phosphate solution and operation parameters of an atomizer.

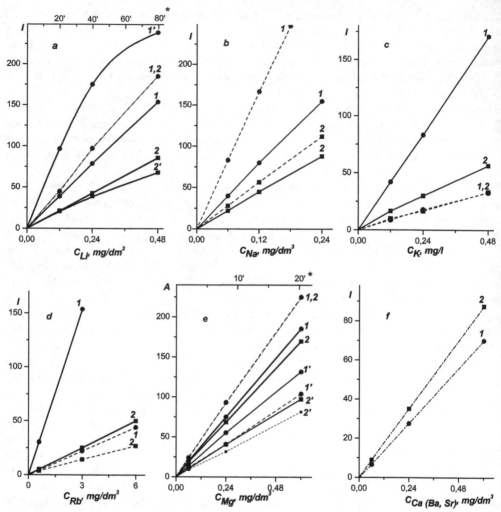

Fig. 5. Effect of H_3PO_4 on emission analytical signals (I) of alkali and alkali-earth elements (a, b, c, d, f) and absorption signals (A) of Mg (e) in acetylene-air (---), acetylene – nitrous oxide (- · - · -), propane- butane – nitrous oxide (- - -). Curves 1, (1') – model solution of element; 2, 2' – element + H_3PO_4. *The burner is perpendicular to the optical axis

2.2.3 Matrix effects

Results of investigation of effect of major elements (Al, W, Ti) of high-melting compounds on analytical signals of analysed elements are presented in Table 3. Concentration of the phosphoric acid was 12.4 mas. %. Measurement conditions matched those given in Table 2. One can see that in comparison with effect of a phosphoric acid, Al increments emission signals of Li and Na. With increase of the Na concentration the rate of this effect decreases. Al shows depressing effect on the emission signal of K, and the rate of this effect depends on redox characteristics of a flame and is incremented with decrease of the content of a combustible gas in a gas mixture. Increase in analytical signals of Na in the yttrium-aluminium garnet matrix is observed, whereas analytical signals of K do not change substantially.

Ele-ment	Concentra-tion, ppm	Flame[2]	Relative intensity of signals					
			H_3PO_4	Al_2O_3	$Y_3Al_5O_{12}$	$MgAl_2O_4$	$CdWO_4$	$SrTiO_3$
Li	0,2...6,0 %	AA	0.22					
	10...500	AA	0.55	1.08			0.98	
	10...500	PBNO	1.00				0.97	
Na	2,0...8,0 %	AA	0.86	1.08		.		
	0,2...1,0 %	AA	0.85	1.14				
	5...200	AA	0.52	1.14	1.12		1.00	
	5...200	PBNO	0.27				1.00	
K	10...500	AA	0.30	0.89	1.01		1.00	
	10...500	PBNO	1.00				1.00	
Rb	10...500	AA	0.17				0.98	
	10...500	PBNO	0.55				0.97	
Cs	10...500	AA	0.14				0.98	
	10...500	PBNO	0.53				1.08	
Mg	0,2...2,0 %	AA	0.74					
	0,2...2,0 %	PBNO	0.81					
	5...50	AA	0.90	0.30			0.55	
	5...50	PBNO	1.00	1.00			0.95	
Ca	5...200	ANO	1.25	0.39	0.78	0.45	1.10	1.55
Sr	10...100	ANO	1.28				1.10	
Ba	10...500	ANO	1.50				1.48	
Ti	200...2400	ANO	1.06	1.30				
V	200...2400	ANO	1.08	1.25		1.25		
Ni	200...2400	AA	0.79	0.50		0.50		
		PBNO	0.76	0.90		0.90		
Co	200...2400	AA	0.81	0.99		1.00		
		PBNO	0.80	1.13		1.12		
Nd	2,0...10,0 %	AA	1.08	2.04				

Table 3. Effect of matrix components on analytical signals of elements

An absorption signal of Mg in the presence of Al decreases slightly and with increase of the Mg concentration the rate of effect of Al increases. Use of a propane-butane-nitrous oxide flame allows analyst to eliminate matrix effect completely.

Aluminium based compounds (Al_2O_3, magnesium aluminium spinel, yttrium-aluminium garnet) show depressing effect on analytical signals of Ca, even in a high-temperature acetylene-nitrous oxide flame. The rate of depressing effect increases in the order yttrium-aluminium garnet < magnesium aluminium spinel < Al_2O_3 which is correlated with aluminium content in these compounds. Increase in analytical signals of Ca in $SrTiO_3$ matrix can be explained by the fact that Sr acts as an ionization buffer (Havezov,1983).

---

[2] AA - acetylene-air, PBNO - propane-butane-nitrous oxide, ANO - acetylene-nitrous oxide

Results of our investigations have shown that absorption signals of hardly volatile elements such as Ti and V, and emission signals of Nd in the presence of Al_2O_3 are increased. Probably, it can be explained by decrease of partial pressure of elemental oxygen and the considerable facilitation of atomisation of oxides of analysed elements in the presence of aluminium (Grossmann, 1987).

In an acetylene-air flame Al demonstrates depressing effect on analytical signals of nickel and cobalt which is more substantial for Ni. On the contrary, absorption signals of Co and Ni in a propane-butane-nitrous oxide flame are increased.

It is known from the literary (Henrion et al., 1979) that W has a depressing effect on emission signals of alkali metals which is incremented in the order Li < Na < K < Rb < Cs. As have shown results of our investigations, in the presence of cadmium and phosphate ions, tungsten acts quite differently. As can be seen from Table 3, a cadmium tungstate matrix does not affect analytical signals of alkali metals in acetylene-air and propane-butane-nitrous oxide flames.

Absorption signals of Mg in a cadmium tungstate matrix in acetylene-air and propane-butane-nitrous oxide flames are reduced by 80 and 30 %, respectively. This matrix has a weak depressing effect on emission signals of alkali-earth elements in an acetylene-nitrous oxide flame.

Thus, character and power of matrix effect in phosphoric acid solutions of examined materials depend on the nature of the element to be analysed, measurement conditions and a composition of a gas mixture. Optimisation of these parameters allows analyst to minimise matrix effect.

2.2.4 Interference effects of analysed elements in phosphate solutions

Avoiding of interference effects is of significant importance in the high-precision determination of a stoichiometric composition of functional materials. Study of interferences of Mg, Al and Y in yttrium-aluminium garnet and magnesium aluminium spinel was performed using solutions prepared by digestion in a phosphoric acid of oxides and mixtures of oxides of these elements. The content of each oxide in a mixture matched to stoichiometric relationships $3Y_2O_3 \cdot 5Al_2O_3$ and $MgO \cdot Al_2O_3$ or by 2-5 times differed from them. The content of dissolved oxides in solutions was 0.5-3 g L^{-1}. It was found that in an acetylene-air flame a strong depressing effect of Al on absorption signals of Mg is observed. In a propane-butane-nitrous oxide flame this effect is weaker, whereas in an acetylene-nitrous oxide flame it is insignificant (Fig. 6). Authors (Havezov, 1983; Lueske, 1992) explained this fact by formation of $MgAl_2O_4$ spinel which stability depends on the composition of a flame and temperature. Magnesium reduces emission and absorption signals of Al that leads to parallel shift of the calibration line without change its bias. In the presence of Y the small increase of emission signals of Al is observed. At the same time Y does not affect absorption signals of Al. Analytical signals of Y decrease in the presence of Al (Fig. 6b).

For elimination of such interferences a lanthanum nitrate modifier is found to be the most effective. Introduction in solutions not less than 1 mas. % of the lanthanum interfering with forming low-volatile compounds at the moment of evaporation of droplets of an aerosol [], allows to eliminate all interferences almost completely (except depressing effect of Al on absorption signals of Mg in an acetylene-air flame). The content of $n \cdot Y_2O_3 \cdot kAl_2O_3$ or $MgO \cdot kAl_2O_3$ (at k <3) in solutions should not exceed 1 g L^{-1} or 0.5 g L^{-1} of $MgO \cdot kAl_2O_3$ (at k >3). In order to prevent precipitation of lanthanum compounds, analysed solutions should contain not less than 1.4 M of HNO_3.

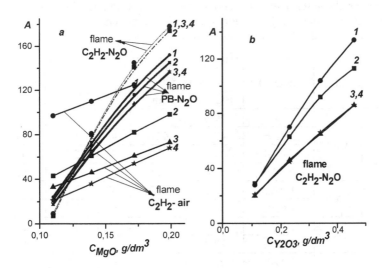

Fig. 6. Influence of Al on absorption signals of Mg (a) in acetylene-air, propane-butane-nitrous oxide and acetylene-nitrous oxide flames; Y (b) in an acetylene-nitrous oxide flame. 1 – Mg (Y) + H_3PO_4; 2 - Mg (Y) + H_3PO_4 + Al; 3 - Mg (Y) + H_3PO_4 + Al + La

It was also found that interferences of Na, K, Ca, Mg during their determination in $CdWO_4$, Al_2O_3, yttrium-aluminium garnet and magnesium aluminium spinel can be eliminated by using of a cesium chloride buffer. In this case, it is possible to determine all these elements from the same solution.

2.2.5 Some practical applications of phosphoric acid digestion of refractory oxide single crystals followed by flame spectrometry analysis

Our investigations have shown efficiency of use of a condensed phosphoric acid for dissolution of different refractory oxide single crystals. The data presented in the previous sections were used for the development of new procedures for determination of main components, dopants and microimpurities in single crystals and raw materials.

For the determination of a stoichiometric composition of yttrium-aluminium garnet and magnesium aluminium spinel, powder reference materials $3Y_2O_3 \cdot 5Al_2O_3$ and $MgO \cdot Al_2O_3$ were used. The content of metals in these compounds was determined precisely by chelatometry. Calibration standards were prepared by dissolving of different masses of a powder reference material in condensed phosphoric acid. All solution contained CsCl as a spectroscopic buffer. An acetylene-nitrous oxide flame was used in the most cases but the determination of Mg is also possible with use of a propane-butane-nitrous oxide flame. Tables 4 and 5 contain data on the methods accuracy check and comparison with the results of chelatometry (in terms of oxides).

As can be seen from the Table 4, high concentrations of Al can be determined by both FAES and FAAS methods but the last one is more preferable. It was found that K and Na can be determined in sparingly soluble tungstates $KGd(WO_4)_2$ and $BiNa(WO_4)_2$ after their dissolving in a condensed phosphoric acid with a confidence interval not exceeding 0.5...0,7 mas. %. Moreover, the absence of matrix effect allowed carrying out the analysis with use of calibration standards containing solvent and analysed elements only (CsCl was used a spectroscopic buffer for the determination of Na).

Material	Oxide	Method	Abundance, mas. %		df[3]	RSD%	t-test (critical value, P=0.95)
			Stated	Found			
Yttrium-aluminium garnet	Al_2O_3	FAES	42.8	42.7	36	1.2	1.19 (2.02)
		FAAS	42.8	42.8	44	1.2	0 (2.02)
		FAES	31.4	31.2	24	2.1	1.67 (2.06)
		FAAS	31.4	31.5	40	2.2	0.98 (2.02)
	Y_2O_3	FAAS	56.8	56.7	44	1.8	0.68 (2.02)
		FAAS	68.8	69.1	42	2.3	1.31 (2.02)
Magnesium aluminium spinel	Al_2O_3	FAES	70.2	69.9	21	1.0	2.19 (2.08)
		FAAS	70.2	70.3	48	1.0	0.98 (2.01)
	MgO	FAAS	27.2	27.3	35	1.1	1.97 (2.03)
		FAAS (propane-butane-nitrous oxide flame)	27.2	27.2	33	0.8	0 (2.01)

Table 4. Accuracy check of the developed procedures

Compound	Oxide	Technical requirements, mas.%	Found, mas.%	
			FAAS	Chelatometry
$3Y_2O_3 \cdot 5Al_2O_3$	Y_2O_3	57.1 ± 1.3	56.4 ± 0.6	56.8 ± 0.2
	Al_2O_3	42.9 ± 1.0	42.6 ± 0.3	42.7 ± 0.1
$Y_2O_3 \cdot Al_2O_3$	Y_2O_3	68.9 ± 1.0	69.2 ± 1.0	69.2 ± 1.0
	Al_2O_3	31.1 ± 0.5	31.2 ± 0.4	31.2 ± 0.4
$MgO \cdot Al_2O_3$	MgO	27.2 ± 0.5	27.2 ± 0.1	27.2 ± 0.1
	Al_2O_3	70.3 ± 1.3	70.3 ± 0.5	70.2 ± 0.2

Table 5. Comparison of results obtained by chelatometry and FAES (n=3; P=0.95)

Accuracy of developed procedures for the determination of dopants in aluminium oxide based materials was checked by comparison with results obtained by an independent method as well as using a different method of sample preparation (Table 6). The proposed procedures are more precise than arc atomic emission analysis and do not require large amounts of samples as is required in X-ray fluorescence analysis.

---

[3] Number of degrees freedom

| Compound | Element | Concentration, mas. % | | Arc atomic-emission spectroscopy |
| | | Flame spectrometry | | |
		Condensed phosphoric acid digestion	Fusion with LiBO$_2$	
α-Al$_2$O$_3$	Ti	$(2.4\pm0.2)\cdot10^{-2}$	$(2.5\pm0.1)\cdot10^{-2}$	
	V	$(8.4\pm0.4)\cdot10^{-2}$		$(10\pm2)\cdot10^{-2}$
	Ni	$(2.5\pm0.2)\cdot10^{-2}$		$(3.0\pm0.3)\cdot10^{-2}$
β-Al$_2$O$_3$	Na	4.4±0.1	4.3±0.2	
		3.0±0.1	2.8±0.2	
		5.0±0.1	5.1±0.2	
		8.0±0.1	8.0±0.2	
Yttrium-aluminium garnet	Nd	0.77±0.04	0.76±0.02	0.85±0.06
		1.24±0.04	1.27±0.03	1.40±0.09

Table 6. Results of the determination of dopants in aluminium oxide based materials (n=3; P=0.95)

Some metrological characteristics of developed procedures are presented in the Table 7. Thus, phosphoric acid digestion is a very effective and universal method of sample preparation of refractory oxide single crystals and raw materials. In some cases this is the only method suitable for the successive determination of alkali metals in single crystals since sample solutions obtained by alkali fusion may have extremely high blank signals.

Compound	Element	Concentration range, мас.%	RSD%	Typical abundance of the element, mas. %
		Dopants		
α-Al$_2$O$_3$, mAl$_2$O$_3\cdot n$MgO	Ti	$1\cdot10^{-2}...2.0$	14...4	$5\cdot10^{-3}...0.4$
	V	$5\cdot10^{-3}...2.0$	10...3	$5\cdot10^{-3}...0.4$
	Co (AA)	$7\cdot10^{-3}...2.0$	15...2	0.01...0.6
	Co (PBNO)	$7\cdot10^{-3}...2.0$	13...2	
	Ni (AA)	$7\cdot10^{-3}...2.0$	12...2	$<7\cdot10^{-3}...0.13$
	Ni (PBNO)	$7\cdot10^{-3}...2.0$	9...2	
β-Al$_2$O$_3$	Li	$1\cdot10^{-3}...10.0$	5...1	$1\cdot10^{-3}...0.7$
	Na	0.2...10.0	4...2	0.7...6.0
	Mg (AA)	0.2...2.0	4...1	0.8
	Mg (PBNO)	0.2...2.0	1	
mAl$_2$O$_3\cdot n$Y$_2$O$_3$	Nd	0.2...6.0	11...5	0.5...4.2
		Impurities		
α-Al$_2$O$_3$	Na	$5\cdot10^{-4}...2\cdot10^{-2}$	14...7	$<5\cdot10^{-4}...0.19$
	K	$1\cdot10^{-3}...5\cdot10^{-2}$	13...2	$<1\cdot10^{-3}...0.06$
	Ca	$1\cdot10^{-3}...2\cdot10^{-1}$	10...2	$1\cdot10^{-3}...0.03$
mAl$_2$O$_3\cdot n$Y$_2$O$_3$	Na	$5\cdot10^{-4}...2\cdot10^{-2}$	12...3	$<5\cdot10^{-4}...7\cdot10^{-3}$
	K	$1\cdot10^{-3}...5\cdot10^{-2}$	13...2	$<5\cdot10^{-4}...9\cdot10^{-3}$
	Ca	$5\cdot10^{-4}...2\cdot10^{-2}$	13...9	$7\cdot10^{-4}...2\cdot10^{-3}$

Compound	Element	Concentration range, мас.%	RSD%	Typical abundance of the element, mas. %
CdWO$_4$	Li	$1 \cdot 10^{-4}...4 \cdot 10^{-3}$	6...3	$<1 \cdot 10^{-4}$
	Na	$1 \cdot 10^{-4}...2 \cdot 10^{-3}$	7...3	$2 \cdot 10^{-4}...1 \cdot 10^{-2}$
	K	$2 \cdot 10^{-4}...2 \cdot 10^{-3}$	6...2	$<2 \cdot 10^{-4}$
	Rb	$5 \cdot 10^{-4}...2 \cdot 10^{-2}$	5...2	$<5 \cdot 10^{-4}$
	Cs	$5 \cdot 10^{-3}...2 \cdot 10^{-1}$	6...2	$<5 \cdot 10^{-3}$
	Mg, Ca	$1 \cdot 10^{-4}...1 \cdot 10^{-2}$	6...2	$<1 \cdot 10^{-4}$
	Sr	$3 \cdot 10^{-4}...1 \cdot 10^{-2}$	2...1	$<2 \cdot 10^{-4}$
	Ba	$1 \cdot 10^{-2}...2 \cdot 10^{-2}$	3	$<5 \cdot 10^{-3}$
SrTiO$_3$	Ca	$5 \cdot 10^{-4}...3 \cdot 10^{-2}$	4...2	$4 \cdot 10^{-4}$

Table 7. Metrological characteristics of procedures of flame spectrometry analysis of single crystals with phosphoric acid digestion at the stage of sample preparation

3. Analysis of aqueous extracts of powders based on aluminium oxide

The traditional methods of sample preparation of aluminium oxide such as fusion with fluxes and selective acid dissolution in autoclaves are time consuming and unsafe.

Digestion of oxide materials in condensed phosphoric acid is express and reliable method. However, a strong depressing effect of phosphoric acid on analytical signals in atomic spectrometry does not allow to provide low detection limits for many elements. Recently, procedures of the direct analysis of oxide materials from aqueous suspension are widely used, but this method (see section 4) requires use of standards identical to analysed samples. This section contains results of the study of aqueous extraction of K, Na, Ca, V, Ni and Co from α-, γ-Al$_2$O$_3$, mAl$_2$O$_3 \cdot n$Y$_2$O$_3$ and mAl$_2$O$_3 \cdot n$MgO used as raw materials for single crystals growing.

3.1 Flame spectrometry determination of alkali metals and Ca in aqueous extracts of α- and γ-Al$_2$O$_3$

Commercially available extra pure γ-Al$_2$O$_3$ powders (Donetsk Plant for Chemical Reagents, Ukraine) were used throughout. a-Al$_2$O$_3$ was obtained by calcination of ammonia alum at 1200 °C during 3 hours (Chebotkevich, 1975), Y$_3$Al$_5$O$_{12}$ was synthesized from Al$_2$O$_3$ and Y$_2$O$_3$ by calcination at 1200 °C (Neiman et al., 1980). a-Al$_2$O$_3$ was powdered in a leucosapphire mortal and sifted through 40, 60, 100 and 160 microns sieves. Particle size of γ-Al$_2$O$_3$ and Y$_3$Al$_5$O$_{12}$ initially was less than 1 μm.

Leaching procedure was carried out in quartz beakers (50 ml) at ambient temperature by mixing of powders with ultrapure water to obtain stable suspension followed by spontaneous sedimentation of a solid phase. Heating of the sample has appeared unacceptable owing to oxide adherence to beaker walls. It was also found that the spontaneous sedimentation of Al$_2$O$_3$ solid phase and Al(OH)$_3$ formed during leaching is an optimum technique of sample preparation. Centrifugation of suspension is found to be unsuitable because of formation of very stable suspension. Especially, this is typical for fine powders of γ-Al$_2$O$_3$ and Y$_3$Al$_5$O$_{12}$.

Effect of a-Al$_2$O$_3$ particle size on the recovery of Na, K and Ca is shown in Fig. 7. One can see that the full extraction of elements is observed up to particle size < 60 microns and for K even to < 100 microns. The average time of the spontaneous sedimentation of these oxide materials depends on the dispersity of powders and is equal to 8 hours for α-Al$_2$O$_3$ and 10-12 hours for γ-Al$_2$O$_3$ and Y$_3$Al$_5$O$_{12}$.

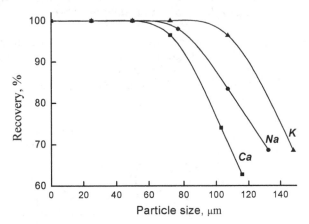

Fig. 7. Effect of a-Al$_2$O$_3$ particle size on the recovery of Na, K and Ca

It is possible to make the following assumptions concerning effects of transferring of impurities of alkali metals and Ca in an aqueous phase during the extraction procedure irrespective of Al$_2$O$_3$ modifications:

- along with increase of specific surface of materials during grinding, a great number of microinhomogeneities and defects are appeared on a surface of particles (Dobrovinskaya et al., 2007; Hanamirova, 1983; Foner, 1984);
- γ-Al$_2$O$_3$ and Y$_3$Al$_5$O$_{12}$ fine powders (<< 1 μm) are partly hydrated and some amounts of Al(OH)$_3$ are always present;
- after calcination of alum at temperature of 1200 °C there is a part of unevaporated sulphates of alkali metals in the obtained aluminium oxide. The full decomposition of suphates takes place at higher temperatures (Chebotkevich, 1975; Hanamirova,1983). Sulphates of alkali metals have a good solubility in water;
- it was experimentally confirmed by other authors (Ermolenko et al., 1971; Hanamirova, 1983; Uksche, 1977 that at temperatures of 600-1200 °C aluminate of alkali and alkali-earth element of spinel type are generated. Owing to a "quasiliquid" state of cationic sublattice ions of alkali metals can migrate freely via interblock clefts and substitute each other in the spinel cluster (Uksche, 1977). Aluminates of alkaline and alkali-earth metals in water are hydrolyzed to Al(OH)3 and hydroxyaluminates (Hanamirova, 1983; Remi, 1973).

Apparently, in aqueous suspensions of Al$_2$O$_3$ and Y$_3$Al$_5$O$_{12}$ all these factors may take place and may have a different contribution depending on the dispersity and "prehistory" of material. It was found that the optimum sample mass/water volume ratio is 0.1 g per 10 ml of water. The time of obtaining of transparent extracts was 10 and 12 hours for a-Al$_2$O$_3$ and γ-Al$_2$O$_3$ (Y$_3$Al$_5$O$_{12}$), respectively. It is a compromise between the sensitivity of analysis and sample preparation time.

Optimum conditions of the flame spectrometry determination of Na, K and Ca in aqueous extracts of Al$_2$O$_3$ and Y$_3$Al$_5$O$_{12}$ have been studied. It has been found that Na, K and Ca increase analytical signals of each other. Al and Y up to concentration of 10 mg L^{-1} do not affect analytical signals of Na and Ca but increase the analytical signal of K. Using of CsCl as a spectroscopic buffer (2 g L^{-1}) has allowed us to eliminate interference effects completely.

The results of analysis of aluminium based oxide materials obtained using the proposed method of sample preparation and digestion in condensed phosphoric acid are shown in Table 8.

One can see that aqueous extraction technique provides lower limits of determination (LOD) of Na, K and Ca and does not yield to more complicated techniques described in (Krasil'shchik et al., 1989; Slovak et al., 1981). The developed technique can be used for the express quality control of fusion mixtures of Al$_2$O$_3$ and Y$_3$Al$_5$O$_{12}$ of different dispersity and for the determination of alkali metals dopants in Al$_2$O$_3$ charge.

Material	Element	Aqueous extracts			Digestion in condensed phosphoric acid		
		Found, ppm	RSD%	C_{min}, ppm	Found, ppm	RSD%	C_{min}, ppm
α- Al$_2$O$_3$	Na	9,1±0,4	3	0.2	11±2	10	5
		67±2	2		66±5	6	
		28±1	2		28±2	4	
γ- Al$_2$O$_3$		14±1	4		14±2	9	
α- Al$_2$O$_3$	K	9,1±0,3	3	0.2	< 10	13	10
		43±2	1		40±4	7	
		130±3	2		140±8	4	
γ- Al$_2$O$_3$		51±3	2		53±4	2	
α- Al$_2$O$_3$	Ca	20±1	3	0.2	17±2	7	5
		29±1	3		28±2	4	

Table 8. Comparison of results obtained using two different methods of sample preparation (n=3; P = 0.95)

3.2 Flame spectrometry determination of dopants Co and Ni in subacid aqueous extracts of α- and γ-Al$_2$O$_3$

The possibility of application of subacid aqueous extraction technique combined with ultrasonic and microwave treatment for the determination of Co and Ni in α- and γ-Al$_2$O$_3$ fusion mixtures was studied.

An ultrasonic dispenser UZDN-A (Selmi, Ukraine) with operating frequency of 22 kHz a microwave digestion system MDS-2000 (CEM, USA) were used for the treatment of subacid aqueous suspensions of Al$_2$O$_3$ powders. Fractions with particle size less than 40 μm were used throughout. 0.1 g of α-Al$_2$O$_3$:Co(Ni) or mAl$_2$O$_3$ nMgO:Co(Ni) was mixed with 10 ml of water and 0.2 ml of HNO$_3$ and then ultrasonic treatment was applied. Solutions obtained after the spontaneous sedimentation of solid phase were used for analysis.

It was found that recovery of Co and Ni is quite similar and increases with augmentation of intensity of ultrasonic irradiation and processing time (Fig. 8 and 9). The maximum in Fig. 8 corresponds to optimum cavitation conditions. However, the maximum recovery at the optimum conditions (I = 78 W, τ =15 min) was only 50 %. It was assumed that only those atoms located on grains boundary (not entered into crystal lattice owing to various ionic radiuses of cobalt (nickel) and aluminium (Dobrovinskaya et al., 2007) have been extracted.

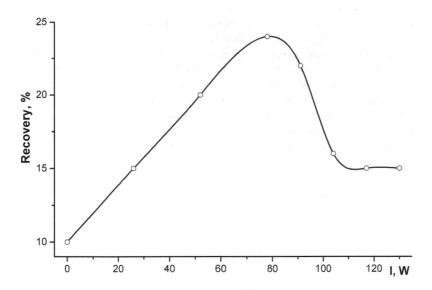

Fig. 8. Dependence of recovery of Co(Ni) on the ultrasonic dispenser driving power (τ =5 min)

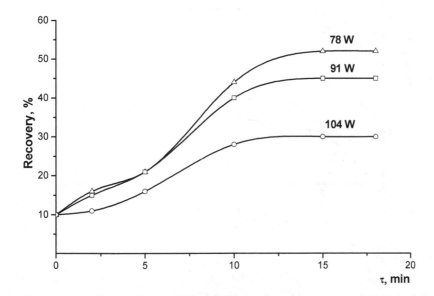

Fig. 9. Dependence of recovery of Co(Ni) on the exposure time

Thus, ultrasonic treatment of aqueous suspensions of Al_2O_3 powders is found to be unsuitable for quantitative analysis of extracts. However, as have shown our further investigations, it is possible to use successfully suspensions of α-Al_2O_3 for direct determination of Co and Ni dopants (see section 4).

Extraction of Co and Ni from subacid aqueous suspensions of Al_2O_3 powders using microwave heating in closed vessels was found to be more effective. Dependence of recovery of Co and Ni on time of exposure (at pressure of 100 psi) and acidity of aqueous suspensions was studied. 0.05, 0.2, 0.5 and 1 ml of HNO_3 were added to 10 ml of aqueous suspensions of Al_2O_3 in order to obtain the samples with different acidity. As can be seen from Fig. 10, the most principal parameter is exposure time. Obviously, the transfer of dopants into solution under microwave heating is caused by deterioration of surface layers and activation of diffusion-controlled transfer processes.

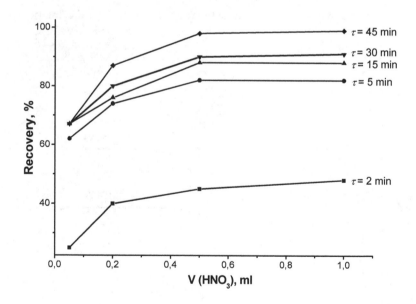

Fig. 10. Dependence of recovery of Co(Ni) on time of exposure and acidity of aqueous suspensions

Accuracy of the developed procedure for the determination of dopants in aluminium oxide based materials was proved by comparison with the results obtained by FAAS after digestion of samples in condensed phosphoric acid (Table 9). The obtained results are in good agreement with each other.

The proposed sample preparation technique has allowed us to improve limits of determination of Co and Ni owing to exclusion of acid solvents from analytical procedure and use of aqueous calibration standards.

Material	Element	Aqueous extracts			Digestion in condensed phosphoric acid		
		Found, ppm	RSD%	C_{min}, ppm	Found, ppm	RSD%	C_{min}, ppm
α-Al$_2$O$_3$: Co(Ni)	Ni	800 ± 20	2	10	810 ± 20	4	50
	Co	190 ± 10	4		210 ± 30	5	
mAl$_2$O$_3$ · nMgO: Co(Ni)	Ni	1600 ± 20	1	10	1610 ± 20	5	50
	Co	790 ± 20	2		810 ± 20	4	

Table 9. Comparison of results of the determination of Co and Ni with use of two different sample preparation techniques ($n = 10$; P = 0.95)

4. Direct flame atomic absorption determination of Co and Ni dopants from aqueous suspensions of aluminium oxide based materials

Whereas Co and Ni dopants cannot be extracted quantitatively to aqueous phase during ultrasonic treatment of Al$_2$O$_3$ suspensions, it is possible to use directly such suspensions in analysis by flame atomic absorption spectrometry. Under ultrasonic irradiation the average size of suspended particles of a powder of Al$_2$O$_3$ decreases and suspension of very good stability and uniformity is formed. This is an important factor which affects degree of atomization and detection limits. For example, analytical signals of Co and Ni in such suspensions are two times higher in comparison with those obtained in suspensions prepared at ambient temperature by mechanical agitation. If there are certified reference materials available, this analysis technique has a number of benefits such us speed and low labor expenditures.

In this work calibration standards were prepared by using melt fusion technique. It was found that for the determination 100...2000 ppm of Co(Ni) it is necessary to have not less than five calibration standards with concentrations of an analyte increased by a factor of 2-2.5. Calibration standards were prepared as follows. Extra pure ammonia alum was melted in platinum crucible on a hot plate and then cobalt (or nickel) sulphate preliminary calcined at 1200 °C was added. The melt was thoroughly stirred and dried under an infrared lamp. The solids was placed into a muffle furnace and heated slowly during 4 hours up to 1100 °C with subsequent holding during 3 hours with subsequent calcination at this temperature during 3 hours. After cooling aluminium oxide was ground to powder in a teflon mortal during an hour and sifted through a 40 μm sieve. Content of dopant in the prepared samples was approved by chelatometry. All subsequent procedures were the same as those described above.

Determined results were consistent with those obtained by analysis of samples digested in condensed phosphoric acid (Table 10).

As can be seen from Table 10, the proposed method has lower relative standard deviations as compared to the acid digestion procedure.

Element	Aqueous suspensions		Digestion in condensed phosphoric acid	
	Found, ppm	RSD%	Found, ppm	RSD%
Co	520 ± 20	4	510 ± 30	5
	990 ± 20	2	1100 ± 50	4
Ni	1100 ± 20	2	1100 ± 50	4
	2100 ± 40	2	2200 ± 70	3

Table 10. Comparison of results of the determination of Co and Ni with use of two different sample preparation techniques ($n = 6$; $P = 0.95$)

5. Conclusion

The results presented in this chapter show that flame atomic spectrometry can be successfully used for the express control of stoichiometric and impurity composition of refractory oxide single crystals such us Al_2O_3, yttrium-aluminium garnet, cadmium tungstate, strontium titanate. Precision and accuracy of analytical procedure as well as metrological characteristics depend strongly on the technique of sample preparation. Use of condensed phosphoric acid as a solvent allows in comparison with traditional techniques of decomposition to unify a sample preparation procedure and to reduce the time of analysis by 5-10 times.

Optimum conditions of digestion of refractory oxide materials in condensed phosphoric acid are established. Influence of solvent, matrix and chemical interference effects on analytical signals of analysed elements with use of different mixed gas flames are also studied. It is shown that in some cases use of a propane-butane-nitrous oxide flame which is preferable and can significantly depress effects of macrocomponents and solvent. Determination of stoichiometric composition of single crystals requires paying much more attention to interference effects as well as matching of analysed materials to calibration standards as compared to impurities determination procedures.

The developed analytical procedures allow to evaluate major elements content such us Mg, Al and Y in different oxide compounds with relative standard deviations (%) 0.8...1.2, 1.0...1.5 and 1.5...2.5, respectively.

The proposed sample preparation methods for water extracts of aluminium oxide powders and yttrium-aluminium garnet allows to carry out an express determination of alkaline metals, calcium, nickel and cobalt in raw materials for single crystals growing. These sample preparation methods ensure low blank signals and better reproducibility of analytical signals as compared to acid digestion technique.

Limits of determination of impurities in all investigated materials after phosphoric acid digestion were 1...5 ppm. Na, K and Ca can be determined in aqueous extracts of Al_2O_3 with the limit of determination of 0.2 ppm.

It was shown that Co and Ni dopants can be determined by FAAS directly from aqueous suspensions of aluminium oxide based materials.

6. References

Bock R. (1979). *A handbook of decomposition methods in analytical chemistry.* Glasgow, International Textbook Co. ISBN 0700202692 [Russian]

Chebotkevich G.V. (1975). Calsination of aluminum ammonium alum. *Zhurnal prikladnoy khimii,* Vol. 48, No.2, pp. 414-417, ISSN 0044-4618 [Russian]

Corbridge D.E.C. (1990). *Phosphorus: an outline of its chemistry biochemistry and technology.* 4-th edition. Amsterdam. Elsevier, ISBN 044-487-4380

Dobrovinskaya E.; Lytvynov L.; Pischik V. (2007). *Sapphire in science and engineering.* Kharkov. STS «Institute for Single Crystals» National Academy of Sciences of Ukraine. ISBN 978-966-02-4589-1 [Russian]

Ermolenko N.F.; Efros M.D.; Oboturov A.V. (1971). *Regulirovanie poristoi struktury okisnykh adsorbentov I katalisatorov.* Minsk. Nauka i teknologia. [Russian]

Foner H.A. (1984). High-pressure acid dissolution of refractory alumina for trace element determination, *Analyt. Chem,* Vol. 56, No. 4, pp. 856-859, ISSN 0003-2700

Grossmann O. (1987). Application of experimental design for investigating the determination of titanium in glass ceramics by flame atomic absorption spectrometry, *Anal. chim. acta.* V.203, No 1, pp. 55-66, ISSN 0003-2670

Haines J.(1988). Sample preparation for AA, *Lab Pract,* Vol. 37, No. 10 pp. 85-86, 88, ISSN 0023-6853

Hanamirova A. A. (1983). *Akumina and methods of reduction of the maintenance to it of impurity.* Er.: AN Arm. SSR [Russian]

Hannaker P.; Qing-Lie H. (1984). Dissolution of geological material with orthophosphoric acid for major-element determination by flame atomic absorption spectroscopy and inductively-coupled plasma atomic-emission spectroscopy, *Talanta,* Vol. 31, No. 12, pp. 1153-1157, ISSN 0039-9140

Haswell S.J.(1991). *Atomic absorption spectrometry.* Amsterdam. Elsevier, ISBN 0444882170

Havezov I.P.; Tsalev D.L. (1983). *Atomic absorption analysis.* Leningrad, Khimia, [Russian]

Henrion G.; Marquardt D.; Lunk H.-J. et al. (1979). Flammenphotometrische bestimmung von kaliumspuren in wolfram, *Z. Chem.,* Vol. 19, No. 10, pp 383-384, ISSN 0044-2402.

Hill S.J. (2005). Atomic absorption spectrometry/ Flame. In: *Encyclohedia of Analytica Science (Second edition)*; Townshend A. (Ed.), Elsevier Ltd. pp. 163-170, ISBN 978-0-12-369397-6

Homeier E.H.; Kot R.J.; Bauer L.J. et al. (1988). Dissolution of alfa alumina for elemental analysis by ICP-AES, *ICP Inf. Newslett.*, Vol. 13, No. 11, p. 716, ISSN 0161-6951

Krasil'shchik V.Z.; Sukhanovskaya A.I.; Voronina G.A. et al. (1989). Atomic-emission and chemical atomic- emission analysis of aluminium compounds using inductive high-frequency discharge, *Zhurnal analiticheskoi khimii*, Vol. 44, No. 10, pp. 1878-1884, ISSN 0044-4502 [Russian]

Krasil'shchik V.Z.; Zhiteleva O.G.; Sokol'skaya N.N. et al. (1986). Autoclave vapour-phase decomposition of some difficultly soluble substances, *Zhurnal analiticheskoi khimii*, Vol. 41, No. 4, pp. 586-590, ISSN 0044-4502 [Russian]

Luecke W. (1992). Systematic investigation of aluminium interferences with the alkaline earths in flame atomic absorption spectrometry. 2. The behavior of magnesium and calcium, *Fr. Z. anal. Chem.*, Vol. 344, No. 6, pp. 242-246, ISSN 0937-0633

L'vov B.V.; Kruglikova L.P.; Polzic L.K. et al. (1975). Theory of flame atomic absorption analysis. Communication 4. Calculated temperature values and equilibrium composition of acetylene- nitrous oxide and acetylene-air flames, *Zhurnal analiticheskoi khimii*, Vol. 30, No.5, pp. 846-852, ISSN 0044-4502 [Russian]

L'vov B.V.& Orlov N.A. (1975). Theory of flame atomic absorption analysis. Communication . Effect of redox characteristics of flames and chemical form of atomized compounds on the degree of aerosol vaporization, *Zhurnal analiticheskoi khimii*, Vol. 30, No.9, pp. 1661-1667, ISSN 0044-4502 [Russian]

Magyar B. (1987). Fundamental aspects of atomic absorption spectrometry, *CRC Crit. Rev. Analyt. Chem.*, Vol. 17, No. 2, pp. 145-191, ISSN 0007-8980

Maitra A.M. (1987). Organophosphorus acid interferences in flame atomic absorption spectrometry, *Anal. chim. acta*, Vol. 193, No. 2, pp. 179-191, ISSN 0003-2670

Mayer C.S., Tinoco C.M., Arraes I.P. (1985). Determination of chromium in geochemical sample by atomic absorption spectrometry, *Atom. Spectroscopy*, Vol. 6, No. 4, pp.121-122, ISSN 0195-5373

Mirenskaya I.I., Shevtsov N.I., Blank A.B. et al. (1994). X-ray spectral determination of the stoichiometry of metal tungstate single crystals, *Zhurnal analiticheskoi khimii*, Vol. 49, No. 5, pp. 522-524, ISSN 0044-4502 [Russian]

Mizoguchi T.; Ishii H. (1978). Analytical application of condensed phosphoric acid – I. determination of ferrous and total iron ores after decomposition with condensed phosphoric acid. 1978, *Talanta*. Vol. 25, No 6, pp. 311-316, ISSN 0039-9140

Morikava H.; Ishizuka T. (1987). Determination of impurities in barium titanate by inductively coupled plasma atomic emission spectrometry, *Analyst*, Vol. 112, No. 7, pp. 999-1002, ISSN 0003-2654

Nagornaya L.; Onyshchenko G.; Pirogov E. et al. (2005). Production of the high-quality $CdWO_4$ single crystals for application in CT and radiometric monitoring, *Nuclear*

Instruments and Methods in Physics Research, Section A, Vol. 537, No. 1-2, pp.163-167, ISSN 0168-9002

Neiman A.Ya.; Tkachenko E.V.; Kvichko L.A. et al. (1980). Conditions and macromechanisms of the solid-phase synthesis of yttrium aluminates, *Zhurnal neorganicheskoi khimii,* Vol. 25, No.9, pp. 1294-1297, ISSN 0044-457X [Russian]

Otruba V.; Navrátil D. (1990). Stanoveni hliniku emisni plamenovou spectrometrii, *Chem. Listy,* Vol. 84, No. 8, pp. 862- . ISSN 1213-7103

Patnaik P.; Dean J.A. (2004). *Dean's analytical chemistry handbook.* 2nd ed. Mc GRAW-Hill New York. ISBN 0-07-141060-0

Pupyshev A.A.; Moskalenko N.I.; Muzgin V.N. et al. (1990). Atomisation of elements I acetylene-nitrous oxide flame, *Zhurnal analiticheskoi khimii,* Vol. 45, No. 12, pp. 2389-2399, ISSN 0044-4502 [Russian]

Remi H. (1973). *Lehrbuch Der Anorganishen Chemie . I und II.,* Akadem Verlagsgesellschaft, Geest Portig.,Leipzig

Rice T. D. (1977). Comparison of dissolution methods for the determination of potassium in rocks and minerals by atomic absorption spectrometry, *Anal. chim. acta,* Vol. 91, No. 2, pp. 221-228, ISSN 0003-2670

Slovak Z., Docekal B. (1981). Determination of trace metals in aluminium oxide by electrothermal atomic absorption spectrometry with direct injection of aqueous suspensions, *Anal. chim. acta.* V.129, No 1, pp. 263-267, ISSN 0003-2670

Smets B. (1980). Vaporisation interference of sulphate and phosphate anions on the calcium flame atomic absorption signal, *Analyst,* Vol. 105, No. 5, pp. 482-490, ISSN 0003-2654

Stratis J.A.; Zachariadis G.A.; Dimitrakoudi E.A. et al. (1988). Critical comparison of decomposition procedures for atomic absorption spectrometric analysis of prehistorical ceramics, *Fr. Z. anal. Chem.,* Vol. 331, No. 7, pp. 725-729, ISSN 0937-0633

Trachevskii V.V; Druzenko T.V; Potapova V.G. et al. (1996). Acidic-alkaline transformation of orthophosphoric acid on heating, *Ukrainskii khimicheskii zhurnal,* Vol. 62, No. 6, Pu pp. 84-88, ISSN 0041-6045 [Russian]

Trachevskii V.V; Druzenko T.V; Potapova V.G. et al. (1996). NMR studying of metal oxide dissolution in phosphoric acids, *Ukrainskii khimicheskii zhurnal,* Vol. 62, No. 6, pp. 84-88, ISSN 0041-6045 [Russian].

Tupitsyna I.A., Grinyov B.V.; KatrunovK.A. et al. (2009). Radiation Damage in CWO Scintillation Crystals With Different Defects, *IEEE Transactions on nuclear science,* Vol. 56, No. 5, pp. 2983-2988, ISSN 0018-9499

Uksche E.A. (1977). Solid electrolytes, Moscow, Nauka [Russian]

Welz B.; Sperling (1999). *Atomic Absorption Spectrometry.* Wiley-VCH, Weinheim, Germany. ISBN 3-527-28571-7

Zolotovitskaya E.S.; Druzenko T.V. & Potapova V.G. (1997). Dissolution of refractory oxide materials in condenced phosphoric acid, *Zhurnal analiticheskoi khimii,* Vol. 52, No. 9, pp. 923-927, ISSN 0044-4502 [Russian]

Zolotovitskaya E.S.; Potapova V.G. (1984). Application of the phosphoric acid as solvent in the analysis of monocrystals based on aluminium oxide, *Zhurnal analiticheskoi khimii*, Vol. 39, No. 10, pp. 1781-1785, ISSN 0044-4502 [Russian]

6

Interference Effects of Excess Ca, Ba and Sr on Mg Absorbance During Flame Atomic Absorption Spectrometry: Characterization in Terms of a Simplified Collisional Rate Model

Mark F. Zaranyika[1], Albert T. Chirenje[1]
and Courtie Mahamadi[2]
[1]Chemistry Department, University of Zimbabwe, Mt Pleasant, Harare,
[2]Chemistry Department, Bindura University of Science Education, Bindura,
Zimbabwe

1. Introduction

Group II elements constitute an environmentally important group of metals. Calcium ranks 5th in relative abundance in nature. It occurs in limestone, dolomite, gypsum and gypsiferous shale, from which it can leach into underground and surface waters. Calcium content of natural waters can range from zero to several hundred milligrams per liter, depending on the source and treatment of the water. Magnesium occurs in nature in close association with calcium. It ranks 8th in abundance among the elements, and is a common constituent of natural waters. As with calcium, concentrations of magnesium in natural water may vary from zero to several hundred milligrams per litre, depending on source and treatment of the water. Calcium and magnesium salts contribute to water hardness. Concentrations of Mg above 125 mg/L can have cathartic and diuretic effects on the water (APHA, 1992). Barium ranks 16th in relative abundance in nature, and occurs in trace amounts in natural waters. Strontium resembles calcium, and interferes in the determination of calcium by gravimetric and titrimetric methods. Although most portable water supplies contain little strontium, levels as high as 39 mg/L have been detected in well water (APHA, 1992).

FAAS and ICP-AES are the preferred methods for determining Gp II elements including Mg. Signal enhancement and/or depression were reported previous when Gp II elements were determined by atomic absorption spectrometry in the presence of other Gp II elements as interferents by several authors (Zadgorska and Krasnobaeva, 1977; Czobik and Matousek, 1978; Kos'cielniak and Parczewski, 1982; Smith and Browner, 1984; Zaranyika and Chirenje, 1999). Our approach to the study of interelement effects in atomic spectrometry involves a technique of probing changes in the number densities of the excited and ground states. Experimental analyte line emission intensity (I) and line absorbance (A) signal ratios, I'/I and A'/A, respectively, where the prime denotes readings taken in the presence of the interferent, are determined and compared to theoretical values derived assuming steady state kinetics. The method was used to follow collisional processes on the excitation and ionization of K resulting from the presence excess Na as interferent (Zaranyika et al., 1991). The approach assumes no change in the rate of introduction of analyte atoms into the excitation source, and no change in the temperature of the torch or flame, upon the simultaneous introduction of an easily ionized interferent element.

Our argument is that according to the local thermodynamic equilibrium (LTE) approach, atomic line absorption and atomic line emission intensities are directly proportional to the population of the ground and excited states, respectively, i.e., $A \propto N_0$ and $I \propto N_j$ whereas N_j and N_0 are related by the Boltzmann equation (Boumans, 1966):

$$N_j = N_o \cdot \left(\frac{g_j}{g_o}\right) \exp(-\Delta E / kT) \qquad (1)$$

where g_j and g_0 are the statistical weights of the excited and ground states, respectively, k is the Boltzmann constant, T is the absolute temperature and ΔE is the difference in the energies of the two electronic states involved in the transition. If the rate of introduction of the analyte atoms into the plasma is kept constant, and we assume no change in the plasma temperature on simultaneous introduction of interfering metal atoms with the analyte, we may write:

$$N_j' = N_o' \left(\frac{g_j}{g_o}\right) \exp(-\Delta E / kT) \qquad (2)$$

where the primes denote the actual populations of analyte ground state and excited atoms in the presence of the interferent. Combining equations 1 and 2, we have

$$\frac{N_j'}{N_j} = \frac{N_o'}{N_o} \qquad (3)$$

Hence,

$$\frac{I'}{I} = \frac{A'}{A} = \frac{n_u'}{n_u} \qquad (4)$$

where n_u is number density of the excited state.

Equation 4 suggests that the effects of collisional processes on the excitation, and line emission of the analyte atoms resulting from the presence of interferent atoms may be followed by measuring the absorption or emission intensities of a given concentration of analyte atoms in the absence and presence of the interferent, and comparing the I'/I and A'/A ratios plotted versus analyte concentration. Applying the approach to absorption spectrometry, the following situations may be identified:

i.   No collisional effects, and therefore no change in the populations of the ground and excited states of analyte atoms: $A'/A=1$
ii. Increase of ground state, e.g. suppression of ionization: $A'/A>1$
iii. Depopulation of ground state, e.g. charge transfer reactions: $A'/A<1$
iv. Increase in excited state population: $A'/A>1$
v.   Depopulation of excited state: $A'/A<1$

The aims of the present work were to investigate and characterize, in terms of a simplified collisional rate model, the interference effects observed when Mg is determined by air-acetylene FAAS in the presence of excess Ca and Sr as interferents.

2. Experimental

2.1 Equipment
An AA-6401 Shimadzu Spectrophotometer with an aberration-corrected Czerney-Turner mounted monochromator, automatic two step gain adjustment beam balance, automatic baseline drift correction using electrical double beam signal processing in peak height and area modes, was used in conjunction with an air-cooled 100-mm slot burner with a stainless steel head with a Pt-Ir capillary nebulizer with a Teflon orifice, glass impact bead and polypropylene chamber, and an air-acetylene flame. The air was supplied by a Toshiba Toscon compressor at 0.35 MPa input pressure, while the fuel gas was supplied from a pressuresed tank (BOC Zimbabwe (Pvt) Ltd, Harare), at 2.0 mL/min. Under these conditions the temperature of the flame was approximately 2573K [3-5], confirmed by a personal communication obtained from Shimadzu Inc., Japan. The spectrophotometer is equipped with automatic fuel gas flow rate optimization for each element.
Experiments were carried out using Hamamatsu Photonics hollow cathode lamps (Hamamatsu, Japan) as source. The lamps was operated at the recommended minimum current of 8 mA for Mg respectively. Measurements were made using the 285.2 nm Mg line. The spectrophotometer employs a high speed 2-wavelength simultaneous measurement Deuterium lamp to correct for background signal.
The nebulization chamber of the spectrophotometer was cleaned with triply distilled water to remove any deposited solids after each set of runs. The instrument was optimized for absorbance measurements, and care was taken not to change the instrumental settings/conditions until all measurements involving a particular interferent were completed. The average absorbance reading was recorded on the instrument computer monitor 5 seconds after aspiration.
Instrumental parameters employed were as follows: Spectral band pass, 0.5 nm; Burner height, 7 mm; Burner angle, 0°; Acetylene fuel flow rate, 2.0 L/min.; Air input pressure, 0.35 MPa. Minimum Hollow cathode lamp current 8 mA. A mean aspiration rate of 3.00±0.06 mL/min. (n = 8), and mean nebulization efficiency of 6.3±1.7% (n = 8) were obtained.

2.2 Materials
The following were used: Calcium chloride AR grade (impurities present: sulphate 0.005%, total nitrogen 0.005%, phosphorous 0.001%, lead 0.001%, iron 0.0005%, magnesium 0.01%, sodium 0.01% and potassium 0.01%); magnesium chloride AR grade (impurities: free acid 0.001%, magnesium oxide 0.0005%, nitrogen compounds 0.0002%, arsenic 0.0001%, barium 0.0002%, calcium 0.00055%, potassium 0.005%, sodium 0.0055%, and zinc 0.0025%); Strontium chloride AR grade (impurities: water insoluble matter 0.005%, sulphate 0.001%, lead 0.0002%, iron 0.0001%, zinc 0.0001%, barium 0.02%, and substances not precipitated by sulphuric acid 0.0002%); Deionized water of conductivity 0.001μSm-1.

2.3 Procedure
Four sets of standard solutions each containing 1 – 30 mg/L Mg were prepared from a freshly prepared stock solution. Three sets were spiked with 1000 mg/L Ca or Sr each

respectively; while one set each was left unspiked. The concentration of the interferent was kept in excess at 1000 mg/L while that of the analyte was kept at 0 -30 mg/L to minimize changes in the physical properties of the test solution. Any such changes in physical properties would affect the set of solutions to be analyzed to the same extent, and this was compensated for by taking blank readings of the solution containing the interferent only.

Absorbance (A) readings were made for the spiked sets of solutions, as well as the unspiked set using distilled water as blank. The readings for the spiked sets of Mg solutions were then adjusted for blank readings of the solution containing the interferent only. A′/A ratios were calculated and plotted against Mg concentration in the test solution in Figure 1. Preliminary experiments were run to determine the aspiration rate and nebulization efficiency for the type of solutions under analysis. A mean aspiration rate (n = 8) of 3.00±0.06 mL/min and a mean nebulization efficiency (n = 8) of 6.3±1.7% were obtained.

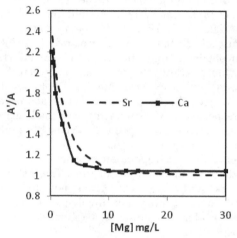

Fig. 1. Interference effects of excess Ca and Sr on Mg absorbance during air-acetylene FAAS.

2.4 Theoretical calculations

Ground state number densities were calculated assuming an aspiration rate of 3 mL/min. and 6 % nebulization efficiency as measured above, and a temperature of 2573 K for the air-acetylene flame as noted above. Ionic number densities were calculated on the basis of the Saha relationship (Allen, 1955). Data obtained are shown in Table 1, in column headed 'n_0 $(cm^{-3}s^{-1})$'.

M	n_0 $(cm^{-3}s^{-1})$	n_+(or n_{m+}) $(cm^{-3}s^{-1})$[a]	a	n_{e^*} $(cm^{-3}s^{-1})$
Mg	$2.5154 \times 10^{12}c$	$1.8252 \times 10^9 c^{1/2}$		$1.8252 \times 10^9 c^{1/2}$
Ca	2.5431×10^{14}	5.6225×10^{11}	0.002	5.5859×10^8
Sr	8.3084×10^{14}	2.5694×10^{12}	0.003	3.8571×10^8

[a]Based on Saha relationship. M = element; α = degree of ionization assuming 2573 K flame temperature. c = analyte concentration in the test solution.

Table 1. Ground state atom, ion and pre-thermal equilibration "hot" electron number densities, n_{e^*}.

3. Results and discussion

The experimental A'/A curves in Fig. 1 show a sharp increase in line absorbance signal enhancement with decrease in the concentration of the analyte in the test solution below about 10 mg/L. Similar results were reported previously in a study of mutual atomization interference effects between Group I elements (Zaranyika and Makuhunga, 1997). Absorbance signal enhancement is attributed to suppression of ionization, which is equivalent to collisional ion-electron radiative recombination. The major processes affecting analyte ground state population in the flame are represented schematically in Fig. 2.

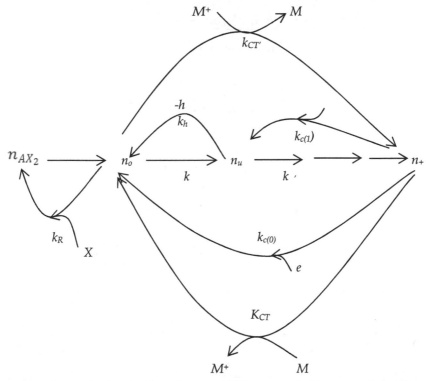

Fig. 2. Proposed kinetic scheme: The subscripts AX2, o, u, and + denote analyte salt, ground state, excited state and ion respectively; M denotes interferent; k_D, k_A, k_A' denote rate constants for thermal dissociation, excitation from the ground state and excitation the excited state respectively; k_{CT} and $k_{CT'}$ denote rate constants for collisional charge transfer involving interferent atoms and ions respectively; $k_{c(o)}$ and $k_{c(1)}$ denote rate constants for collisional radiative recombination to the ground state and excited state respectively; k_{hv} and k_R denote rate constants for radiative relaxation and analyte atom and counter- atom recombination respectively.

Assuming a steady state with respect to the analyte ground state and the excited state:

$$\frac{dn_A}{dt} = k_D n_{AX_2} + k_{c(o)} n_u n_e' + k_{CT} n_+ n_{Mo} + k_{hv} n_u' - k_\varphi n_o' - k_{CT}' n_o n_{M+} - k_R n_o' n_X = 0 \qquad (5)$$

$$\frac{dn_u}{dt} = k_\phi n'_o - k_{h\upsilon} n'_u - k'_\Delta n'_u = 0 \tag{6}$$

Rearranging Eq. 6, we have

$$k_{h\upsilon} n'_u = \left(\frac{k_{h\upsilon}}{k_{h\upsilon} + k'_\Delta} \right) k_\phi n'_o \tag{7}$$

Substituting into Eq. 5 and rearranging:

$$n'_o = \frac{k_D n_{AX_2} + k_{CR(o)} n_+ n'_e + k_{CT} n_+ n_{mo}}{k'_{CT} n_{m+} + k_R n'_X + k_\phi \left(1 - \dfrac{k_{h\upsilon}}{k_{h\upsilon} + k'_\Delta} \right)} \tag{8}$$

$$n'_o = \frac{k_D n_{AX_2} + k_{CR(o)} n_+ n'_e + k_{CT} n_+ n_{mo}}{k'_{CT} n_{m+} + k_R n'_X + k'_\phi} \tag{9}$$

Where

$$k'_\phi = k_\phi \left(1 - \frac{k_{h\upsilon}}{k_{h\upsilon} + k'_\Delta} \right) \tag{10}$$

Therefore

$$\frac{n'_o}{n_o} = \left[\frac{k_D n_{AX_2} + k_{CR(o)} n_+ n'_e + k_{CT} n_+ n_{mo}}{k_D n_{AX_2} + k_{CR(o)} n_+ n_e} \right] \left(\frac{k'_\phi + k_R n_X}{k'_\phi + k'_{CT} n_{m+} + k_R n'_X} \right) \tag{11}$$

Two major limiting cases can be defined for Eq. 11, thus:
Limiting Case I (LC I):

$$k_D n_{AX2} \gg k_{CR(o)} n_+ n'_e + k_{CT} n_+ n_{mo}$$

$$\frac{n'_o}{n_o} = \frac{k'_\phi + k_R n_X}{k'_\phi + k'_{CT} n_{m+} + k_R n'_X} \tag{12}$$

Since $k_R n_X \gg k_R n_X$, n_o/n_o will be less than unity, i. e., a depression of absorbency signal is expected in this case, contrary to the experimental results in Fig. 1. Although the present work reports analyte line absorption signal enhancement, analyte line absorption signal depression has been reported by several workers (Herrmann and Alkemade, 1963; Brown et al., 1987). Analyte line absorption signal depression conforming to Eq. 12 was also reported previously at low flame temperature when 0.2 – 1.0 mg/L K solutions were determined in the presence of 1000 mg/L Na in the secondary reaction zones of the air-acetylene flame (Zaranyika et al., 1991).

Limiting Case II (LC II):

$$k_D n_{AX2} \ll k_{CR(o)} n_+ n_e$$

$$\frac{n'_o}{n_o} = \left[\frac{k_{CR(o)}n_+n'_e + k_{CT}n_+n_{mo}}{k_{CR(o)}n_+n_e} \right] \left(\frac{k'_\phi + k_R n_X}{k'_\phi + k'_{CT}n_{m+} + k_R n'_X} \right)$$ (13)

Two further limiting cases can be defined for Eq. 13, thus:
Limiting Case IIA (LC IIA):

$$k'_\phi \gg k'_{CT}n_{m+} + k_R n'_X$$

$$\frac{n'_o}{n_o} = \frac{k_{CR(o)}n_+n'_e + k_{CT}n_+n_{mo}}{k_{CR(o)}n_+n_e}$$ (14)

Since $n_e' = n_+ \, \Delta n_e = n_+ + \alpha n_{mo}$, where Δn_e represents the change in the electron number density in the presence of the interferent, and α is the degree of ionization of the interferent, Eq. 14 becomes

$$\frac{n'_o}{n_o} = 1 + \left(\frac{\alpha k_{CR(o)} + k_{CT}}{k_{CR(o)}} \right) \frac{n_{mo}}{n_+}$$ (15)

Or

$$\frac{n'_o}{n_o} = 1 + \frac{k n_{mo}}{n_+}$$ (16)

i.e., absorbance signal enhancement is directly proportional to interferent number density and inversely proportional to analyte number density. In addition, if we assume that $A'/A = n_{o'}/n_o$, a plot of A'/A versus n_{mo}/n_+, should be linear with an intercept of unity. At constant interferent concentration in the test solution, a plot of A'/A versus $1/n_+$, should also be linear with an intercept of unity.
Limiting Case IIB (LC IIB):

$$k'_\phi \ll k_R n_X$$

It can be shown that

$$\frac{n'_o}{n_o} = \left\{ 1 + \left(\frac{\alpha k_{CR(o)} + k_{CT}}{k_{CR(o)}} \right) \frac{n_{mo}}{n_+} \right\} \left(\frac{k_R n_X}{\alpha k'_{CT}n_{mo} + k_R n'_X} \right)$$ (17)

i.e., although the absorbance signal enhancement is still dependent on interferent number density and inversely proportional to analyte number density, the enhancement observed will be reduced by a factor of at least n_X/n_X'. In addition, if we assume the $A'/A = n_o'/n_o$, a plot of A'/A versus n_{mo}/n_+ should give a non-linear slope dependent on the concentration of interferent metal atom and counter atom, and with an intercept that is less than unity and equal to at least n_X/n_x'.
Table 2 shows the slope, intercept and R^2 values for the regression plots of A'/A versus n_{mo}/n_+ obtained when the enhancement factor A'/A for Mg is determined in the presence and absence of Ca and Sr as interferents. It is apparent from the data in Table 2 that the

intercept values of 1.09 and 1.22 obtained are close to unity, in close agreement with Eq. 11. These data suggest that the signal enhancement obtained when Mg is determined in the presence of excess of Ca and Sr as interferents conforms to LC IIA.

Interferent	Slope	Intercept	R²
Ca	2.34×10^{-4}	1.08714	0.909
Sr	3.38×10^{-3}	1.21649	0.732

Table 2. Regression data: A'/A vs n_{mo}/n_+ for Mg

Signal enhancement in the presence of easily ionizable elements (EIEs) in atomic absorption spectrometry is commonly attributed to suppression of ionization (Foster Jr. and Hume, 1959; Smit et al., 1951). Supression of ionization is in effect collisional radiative recombination, and assumes that in Eq. 14 $k_{CR(o)} n_+ n_e' \ggg k_{CT} n_+ n_{mo}$ so that Eq. 14 reduces to

$$\frac{n_o'}{n_o} = \frac{k_{CR(o)} n_+ n_e'}{k_{CR(o)} n_+ n_e} = 1 + \frac{k_{CR(o)} n_+ \Delta n_e}{k_{CR(o)} n_+ n_e} \tag{18}$$

Or

$$\frac{n_o'}{n_o} = 1 + \frac{k_c n_+ \Delta n_e \exp(-E_a' / kT)}{k_c n_+ n_e \exp(-E_a / kT)} = 1 + \frac{\Delta n_e \exp(-E_a' / kT)}{n_e \exp(-E_a / kT)} \tag{19}$$

where Δn_e is the change in electron number density upon the addition of the interferent, i.e., $\Delta n_e = n_{m+}$, E_a' and E_a are the activation energies for the electrons from the ionization of the interferent and analyte respectively, and. k_c is the collisional rate constant given by (Weston and Schwarz, 1972):

$$k_c = Q_{12} \left(\frac{8kT}{\pi\mu} \right)^{1/2} \tag{20}$$

where Q_{12} is collision cross-section between particles 1 and 2, and μ is their reduced mass. If we assume thermal equilibrium conditions, then all the electrons will require the same activation, i.e., $E_a' = E_a$, and Eq. 19 reduces to

$$\frac{n_o'}{n_o} = 1 + \frac{\Delta n_e}{n_e} \tag{21}$$

where $\Delta n_e = n_{m+}$ and $n_e = n_+$. It is apparent that substitution for Δn_e and n_e from Table 1 will yield values of n_o'/n_o up to three orders of magnitude greater than experimental A'/A values at low Mg concentrations. If however we assume pre-thermal equilibrium collisional radiative recombination, then the electrons from the ionization of the interferent would require further activation by an amount of energy equal to the difference between the ionization potentials of the analyte and that of the interferent, i.e., in eq. 19,

$$E'_a = IP_a - IP_m \tag{22}$$

And

$$E_a = IP_a - IP_a = 0 \tag{23}$$

where the subscripts a and m denote analyte and interferent respectively. If we make this assumption, then Eq. 15 becomes

$$\frac{n'_o}{n_o} = 1 + \frac{\Delta n_e}{n_+} \exp(-E'_a / kT) \tag{24}$$

Table 3 shows the fraction of such electrons having the appropriate energy, and Table 1 shows the corresponding pre-thermal equilibration number density of "hot" electrons, n_e^*, arising from the analyte, Mg ($n_e^* = n_+$), and the interferents Ca and Sr ($n_e^* = n_+\exp(-E_a'/kT)$). $^*E_a' = IP_a - IP_m$.

Element	IP$_a$ (eV)	IP$_m$ (eV)	E$_a$' (eV)*	Exp(-E$_a$'/kT)
Mg	7.644			
Ca		6.111	1.533	9.9349 x 10^{-4}
Sr		5.692	1.952	1.5012 x 10^{-4}

Table 3. Pre-thermal equilibration fraction of electrons having the requisite activation energy.

Substitution of the appropriate quantities from Table 1 and 4 into Eq. 24 yields Eqs. 25 and 26 for the absorbance signal enhancement factor, n_o'/n_o, for the presence of 1000 µg/ mL excess Ca and Sr as interferents:

$$\left(\frac{n'_o}{n_o}\right)_{Mg/Ca} = 1 + \frac{3.0604x10^{-1}}{\sqrt{c}} \tag{25}$$

$$\left(\frac{n'_o}{n_o}\right)_{Mg/Sr} = 1 + \frac{2.1132x10^{-1}}{\sqrt{c}} \tag{26}$$

A major objective for kinetic modeling of interference effects is to be able to predict the interference observed experimentally. If assume that

$$\frac{A'}{A} = \frac{n'_o}{n_o} \tag{27}$$

then the calibration curve obtained in the presence of the interferent can be predicted. Theoretical calibration curves predicted on the basis of Eqs 25 and 26 for the effects of 1000 mg/L excess Ca, Br and Sr respectively on the absorbance signal of Mg in the air-acetylene flame are shown in Fig. 3, together with the experimental A' calibration curves for comparison.

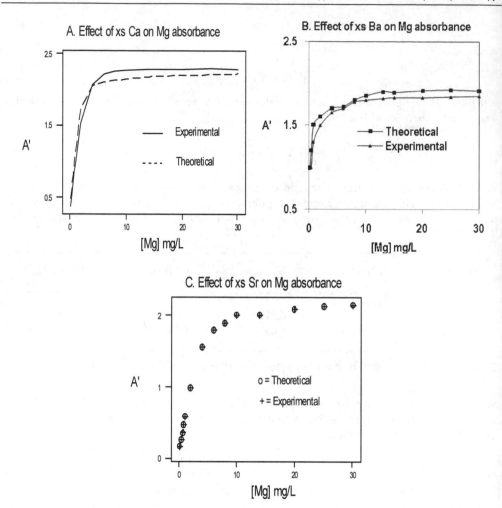

Fig. 3. Theoretical and experimental A' calibration curves: Effect of excess Ca (A), Ba (B) and Sr (C) on Mg absorbance during air-acetylene FAAS.

It is apparent from Fig. 3A that Eq. 24 leads to close agreement between theory and experiment at Mg concentrations below about 5 mg/L when Mg is determined in the presence of excess Ca, while the theoretical curve deviates slightly from the experimental curve above this concentration. In contrast to the case when Mg is determined in the presence of excess Ca, exact agreement between theory and experiment is obtained throughout the range of Mg concentrations studied when determined in the presence of excess Sr. Although the work presented in this paper is rather limited in scope, the remarkable success of the model in predicting the interference of excess Sr on the absorbance signal of Mg, confirms the potential of the model as represented by Eq. 24 in simulating the absorbance signal enhancement interference effects during flame atomic spectrometry. Further work is underway to test the model on other systems.

Interference Effects of Excess Ca, Ba and Sr on Mg Absorbance During Flame Atomic Absorption Spectrometry:
Characterization in Terms of a Simplified Collisional Rate Model
125

4. Conclusions

From the foregoing discussion we conclude that the interference effects between Group II elements can be characterized using a simplified rate model that takes into account collisional radiative recombination, charge transfer between analyte and interferent species, and collisional recombination between analyte atom and counter atom. The model predicts that, depending on the specific experimental conditions employed, interference effects in flame atomic spectrometry can manifest themselves as enhancements (Eq. 15) or depressions (Eqs. 12 and 17) of the analyte line absorbance signal. Data relating to the signal enhancement interference effects of excess Ca and Sr on Mg absorbance signal during air-acetylene flame absorption spectrometry suggest that the signal enhancement can be simulated on the basis of a simplified rate model that assumes pre-LTE ion-electron radiative recombination.

5. Acknowledgements

This work was supported by a grant from the Research Board of the University of Zimbabwe (UZ).

6. References

American Public Health Association (APHA), American Water Works Association (AWWA) and Water Environment Federation (WEF), in Standard Methods for the examination of water and wastewater, 18th Ed., A .E. Greenberg (APHA), L. S. Clesceri (WEF) and A. D. Eaton (AWWA), eds. American Public Health Association, Washington DC, 1992.

Czobik, E.J., and Matousek, J.P., (1978). Interference effects in furnace atomic absorption spectrometry. Analytical Chemistry, 50, 1, 2-10.

Smith, D.D., and Browner, R.F., (1984). Influence of aerosol drop size on signals and interferences in flame atomic absorption spectrometry. Analytical Chemistry, 56, 14, 2702-2708.

Zaranyika, M.F., and Makuhunga, P., (1997). A possible steady state kinetic model for the atomization process during flame atomic spectrometry: Application to mutual atomization interference effects between group I elements. Fresenius Journal of Analytical Chemistry, 357, 249-257.

Allen, C.W., (1955). Astrophysical Quantities, 2nd Ed., Athlone Press, ISBN: 13: 9780387987460, London.

Herrmann, R., and Alkemade, C.T.J., (1963). Chemical analysis by flame photometry. 2nd ed., Interscience Publishers, New York.

Brown, A.A, Roberts, D.J., and Kahokola, K.V., (1987). Methods for improving the sensitivity in flame atomic absorption spectrometry. Journal of Analytical Atomic Spectrometry, 2, 201-204.

Zaranyika, M.F., Nyakonda, C., and Moses, P., (1991). Effect of excess sodium on the excitation of potassium in an air-acetylene flame: a steady state kinetic model which takes into account collisional excitation. Fresenius Journal of Analytical Chemistry, 341, 577-585.

Weston, R.E., Schwarz, H.A., (1972). Chemical Kinetics, Prentice-Hall, ISBN: 9780131286603, New Jersey.

Foster, W.H Jr., and Hume D.N., (1959). Mutual cation interference effects in flame photometry. Analytical Chemistry, 31:2033-2036.

Smit J., Alkamade C.T.J., and Verschure, J.M.C., (1951). A contribution to the development of the flame-photometric determination of sodium and potassium in blood serum. Biochimica ET Biophysica Acta, 6, 508-523.

Zadgorska, Z and Krasnobaeva, N (1977), Influence of the anionic component of the additive on the relative intensity of atomic spectral lines in the analysis of dry residues from solutions, Spectrochim. Acta. 32B, 357-363.

Kos'cielniak, P and Parczewski, A (1982), Theoretical model of alkali metals interferences in flame emission spectrometry, Spectrochim. Acta 37B, 881-887.

Zaranyika, M. F and Chirenje, A. T (1999), A possible steady-state kinetic model for the atomization process during flame atomic spectrometry: Application to atomization interference effects of alumimiun on Group II elements. Fresenius J. Anal. Chem, 364, 208-214.

Boumans, P. W. J. M (1966), Theory of Spectrochemical Excitations, Higler and Watts, London.

7

Nutritional Metals in Foods by AAS

Mary Millikan

Sch. of Engineering & Science, FOHES & ISI
Victoria University
Australia

1. Introduction

It is well known that a balanced diet is essential in maintaining good health. Hence, the nutritional value of foods is an important aspect that should be considered especially with respect to metal intake such as iron, calcium, magnesium, potassium, sodium, selenium, manganese, copper, chromium and zinc. Iron being required for the haemoglobin; calcium for relaxing the central nervous system; magnesium to prevent muscle spasms; potassium and sodium for electrolyte balance; selenium has a number of functions including deactivating heavy metals from external exposure; manganese and copper are linked to superoxide dismutase (SOD): chromium stabilizes blood sugar and zinc is important in the healing of wounds.

An overview of the literature will be given of applications of Fame AA together with some references to Graphite Furnace AA to the analyses of foods such as: meat the main source of iron; dairy products, the source of calcium and fruit and vegetables for a range of metals. Comparisons will be given of metal content in these products particularly in meat and dairy products. Of the metals listed above, not all of these will be considered in every product: only where they are the metal of highest concentration. The aim of this chapter is to give a general comparison of the metal content in these products, which will not be exhaustive, particularly, with respect to fruit and vegetables but the ones most commonly consumed. The emphasis is on nutrition and to give the general reader and health professional a concise view of the metal content of these food products. From scientific aspect the methodology for Flame AA is relative straight forward, as is the work up for instrument presentation but there are often extra procedures that are required depending on the matrix that are essential for obtaining a valid result.

Flame Atomic absorption spectroscopy even though it is a well-established technique that was discovered about fifty years ago is still used extensively today for trace metal analyses in industry, commercial laboratories and universities.

2. History of the discovery of flame atomic absorption spectroscopy

The method for the analyses of metals in a wide range of samples from food, agriculture, mining, environmental, pharmaceutical and biochemical industries was made possible by the discovery of a new technique in the early 1950's by Dr. Alan Walsh. Like many scientists before him, he was not in his laboratory when the idea came to him but in his vegetable garden one Sunday morning in 1952. The idea proved successful and was the basis of the atomic absorption spectrophotometer, an instrument for quantitative chemical analyses of metals (Hannaford 2002) that did not require slow wet chemical procedures. Techtron Pty. Ltd., Melbourne, Australia, manufactured the first instrument based on his design in the mid 1960's. The company was taken over many years later by Varian Australia Inc. and several months ago by Agilent Technologies Pty. Ltd, Melbourne who still manufactures an instrument based on this early technology. (Hannaford 2002)

The scientific importance of Alan's idea was that he realized in his attempt to measure the concentration of metals in solution by spectroscopic means, he had been trying to measure the incorrect parameter. Rather than measuring emission he should have been measuring absorbance. When he mentioned this to his colleague John Willis (CSIRO, Division of Chemical Physics) he said that they had considered this aspect before and that it would not work because of the emitted light at the same wavelength. The reply Alan gave was that this could be overcome by having a chopper to eliminate this emission and to use an amplifier. Several days later Alan measured the absorbance of sodium but his colleague at this time did not appreciate the significance of this major scientific breakthrough that was the basic principle of the instrument (Hannaford 2002) .

Fig. 1. Photograph of prototype Atomic Absorption Spectrometer build by Dr. Alan Walsh, with permission from Agilent Australia, Mulgrave, Victoria., where the instrument is located.

3. Nutritional significance of minerals in the diet

Minerals are divided into two groups Essential and Trace minerals, which is related to the quantity required and found in the body, the former being present in the largest amounts. These minerals will now be discussed briefly in this order.

3.1 Essential minerals
The essential metals are the macro metals:
- Calcium
- Magnesium
- Potassium
- Sodium

Calcium is responsible for strong bones and teeth and accounts for ninety percent of the calcium in the body whereas the other one percent is circulating in fluids in order to ionise calcium. The metal's function is related to transmitting nerve impulses; contractions of muscles; blood clotting; activation of some enzyme reactions and secretion of hormones Magnesium has many roles including supporting the functioning of the immune system; assists in preventing dental decay by retaining the calcium in tooth enamel; it has an important role in the synthesis of proteins, fat, nucleic acids; glucose metabolism as well as membrane transport system of cells. Magnesium also plays a role in muscle contraction and cell integrity. Potassium and sodium work together in muscle contraction nerve transmission. Sodium is important in muscle contraction and nerve transmission Sodium ions are the main regulators of extra cellular fluid and volume (Whitney and Rofles 2002).

3.2 Trace minerals
These are particularly important for health promotion and prevention of disease. Trace metals being considered in this work are:
* copper,
* chromium,
* iron,
* manganese,
* molybdenum
* selenium
* zinc.
The non-metals also in the group are iodine and fluorine that will not be discussed. Copper has the role of assisting in the formation of haemoglobin, helping to prevent anemia as well as being involved in several enzymes. Chromium function is related to stabilising blood sugar levels with respect to insulin required for release of energy from glucose. Iron is the central metal in the haemoglobin molecule for oxygen transport in the blood and is portion of myoglobin located in muscles. Manganese is one of the co-factors in a number of enzymes as is molybdenum. Selenium has several roles such as regulating the thyroid hormone as well as being part of an enzyme that protects against oxidation (Whitney and Rofles 2002). Selenium has also been reported as assisting in deactivating heavy metals.

3.3 RDI of minerals according to age and gender
The Recommended Daily Intake, RDI of metals is related directly to age, and gender. The requirements for babies,, toddlers,, children , adolescents, and elderly vary with gender and country due to soil type. These requirements are continually being reviewed in the light of more research that is undertaken by food regulating bodies such as Food Standards Australia and New Zealand, FSANZ, United Stated of America, Food and Drug Administration, FDA, and European Authorities to name three such groups. The work done by these bodies includes all food groups in addition to vitamins, minerals: cereals, fat, protein, carbohydrates, sugars and so on, as well as research on different age groups in particular locations in many countries, to assist in maintaining and improving the health of the various groups and the population in general.

Major minerals	Recommended Daily Intake , RDI
Calcium	1000 mg
Magnesium	350 mg
Potassium	3500 mg
Sodium	2400 mg
Trace minerals	
Chromium	120 µg
Copper	2 mg
Iron	15 mg
Manganese	5 mg
Molybdenum	75 µg
Selenium	35 µg
Zinc	15 mg

Table 1. The table above represents RDI values recommended by experts and agencies for a normal adult population. http:// lenntech.com/recommended-daly-intake.htm

3.4 Analyses of foods with respect to safety and toxicity

Fresh foods and others are monitored regularly for safety and to be sure that the level of undesirable metals is below the safe limit or not present at all. Similarly, the dietary surveys test for these as well as the nutritional value of the various food groups.

4. Variation of mineral content according to soil and country

According the age of the rocks that contain the minerals and type of rocks and soil with respect to their geological age, the mineral content will vary considerably, as well as different minerals being found in the respective rocks. There can be similar soil types that occur in countries that are not near each other. For example, gold was found in California and in Australia in the mid 1800's when the gold rush took place and prospectors came from many countries to make their fortunes. Soil can vary considerably within a particular region, state, territory or through out a particular country with respect to minerals found in the soil. Soils are studied by agriculturalists and farmers, so they can add certain minerals when they are in low levels in order to increase the yield and quality of crops. Even within a particular paddock or field the soil can vary so farmers need to add fertilizers and minerals appropriately in order to obtain a uniform yield of the crop (Dundas and Pawluk 1977). Hence mineral levels for the countries will vary accordingly to the soils and the additives required for maximum crop yield and the fortification of crops, cereals or grains to ensure that the products manufactured would still give sufficient portion of the RDI to maintain the health of the population . RDI allowances of minerals will essentially be very similar in most countries for the various age and gender groups but in some cases extra fortification of foods will be required where there is a low level of an essential or trace mineral.

5. Metals in meat

The principal source of iron in the diet is mainly from meat, particularly, red meat. Other metals are also important such as calcium, magnesium, sodium and potassium, but are generally in lower amounts. Trace metals include: copper, manganese, zinc and chromium. Metal analyses in meat including beef, pork, poultry and fish will be discussed with some comments on sample preparation and values of the metals obtained in these samples

5.1 Selected metals in some meat

A comprehensive study of the analyses of iron, calcium, magnesium, potassium, sodium, manganese, copper, zinc and manganese in foods including meat was undertaken by Maurer (Maurer 1977) here he compared three extraction methods using HNO_3 or HCl/HNO_3 or dry ashing at 450°C. The main focus of this research was to thoroughly evaluate each of the three extraction procedures for the metals listed on a wide variety of food, to determine the daily consumption per person over given time intervals. Recoveries were also determined for the different extraction methods. Results obtained indicated that extraction with the combined acids was the most suitable procedure for all metals, particularly for copper and zinc when compared with dry ashing, which was markedly, affected by the matrix. Recoveries were high for all metals: in the high nineties except for zinc that was only 87%. A Perkin-Elmer (model 300) instrument was employed where calcium and magnesium were analysed with nitrous oxide/acetylene while air/acetylene was used for the other elements. Siong (Siong, Khor Swan, and Siti Mizura 1989) compared the analysis of iron in meat and other foods using AA and the phenanthroline colorimetric method. The two methods compared well for the foods tested and also gave satisfactory recoveries. Values iron obtained for some meat products are as follows: beef extract 10.66; canned beef liver rendang 4.20; canned chicken curry 2.82; chicken heart 2.05; corned beef 1.67; duck 0.69; roast4ed duck 0.84; canned mutton curry 3.67 where all values are expressed as mg Fe/100 g.

	Cu			Mn			Se			Zn		
PIG												
meat	**0.90**	±	0.61	**0.12**	±	0.052	**0.044**	±	0.017	24	±	11
liver	**9.0**	±	4.0	**3.0**	±	0.52	0.50	±	0.062	**74**	±	2
kidney	6.1	±	2.0	1.5	±	0.24	**1.9**	±	0.35	**22**	±	3.3
CATTLE												
meat	**0.87**	±	0.12	**0.093**	±	0.044	0.030	±	0.020	**49**	±	18
liver	**39**	±	27	**3.2**	±	0.67	0.030	±	0.035	40	±	8.5
kidney	3.7	±	0.59	1.1	±	0.24	**0.86**	±	0.28	16	±	1.5

Table 2. Portion of data extracted from (Jorhem et al. 1989) showing in a horizontal row the concentrations in mg/kg of the four elements per product.

A study was conducted by (Jorhem et al. 1989), on the levels of the trace metals aluminium, chromium, cobalt, copper, manganese, nickel and selenium in the kidney, liver and meat of Swedish cattle and pigs at the slaughter houses and verified using standard reference materials. Data for aluminium and nickel will not be discussed, only the other beneficial metals. In general, all the samples were prepared for analysis by dry ashing at 450°C with some extra procedures required for some metals before ashing. Copper, manganese and zinc were analysed by Flame AA employing a Vanian AA-6 instrument equipped with a H_2 lamp for background; correction using an air/acetylene flame. The method of standard addition was used for manganese, hydride/generation AAS for selenium whereas chromium and cobalt were analysed by graphite furnace AA.

In Table 2, it can be seen in at a glance, the metal content in meat, liver and kidney from pigs and cattle. From the results shown, it is clear that liver and kidney are good sources of dietary copper. The manganese levels in this study are low and the authors attribute it the method of standard additions used in the AA analyses. Reported data shows the best source of selenium is pig kidney but the results are generally low in Scandinavian due to the soil. Meat products are an excellent source of zinc, especially liver. Data for chromium is not shown but the results indicated that these products are a poor source of this metal in the diet. Cobalt data also is not shown was found to be four times higher in cattle liver than in pig's liver being 0.043 ± 0.028. The authors compared their results with the published literature for the above metals, as well as others, not stated here, for the three parameter in Table 2 for pigs and cattle and found there was good agreement with other countries that included Austria, Australia, Finland, FRG, Italy, Norway, Sweden and USA, however, not all metals studied had data from all of these countries.

Tinggi'group (Tinggi, Reilly, and Patterson 1997) have reported chromium and manganese levels in some meat products. The samples that were anlysed included the following: chicken (cooked); beef steak; ham; lamb chops and sausages. Results obtained are given below:

Meat	Cr (ng/kg)	Mn (mg/kg)
Chicken (cooked)	12.2 ± 5.0	0.19 ± 0.01
Beef steak	49.0 ± 1.8	0.52 ± 0.2
Ham	26.4 ± 2.7	2.0 ± 0.2
Lamb chops	30.0	0.36
Sausages	32.0 ± 1.4	1.2 ± 0.3

Table 3. Chromium (ng/kg) and Mn (mg/kg) concentrations in meat products.

It is interesting to note in Table 3 that the highest level of chromium is in beef steak 49.0 ng/kg and since it is related to blood sugar levels would account for the feeling of satiety and sustaining of energy after consuming beef in comparison to chicken with the low level of 12.2 ng of chromium. Manganese in contrast is highest in ham 2.0 mg/kg but lowest in chicken .12 mg/kg.

Siong et al (Siong, Khor Swan, and Siti Mizura 1989) analysed the calcium content of eight food groups that included meat and fish. Samples were from local markets and stores, they were homogenized, oven dried in air then the charred and ashed in a muffle furnace at 550°C. A Varian instrument was used employing an air/acetylene flame. Some selected values from the large range of products analysed reported in mg Ca/100g included: de-bonded chicken feet 25.1; chicken heart 6.0; canned beef rendang 31.1; canned mutton curry 16.1 and beef extract 40.4. They compared the AA analyses with potassium permanganate titration where both methods were satisfactory using the paired t-test (p< .0.05). For only two groups legumes and vegetables there was a significant difference in the two methods. Analyses of mechanically de-boned poultry from five Dutch processors was undertaken by (Germs and Stennenberg 1978) employing AA and oxidimetry procedures. The mean AA calcium content was found to be 2.36 g/kg. Of the two oxidimetry methods used, the AOAC procedure gave the best agreement with the AA method where the coefficient of correlation was equal to 0.9996 and the standard error of regression, 0.05. Nakamura (Nakamura 1973) determined the water-extractable calcium content of chicken breast during postmortem aging by AA where he addressed the anionic and cationic interferences. In order to eliminate the anionic interferences, he added a 1% solution of the di-sodium salt of EDTA to the test solutions. In contrast, cationic interferences were overcome by adding a known amount of calcium to the test solutions to being the concentration to 0.1 mM. Results showed that the calcium content increased during the postmortem aging and reached a maximum after twenty-four hours. Atomic absorption spectroscopy and a modified AOAC fibre method were used to determine magnesium and manganese in meat –soy blends. It was found that in regular ground beef the magnesium and manganese levels were 151 mg/kg and 7.4 mg/kg, 4.9% soy flour, respectively (Formo, Honold, and MacLean 1974).

Copper and other metals in meat were analysed by (Ybanez, Montoro, and Bueso 1983) using both dry ashing and wet digestion with HNO_2 and H_2O_2, essentially there was not any significant difference in the two methods of sample preparation. Samples analysed included both cold and cooked ham; mortadella, Frankfurters and liver paste. Recoveries for added metal for copper was found to be 100%. Copper determined in bovine liver standard was 193 ± 10 µg/g. The limit of detection for copper was determined as 0.17 µg/g. An analysis of copper and zinc in meat and meat analogues was done by Schaefer *et al* (Schaefer et al. 1979). Sample work up was carried out with nitric/perchloric acid digestion followed by flame AA. The mean zinc content found in meat and meat analogues was 3.36 and 1.07 mg/100g respectively while the copper content was 0.09 and 0.32 mg/100 g in the same product order. It can be seen that there is a contrast between the two metals in meat products: zinc being higher is meat whereas copper is higher in the analogues. Work undertaken by (Dalton and Malanoski 1969) determined copper and lead in meat ad meat products but used dry ashing at 500ºC for sample preparation unlike Schaefer's group who used wet digestion. Copper was found to be in the 1–5 ppm range.

Zinc and magnesium were analysed in a range of Philippine foods by AAS. Samples were prepared for analyses by hydrochloric acid extraction of the metals. One of the major advantages of the technique was the short sample preparation time for AA compared to other procedures such as gravimetric and colorimetric methods. A variety of foods were analysed that included meat and poultry (Lustre and Lacebal 1976)

Selenium is not an easy metal to analyse and reports in the literature are few. Research by (Hoenig and van Hoeyweghen 1986) determined selenium and arsenic in animal tissues and addressed the spectral interferences that are due to calcium and magnesium phosphates. Their method employed platform furnace atomic absorption spectrometry and deuterium background correction. The interferences just mentioned cannot be corrected using the deuterium arc. To overcome these interferences in animal tissues, they added nickel nitrate to the samples. The quantity added is critical as less than 20 mµg nickel nitrate added to a 10 µl sample does not allow correct development of analyte absorbance signals but more than this causes a loss of sensitivity. It was found that the measurement of peak height was the most suitable approach as integrated absorbance was partly influenced by the matrix. For different animal matrices the slope constants of working curves obtained were very close hence direct calibration was possible. Results were verified by analyses of a number of reference materials.

Investigation of the effect of adding selenium to chicken feed on the amount found in chicken meat and eggs was done by (Turker and Erol 2009). The method used was hydride-generation atomic absorption spectrometry. Optimisation of conditions for the technique such as HCl and $NaBH_4$ concentrations; flow rate of carrier gas; analytical parameters and the effects of digestion procedures on the analyses were all determined. The limits for detection and quantification were 0.78 and 2.35 mg/l respectively. Validation of the method was done by means of certified reference materials. Fortification of the chicken feed increased the levels of selenium in the meat and eggs. Four digestion methods: two dry ashing and two wet ashing procedures for trace metal analysis of Pb, Cd, Cu and Zn were evaluated for pork meat and fish (Zachariadis et al. 1995). Results showed that the wet ashng with HNO_3/HCl mixture gave the best results: For pork meat the copper level was determined to be 0.208 mg/kg and zinc 7.63 mg/kg. The data was collect on a Perkin-Elmer Model 2380 instrument with a HGA-400 graphite furnace that included a Deuterium lamp background corrector.

6. Metals in fish

Most of the analyses of metals in fish are related to heavy metals analysis to monitor pollutants such as zinc, cadmium, mercury and arsenic. Such metals find their way into the marine life in estuaries and rivers due to effluent from heavy industry accidentally leaking into the water ways or by faulty filtering systems and practice that do not comply with the Environmental Protection Agency in that particular country. Most commercial fishing is undertaken in deep-sea waters where these metals would not be a problem, however, fish caught in rivers and estuaries that are many kilometers from heavy industry can sometimes still be affected by metals carried by currents that cause pollution where it is not expected. Reports of beneficial metals analysed by AA in fish are few. Salmon, for example is a good source of calcium since in canned products the bones are not usually removed, except in specialty products. In this discussion emphasis will be given to t nutritional metals and the other metal pollutants will not be mentioned. In their paper, (Carvalho, Santiago, and Nunes 2005) stated that fish are an important source of lipids, proteins, liposoluble vitamins and polyunsaturated fatty acids that are important in assisting to reduce hypertension, cancer risk and coronary heart disease.

6.1 Limited references to some beneficial metals in fish

An analysis of essential and heavy metal levels in edible fish muscle was undertaken by (Carvalho, Santiago, and Nunes 2005) who employed two different techniques: energy-disperive X-ray fluorescence (EDXRF) and flame AAS. Samples of nine fish were analysed, namely, Forkbeard (For), Meagre (Mea), White sea bream (WSB), Axillary sea bream (ASB), Red sea bream (RSB), Common sea bream (CSB), Rockfish (Roc), Common sole (Cso),

Anglerfish (Ang) and Octopus (Oct) that came from coastal markets in Portugal. Flame AAS was used to determine the content of Cu, Cr, Ni, Hg, Pb and Cd employing a Varian (Australia) Spectr AA20 spectrometer. Sample preparation was achieved by means of incineration and dissolution in nitric acid. An EDXRF spectrometer was used for the determination of K, Ca, Fe, Rb, Se and Zn. Of the elements studied, it was fond that calcium and potassium were the most abundant elements detected in the fish samples. The highest potassium level detected was found in octopus with an average value of 12,660 μg/g dry weight and the lowest in Axillary sea bream being 6.170 μg/g. Common sea bream had the highest calcium content of 788 μg/g whereas White sea bream had the lowest value of 444 μg/g. The highest iron concentration in these fish was observed for Octapus at 109 μg/g and the lowest for Forkbeard at only 6.4 μg/g. In a Turkish study, nine fish were studied for their trace metal levels. The fish from the Black Sea were: European anchovy; Whiting; Red mullet; Bluefish; Atlantic horse mackerel; Flathead mullet and Atlantic bonito. Two species from the Aegean Sea were Black swordfish and Gilthead sea bream. A Perkin Elmer AAnalyst 700 AAS instrument equipped with a HGA graphite furnace and deuterium background correction was used for the analyses of chromium, copper, lead and nickel. Flame AA with an air/acetylene flame was employed by (Uluozlu et al. 2007) to determine the copper, iron, manganese and zinc content of fish samples.. Of the fish studied, Bluefish had the highest copper content of 1.83 ± 0.10 μg/g and the lowest level for European anchovy ± 0.08 μg/g. For chromium, the highest level was again for European Anchovy at 1.98 ± 0.10 μg/g but the lowest value for Atlantic horse mackerel 0.95 ± 0.07 μg/g; red mullet has the highest iron level 163 ± 12 μg/g and Bluefish the lowest 698.6 ± 5.3 μg/g; manganese gave the highest level of .54 ± 0.50 μg/g and the lowest for Bluefish 1.28 ± 0.10 μg/g and lastly zinc had the highest level for Red mullet of 106 ± 9.1 μg/g and the lowest level for Bluefish at 35.4 ± 3.2 μg/g. Hence, it can be seen that Red mullet had the highest zinc and iron levels, which were considerably higher than any of the other metals just mentioned in this analysis.

7. Metals in some dairy products

Traditionally, dairy foods are the main source of calcium in the diet where milk, yogurt and cheese are the foods commonly consumed. Calcium and also magnesium are the most important metals for building strong bones and teeth and to prevent rickets and osteoporosis in older citizens particularly women. Vitamins A and D as well am magnesium are usually added to over the counter dietary supplements to ensure adequate absorption of the calcium. In addition, magnesium is also present in milk and dairy products but ii is in larger quantities in leafy green vegetables in the form of chlorophyll being the central metal. Although magnesium is present in dairy products it is in considerably lower concentrations so the following discussion with be focusing mainly on calcium.

7.1 Calcium and some other metals in milk and yogurt

A method for the analysis of calcium and magnesium in dairy products has been reported by (Brandao, Matos, and Ferreira) employing a high resolution continuum source flame atomic absorption spectrometer (HF-CS-FAAS) using secondary lines. The advantage of using these secondary lines is that samples with high concentration of these elements did not need substantial dilution. Samples of milk powder, cow milk and yogurt were obtained from local supermarkets in Salvador City Brazil. In order to prepare the samples for analysis two methods were used: slurry sampling and digestion. Data obtained indicated that there was not any significant difference in the results for the two procedures. The instrument used was an Analytik jenna Model ContrAA 300 High Resolution-Continuum Source Flame Atomic Absorption Spectrometer (GI.E. Berlin, Germany) equipped with a xenon short-arc lamp XBO 301 with a nominal power of 300 watt. A nitrous oxide/acetylene flame was used for the for the analysis of calcium and magnesium. For yogurt samples the calcium and magnesium levels were found to be 1.40 and 0.13 mg/g respectively. Values of calcium and magnesium in whole milk were 1.23 and 0.12 mg/mL respectively while for skim milk the concentrations were almost the same. In milk powder, however, the values were higher, namely, 8.92 and 0.83 mg/g for calcium and magnesium in this order. In contrast for skim milk the calcium and magnesium concentrations were 1.21   and 0.118 mg/g. A detailed study by (Miquel et al. 2005) for calcium, iron and zinc in toddler milk-based formula were 861 ±, 27; 12 ± 5; 7 ± 5 mg/L and for soluble fraction in the same product the values obtained were 704 ± 24; 7 ± 1 and 5 ± 1 mg/L, respectively. The chromium content of *acidaphilus* milk culture was reported by (Larsen and Rasmussen 1991) as 0.76 ng/mL: for cream 0.67 ng/mL; low fat cream, 13% fat, 0.77 ng/mL and yogurt 2.2 ng/mL.. A value of 14.3 ± 4.0 mg/kg for chromium and 0.27 ± 0.05 mg/kg for manganese in yogurt was reported by (Tinggi, Reilly, and Patterson 1997).

7.2 Manganese and chromium levels in milk powders

A procedure for the determination of manganese in dried milk employing a Zeeman furnace AAS was reported by (Koops and Westerbeek 1993). Samples were prepared for analysis by boiling them in nitric acid for 15 minutes then adding magnesium and palladium as matrix modifiers. Manganese levels were determined in 36 dried whole milk powder samples. It was found that the average manganese concentration of the samples was 0.21 mg/g of total solids present. Hence, these results showed that it was possible to calculate the manganese content for reconstituted milk as approximately 25 µg/kg. An analysis of the chromium and manganese levels in skim milk was determined by (Tinggi, Reilly, and Patterson 1997) who found they were 6.7 ± 1.5 ng/kg and 0.40 ± 0.20 ng/kg respectively.

7.3 Molybdenum in milk products

Although molybdenum is a essential trace metal, there are few references in the literature of atomic absorption analyses of molybdenum in foods, this may be because it is not a straight forward metal to analyse, as matrix modifiers are required. Determination of the concentration of molybdenum in skimmed milk based drinks and infant food samples was done by (Regina de Amorim et al.) using a graphite furnace AAS. A considerable amount to time was spent in evaluating matrix modifiers and temperature programming to optimize the procedure. Samples were obtained from a local store in Brazil: they were dissolved in ultra pure water, sonicated then used without further pre-digestion. The analyses were performed on a Perkin Elmer AAnalyst 300 atomic absorption spectrometer (Norwalk, CT,USA) fitted with an HGA 800 graphite furnace and AS-72 autosampler using argon purge gas. Values obtained for the molybdenum levels in the milk products are given below:

Milk products	Mo level (g/g)
• Skimmed milk	0.034 ± 0.01
• Whole milk:	
• Sample (A)	0.37 ± 0.05
• Sample (B)	**0.22 ± 0.03**
• Sample (C)	0.27 ± 0.02
• Sample (D)	0.30 ± 0.04
• Sample (E)	**0.40 ± 0.02**
Infant formula	
• Sample (A)	**0.43 ± 0.03**
• Sample (B)	**0.36 ± 0.01**
• Sample (C)	0.37 ± 0.06
• Milk based drinks	
• Sample (A)	**1.57 ± 0.28**
• Sample (B)	**0.040 ± 0.006**
• Sample (c)	0.32 ± 0.01

Table 4. Mo concentrations in µg/g for milk products determined by (Regina de Amorim et al.).

From Table 4, it can be seen that the range of Mo in whole milk samples was in the range 0.22 to 0.40 µg/g while the level in skimmed milk was considerable lower, 0.034 µg/g; infant formula 0.36 to 0.43 µg/g and milk based drinks 0.040 to 1.57 µg/g.

7.4 Analyses of some metals in butter

A Perkin-Elmer–303 Atomic Absorption Spectrophotometer was used to analyse copper, iron, manganese, magnesium, potassium, sodium and calcium in butter samples. Samples were prepared by two methods, either by direct dry ashing or by dry-ashing of the HNO_3 extract. Dry ashing was not suitable for copper, iron and manganese due sample loss and a wide range of results as well as being a very time consuming procedure. In contrast, for the other metals both methods gave satisfactory results. Three factories were included in the study, namely, 58 butter samples were obtained from Latvian dairy factories (LD); 33 from the Vladimir region factories (VR) and 58 from Krasnodar district factories.(KD). Results from these three factories in mg/kg can be seen in the Table 5 below.

Metal in mg/kg	Latvian Distract	Valadimir Region	Krasnodar District
Calcium	144	168	172
Copper	0.82	1.45	1.26
Iron	1.57	1.84	2.15
Potassium	173	197	144
Magnesium	14.4	19	21.2
Manganese	0.05	0.1	0.004
Sodium	74	151	146

Table 5. Results from (Lovachev et al. 1972) comparing the seven metal concentrations from the three regions.

It can be seen from Table 5 that the highest metal concentration was potassium in the Valadimir Regiion, 197 mg/kg followed by 173 mg/kg in the Latvian District. Calcium was the metal with the highest concentration of 172 mg/kg in the Krasnodar District. In the Valadimir Region the calcium level was 168 mg/kg but in the Latvian District it was 144 mg/kg. The metal with the lowest concentration in these three regions was for manganese: 0.05 mg/kg; 0.1 mg/kg and 0.004 mg/kg for Lativan, Valadimir and Krasnodar areas, respectively.

7.5 Chromium levels in cheese

Results are given below for analyses obtained by (Larsen and Rasmussen 1991) using a Zeeman graphite furnace atomic absorption spectrometry to determine chromium, cadmium and lead in some Danish dairy products including cheese. Only data for chromium in cheese samples will be presented here. A Perkin Elmer model 5000 Zeenam atomic absorption instrument with a HA 500 graphite furnace and autosampler AS40 were employed for the analyses. Cheese samples were cut into small cubes then ashed in a bomb with nitric acid for four hours then ashed overnight. Chromium concentrations for six cheese samples that the authors state are only indicative, due to the small sample set, will now be given: Brie 45, 20 ng/mL; Camembert 30, 13 ng/mL; Danbo 45, 8.6 ng/mL; Danbo 45, 19 ng/mL; Havarti 60, 4.6 ng/mL and Maribo 45, 20 ng/mL. The authors state that the results obtained in this work compare well with other countries. Chromium and manganese levels in cheese were also reported by (Tinggi, Reilly, and Patterson 1997) as 95.0 ± 29.2 ng/kg and 1.1 ± 0.2 mg/kg, respectively.

8. Selected metals in fruit

Some of the metal analyses on fruit products are, in fact, related to metal analyses in fruit juices and purees that are often related to authenticity and country or region of origin. This is to detect adulteration with an inferior juice since metal analyses are indicative of the soil type and hence the location. Authenticity studies including metal analyses are usually in conjunction with other techniques such as GC/MS as well as AAS. Some examples will be given on juice and fruit with a limited number of metals. Other studies are related to contaminants such as tin from canned fruit and juices. Even though this article is directed to nutritional metals a few examples will be given of tin concentrations in canned fruit products for interest.

8.1 Examples of metal analyses in fruit juices and purees

Selenium levels in wild fruit juice from the mountainous area in north China, Lantingguo, was reported by (Yongming et al. 1996) using a graphite furnace AAS. Electrothermal atomic absorption spectriion spectrometry using a Perkin-Elmer 5000 atomic absorption spectrometer linked to a Model HGA 500 graphite furnace with a selenium HCL was the instrument set up employed for this work. This analysis was complicated by the fact that there were other interfering ions present in the juice, hence matrix modifiers were necessary. It was found that the modifier consisting of 10 µg of platinum and 200 µg of nickel gave the most satisfactory results. As already mentioned the metal content in the juice are related to the different fruit growing regions. Other metals found in these locations that interferes with the Se analyses include: potassium greater that 1000 mg/L followed by calcium, magnesium, manganese, iron, phosphorous and zinc in lower amounts 100 – 1000 mg/l. There were also trace amounts of cadmium, copper, lead, nickel, silicon, strontium and selenium. The average values found for eight determinations of three juices from fruit grown in different locations in the above region were: 0.20, 0.23 and 0.10 mg/L selenium.

Tin is not a desirable metal but a contaminant and was analysed in a number of juices, purees and fruit by the Comite Europeen de Normalisation (Foodstuffs. Determination of trace elements. Determination of tin by flame and graphite furnace atomic absorption spectrometry (FAAS and GFAAS) after pressure digestion 2009) in a collaborative study.. Some of the products analysed included: carrot puree, tomato puree, pineapple, mixed fruits, powdered peach and tomato. Samples were prepared by pressure-assisted digestion then analysed by flame AAS or graphite furnace AAS. Data obtained for the products analysed were in the range of 43 – 260 mg/kg for AAS and or 2.5 – 269 mg/kg for graphite furnace AAS.

8.2 Beneficial metal concentrations in some fruits

Slurried fruit samples were tested by (Cabrera, Lorenzo, and Lopez 1995) for the levels of cadmium, copper, iron, lead and selenium by Electrothermal AAS. Only results for copper, iron and selenium will be mentioned, as these are the nutritional metals as distinct from the others that are contaminants. In addition, to the sample preparation of slurries, the samples were also mineralized in a microwave acid-digested bomb and the data compared for accuracy and precision. A total of 40 samples comprising 8 types of fruit that are regularly consumed were tested. These samples were: banana; custard apple; kiwifruit; mango, medlar, papaya; pineapple and strawberries. For these fruit samples the mean range of the metals copper, iron and selenium were 2.00 – 5.50 µg/g; 0.050 - .0.396 µg/g and 0.010 – 0.020 µg/g, respectively. Chromium and manganese were analysed (Tinggi, Reilly, and Patterson 1997) in fruit by AAS, after wet digestion but found that these two metals were relatively low in fruit when compared to other foods. Fruit samples tested were: apple; banana; grapes; orange; pear; pineapple (canned) and rock melon.

Fruit	Cr (ng/kg)	Mn (mg/kg)
Apple	19.3 ± 3.3	0.5 ± 0.1
Banana	5.2 ± 1.3	3.3 ± 0.9
Grapes	4.3 ± 1.2	0.6 ± 0.1
Orange	6.3 ± 1.2	0.4 ± 0.1
Pear	12.6 ± 1.8	0.8 ± 0.1
Pineapple (canned)	21.3	1.5
Rock melon	9.8 ± 1.5	0.4 ± 0.1

Table 6. Concentrations of chromium (ng/kg) and manganese (mg/kg) in selected fruit samples.

It can be seen in the Table 6 that the chromium levels in these fruit samples ranged from 4.3 ng /kg in grapes to 21.3 ng/kg in pineapple. In contrast the manganese concentrations were higher than those for chromium in these same samples where the range was from 0.4 mg/kg for both orange and rock melon up to 3.3 mg/kg for banana.

Fifteen elements were analysed by flame AAS after microwave-assisted digestion of the cultivar citrus reticulate Blanco CV. Ougan fruits. The analyses gave high concentrations of these metals at both of the ripeness stages (Mojsiewicz-Pienkowska and Lukasiak 2003)

8.3 Analyses of tin a contaminant in canned fruit and fruit juices

A report by (Dogan and Haerdi 1980) on the analysis of the tin content in peaches, pears, pineapple, mandarin, peeled tomato and fruit cocktail by a number ot techniques including AAS will now be presented. Sample preparation was achived by using Lumatom, which is a trade organic chemical that contains quaternary ammonium hydroxide suspended in isopropanol. After this sample preparation procedure the.fruit and juice samples were introduced directly into the graphite AAS instrument. The tin concentrations levels were: quartered mandarin 68 ppm; peeled tomato 57 ppm and fruit cocktail 57 ppm. Rigin (Rigin 1979) used flameless AAS to determine the tin levels in canned tomato, apple and orange juice. The results for 5 replicates of each sample after three months storage for canned tomato, apple an orange juice were 12.1, 2.75 and 30.5 µg/ml respectively. After twelve months, however, the values had increased to 76.3, 4.26 and 45.4 µg/ml for the samples in the same order as for three months storage. Wehrer *et al* (Wehrer, Thiersault, and Laugel 1976) used AAS with wet and dry ashing to determine tin content of canned samples including stewed apples. The concentration of tin in the stewed apples for ten replicates was 52.4 mg/kg after 2N HCl digestion and 57.5 mg/kg after dry ashing that involvd calcinations using magnesium nitrate. Vijan and Chan (Vijan and Chan 1976) determined the tin content of a number of different types of samples including apple, apple/cherry, apple/pineapple and tomato juices. For the first three products the tin level was less than 0.1 µg/ml but for tomato juice it was considerably higher at 90 µg/ml.

9. Metal levels in vegetables

Detailed analyses of chromium and manganese on a wide range of food groups has been undertaken by (Tinggi, Reilly, and Patterson 1997) at this point in time the discussion will be limited to vegetables. The results show that the chromium levels are lower (ng/kg) in the vegetables analysed than in manganese (mg/kg) where the sample sizes ranged from 3 to 5 per vegetable. Vegetables studied included: beans (boiled); broccoli (boiled); carrot (boiled); cauliflower (boiled); lettuce; peas (frozen); potato (roasted); pumpkin (boiled);); tomatoes and zucchini. Portion of the data for these two elements have been extracted from the study under consideration in this section and will be presented below in Table 7.

It can be seen from Table 7 that the highest level of chromium is in tomatoes, 30 ng/kg, and the lowest level in boiled bean 5.3 ng/kg. Whereas for manganese the highest level was for frozen peas, 6.0 mg/kg, while the lowest value was for potato and pumpkin that were equal at 0.9 mg/kg. Hence tomato is a good source of chromium, which stabilizes the blood sugar levels and peas are a good source of manganese.

A study on the mineral levels in some Slovenian foods that are regularly consumed was undertaken by (Zuliani et al. 2005) After sample workup employing microwave-assisted digestion the samples were analysed by flame and electrothermal atomic absorption spectrophotometer. The minerals content investigated in this study were, Zn, Cu, Cd, Pb and Ni. It was found that the samples tested did not contain any Cd, Pb or Ni contamination. Zinc levels reported for cabbage and tomatoes were less than 50 mg/kg. Copper content, in contrast, was between 2 and 3 mg/kg in the majority of the samples while the chromium content was below 0.05 mg/kg. An analysis of Fe, Zn and Cu in some foods consumed in Mexico including vegetables, legumes, fruits, cereals and animal foods was reported by Lopez and co-workers (Lopez et al. 1999). Considering these products, it was found that the zinc level had a range of 0.018 mg/100g to 9.193mg/100 g for strawberry and beef. The iron concentrations were from 0.113 mg/100g to 19.82 mg/10g for yogurt and commercial cereal but the latter was fortified with minerals. Copper was not detected in all foods but was found to be the highest in beef liver, namely 3.371 mg/100g. Selenium levels in some Egyptian foods were determined by electro-thermal (ETAAS) and hydride generation (HGAAS) atomic absorption spectrometry by (Hussein and Bruggeman 1999). They found the metal was only in trace amounts: in the range of 1-33 μg/kg. Other products were tested but are not reported here.

Vegetable	Chromium (ng/kg)	Manganese (mg/kg)
Bean (boiled)	5.3 ± 1.6	3.4 ± 0.4
Broccoli (boiled)	8.0 ± 2.0	1.2 ± 0.1`
Carrot (boiled)	13.0 ± 2.3	1.5 ± 0.1
Cauliflower (boiled)	6.3 ± 1.6	1.1 ± 0.2
Lettuce	9.2 ± 1.3	1.3 ± 0.1
Peas (frozen)	28.3 ± 3.5	6.0 ± 1.2
Potato (roasted)	19.0 ± 2.2	0.9 ± 0.06
Pumpkin (boiled)	16.0 ± 2.2	0.9 ± 0.2
Tomatoes	30.0 ± 2.1	2.0 ± 0.1
Zucchini	6.3 ± 1.2	1.3 ± 0.3

Table 7. Chromium and manganese concentrations in some vegetables in ng/kg and mg/kg respectively (mean ± SD)

Vegetable	mg Ca/100 g
Asparagus (fresh)	13.9
Asparagus (canned)	14.7
Leek	16.2
Mushrooms (fresh) - grey oyster	1.0
Peak (fresh) – garden	62.5
Radish (pickled) - Chinese	94.9
Rhubarb (petioles) - ;pie ;plant	268.9
Seaweed (agar)	510.2
Spinach - Ceylon	116.2
Spinach -Bayam pasi	319.4
Tomato – tree	11.2
Yam bean	12.4

Table 8. Results of AAS Ca analyses, selected examples in vegetable, edible portion, from part of a table by Siong.

Calcium levels in many foods have been reported by (Siong, Khor Swan, and Siti Mizura 1989): some results will now be given for a selection of the vegetables tested. Two methods were compared AAS and potassium permanganate titration. A Varian Atomic Absorption spectrophotometer, Model 175 using an air/acetylene flame was used for the analyses. Samples were prepared for introduction into the instrument by ashing. Data for the two methods were in good agreement, however, only AAS results will be given for some vegetables reported by Siong's group.

For the vegetables listed in Table 8, seaweed contains the highest concentration of calcium being 510.2 mg/100 g. Spinach followed next with a value of 319.4 mg/10g.whereas fresh mushrooms contain only 1 mg/100 g of calcium, the lowest concentration reported in this group. Atomic absorption spectroscopy was used to determine the mineral content of hummyad (*Rumex vesicarius*) leaves that are grown in both the northern and central areas of Saudi Arabia.. Elements analysed included calcium, copper, iron, magnesium, potassium, sodium and zinc. The range of values (Alfawaz 2006) obtained for these metals are as follows: calcium 1790 – 2680 mg/100 g; copper 24.1 – 43.5 mg/100 g; iron 1320 – 2270 mg/100 g; potassium 2710 – 3230 mg100 g; sodium 846 – 1100 mg/ 100 g and zinc 3.7 – 8.8 mg/100 g. Hence, it is clear that this plant is very rich in calcium, iron and potassium, the highest mineral content being for potassium followed by calcium and iron indicating that these leaves are a nutritious source of these essential metals The germanium content in different foods including vegetables was determined by (McMahon, Regan, and Hughes 2006) using a Graphite furnace atomic absorption instrument. Sample workup was via drying and ashing. Values for several vegetables are as follows: carrot 0.60 µg/g; potato 1.85 µg/g; garlic 2.79 µg/g. and soy mince 9.39 µg/g. The latter sample having the highest germanium content of the vegetables tested. Food, crops and soils in Taiwan were analysed by (Huang, Wen, and Chern 1987) for the selenium content. Metals found in soils are directly related to the uptake of minerals in plants. Considering soil in the Taiwan region, the selenium level was determined to be in the range 0.03 – 0.23 ppm. Selenium content in crops, fruit and vegetables, in contrast, was reported to be approximately 0.1 ppm, however, for mushrooms the level was higher, namely, 0.55 ppm.

An analysis of two cultivars of onions grown in Venezuela: Yellow Granex PRR 502 and 438 Granex were tested for the concentrations of calcium, copper, iron, manganese, potassium and zinc, by total reflection X-ray fluorescence (TXRF), then the results compared with those from FAAS. A more efficient sample preparation was employed where the samples were acid extracted from the crude products using an ultrasonic bath, avoiding time consuming digestion. Sample work up was also compared with wet and dry ashing. The mineral content of the onions is important so the soils can have more elements added if, required, to improve their nutritional value. It was found that the ultrasound work up and dry ashing gave similar results. Levels of calcium copper and iron were found to be significantly greater in the Yellow Granex cultivar while potassium, manganese and zinc were significantly higher in 438 Granex. Levels of calcium and potassium were very much greater than the concentrations of the other elements: potassium being slightly higher than calcium, irrespective of the work up procedure: ultrasonic extraction or wet or dry ashing methods (Alvarez et al. 2003). A thorough analyses of a wide range of metals in Jamaican foods has been reported to (Howe et al. 2005) for legumes, leafy and root vegetables, fruit and other root crops. Only some metals levels will be mentioned here for the first three products listed above. Data will be given for calcium, chromium, copper, iron, magnesium, manganese, sodium and zinc. Comparison data was also undertaken with other countries but will not be given here.

Element	Legumes	Leafy vegetables	Root vegetables
Calcium	514	2580	390
Chromium	0.07	0.08	0.09
Copper	2.28	0.6	0.7
Iron	30.02	15.5	12.9
Potassium	0.67	0.39	0.41
Magnesium	790.00	359.0	157
Manganese	10.6	8.3	4.3
Sodium	4.16	90.1	727
Zinc	16.30	4.2	3.4

Table 9. Mean concentrations of metals, mg/kg in legumes, leafy and root vegetable grown in Jamaica

In addition to the analyses on calcium in foods, (Siong, Khor Swan, and Siti Mizura 1989) also undertook a similar study for iron levels, again employing a Varian Atomic Absorption Spectrophotometer model 175 using an air/acetylene flame. Analyses of iron in foods are important as it is a mineral that is often lacking in diets low in red meat, which is necessary to prevent anaemia Readily available information assists consumers and health care professionals to advise on foods that contain this element where the total RDA for that metal is kept in mind. The edible portions of the samples were homogenized, oven dried, charred then ashed in a muffle furnace. The AAS method was compared with a colorimetric phenanthroline procedure where the results indicated that both methods were found to be satisfactory for iron analyses although AAS would be a less time consuming. Only some data for vegetables by AAS will b e given below.

Vegetable	mg Fe/100 g
Asparagus (canned)	7.06
Asparagus (fresh)	0.55
Broccoli	0.47
Cucumber (hairy)	0.15
Leek	0.33
Mushrooms (fresh) grey oyster	0.84
Mustard leaves (Chinese)	1.35
Mustard leaves (Indian)	1.46
Peas (fresh) garden	0.75
Seaweed (agar)	5.33
Seaweed (dried)	22.94
Spinach (Ceylon)	0.88
Spinach (red)	2.64
Spinach (Bayam duri)	1..69
Yam bean	0.26

Table 10. Concentration of iron in mg/100 g of edible portion of vegetables

It is interesting to note from Table 10 that the highest level of iron is in dried seaweed at 22.94 mg Fe/100 g, followed by canned Asparagus, 7.06 mg Fe/100 g and then seaweed (agar), 5.33 mg Fe/100 g. In contrast the lowest iron content was found in hairy cucumber being 0.15 mg Fe/100 g. Sodium and potassium content of a large number of foods and composite foods consumed in Canada was carried out with respect to a Total Diet Survey. Generally only some unprocessed vegetable data obtained by (Tanase et al.) will be given in this Section. Potassium samples were analysed by AAS, Perkin Elmer AAnalyst 400, but sodium analyses were performed employing atomic emission spectroscopy. The mean potassium and sodium levels in mg/kg of the various unprocessed vegetables will now be quoted in this order. Samples were mechanically homogenized, filtered with Whatman 541 filter paper then diluted with dilute nitric acid that contained $CsCl_2$ (1000ug/ml) , a matrix modifier.

Vegetable	Potassium mg/kg	Sodium mg/kg
Asparagus (fresh)	2081	40
Baked beans (canned)	2542	**2892**
Beans string (fresh: canned)	1550	**1704**
Broccoli	1647	153
Brussels sprouts (fresh)	2946	64
Cabbage	1694	95
Carrots	2182	163
Cauliflower	1266	148
Celery	1627	626
Lettuce iceberg; romaine 3:1	1520	127
Mushrooms (button)	1844	38
Onions	825	33
Peas (frozen: canned)	**822**	1083
Peppers, green	1698	**4**
Potatoes (peeled; boiled)	2032	20
Potatoes (baked: skin)	**4921**	41
Spinach	**4007**	619
Tomatoes (fresh)	2599	26

Table 11. Selected vegetables (mean values, mg/kg) generally fresh but with some canned or cooked.

It can be seen in Table 11 that all of the vegetables listed have high potassium content with baked jacket potatoes having the highest potassium level of 4921 mg/kg followed by Spinach at 4007 mg/kg. The lowest potassium content is this group is for onions at 822 mg/kg. In contrast, the highest sodium concentration was found to be in canned baked beans at 2892 mg/kg followed by string beans either fresh or canned at 1704 mg/kg. The lowest sodium level was found be in green peppers at 4 mg/kg. Mineral concentrations in a range of vegetables have been reported by (Howe et al. 2005). Calcium, copper, chromium, iron, magnesium, manganese, potassium, sodium and zinc values will be given for some vegetables for the highest observed values expressed on a fresh weight basis. See Table 12 on the next page.

Metal	Vegetable	Level mg/kg
Calcium	Cabbage	20160
Magnesium	Cow peas	1621
Potassium	Corn	3.8%
Sodium	Carrot	1920
Copper	Cow peas	5.0
Chromium	Sweet potato	1.1
Iron	Red kidney beans	76.5
Manganese	Cow peas	27
Zinc	Daheen	76

Table 12. Minerals in selected vegetables in mg/kg

In Table 12, interesting to note that cabbage has such a high level of calcium at 20160 mg/kg and carrots contain 1920 mg/kg of sodium. Red kidney beans are good source of iron having 76.5 mg/kg that is an excellent source of this metal for vegetarians. Cow peas, in addition to containing a very high level of magnesium, 1620 mg/kg also have a concentration of manganese, 27 mg/kg. Dasheen is rich in zinc, 76 mg/kg.

9.1 Mineral and trace mineral content in tomatoes

Essential and trace mineral content in a number of tomato fruit cultivars was determined by (Ruiz et al. 1995) Their work is reported here under the section on vegetables as in Australia and some other countries, tomatoes are regarded as a vegetable. Essential minerals studied were calcium, magnesium, potassium and sodium while the trace metals were copper, iron, manganese and zinc. Data on samples reported here are new cultivars that were being developed in order to increase their health benefits. The four cultivars under consideration are: S. *lycopersicum*; S. *pimpinellofilium*; S. *cheesmaniae* and S. *harbrochaites*. Samples were grown in a greenhouse during the spring-summer time in Valencia, Spain. Preparation of the sample was firstly by freeze-drying,, ashing at 450°C and finally making up in acids: 50% HCl and HNO_3. A Perkin-Elmer 2280 spectrophotometer was employed using an air/acetylene flame for the analyses. Results from the study will now presented in the table below and cam be compared with other data for tomatoes reported by other researchers in this section on vegetables.

Sample	Ca	Mg	K	Na	Cu	Fe	Mn	Zn
A	23.57 ± 0.86	6.40 ± 0.52	78.47 ± 2.54	36.89 ± 1.48	0.43 ± 0.02	0.55 ± 0.3	0.1 ± 0.00	0.26 ± 0.01
B	37.69 ± 2.25	12.93 ± 0.44	212.83± 5.88	54.94 ± 2.65	0.74 ± 0.02	0.96 ± 0.05	0.17 ± 0.01	0.33 ± 0.01
C	45.78 ± 0.46	14.37 ± 0.25	222.51± 2.34	84.4 ± 1.49	0.72 ± 0.01	1.17 ± 0.11	0.19 ± 0.01	0.70 ± 0.01
D	46.88 ± 2.73	20.49 ± 0.16	194.27± 8.51	82.61 ± 2.89	0.62 ± 0.03	1.02 ± 0.06	0.15 ± 0.01	0.17 ± 0.01

Table 13. The concentrations of the essential and trace metals for the four cultivars A, B, C and D are given in mg/100 g Where A, B, C and D represent: S. *lycopersicum*; S. *pimpinellofilium*; S. *cheesmaniae* and S. *harbrochaites respectively.*

In Table13, it can be seen that cultivar C has the highest K level at 222.51 mg/100g; as well as highest Na, Fe, Mn and Zn levels: 84.4 mg/ 100g; 1.17 mg/100g , 0.19 mg.100g and 0.70 mg.100g, respectively. Cultivars B and D following closely behind C with Ca levels of 212.83 mg/100g and D 194.27 mg/100 g.. Cultivar D has the highest Ca and Mg concentrations at 46.88 mg/100g and 20.49 mg/100 g respectively.

10. Mineral content in some herbs and spices

Much of the literature on herbs and spices is related to contamination of heavy metals and analyses are undertaken to ensure that they are safe to consume and have not been grown in polluted areas, such as near motorways where there is the pollution from lead in old style cars. Possibly, they have been grown in soil near mining sites where there is pollution from heavy metals. Crops can be grown near polluted waterways where they are down stream from heavy industry factories so heavy metals seep into the soil and/or contaminate the ground water. In this section metals that are of health benefits will be emphasised with only a brief mention of those that are considered to be pollutants. Metals such as iron and calcium and others import in maintaining good health. Studies have also been reported on the metal content of Chinese herbs that have specific health benefits for certain disease states.

10.1 Minerals related to kidney function

Selected metals in Chinese medicinal herbs that are used in order to improve kidney function has been reported by (Kolasani, Xu, and Millikan 2011) Dried, unprocessed herbal samples were purchased from local importers of Chinese herbs in Melbourne. These are herbs that the Chinese medical practitioners would prescribe to their patients. Samples came from different parts of the plant: such as leaves, whole plant, stem, twig, bark or roots. Seven metals were tested by atomic absorption spectroscopy that included: calcium, iron, magnesium, manganese, sodium, potassium and zinc. A Varian spectra-400 Atomic Absorption Spectrophotometer (Varian Inc. Mulgrave, Australia) an air/acetylene flame was employed for the metal analyses. Samples were ground then digested in concentrated nitric. The range of metals found in these samples were as follows: Ca (130 – 560940 µg/g); Fe (20 – 8020 µg/g); Mg (90-5520 µg/g); Mn (20-140 µg/g); K (270-90260 µg/g); Na (30-4500 µg/g) and Zn (10-1010 µg/g); Na (30-4500 µg/g). Results indicated that calcium and potassium levels were the highest elements detected in all sample compared to the other metals. It was also found that the calcium concentration was greatest in fossils, then in plants whereas iron, potassium, manganese and zinc levels were highest in plants. Roots contained the highest magnesium concentration while flowers contained the highest sodium values.

10.2 Iron levels in some commonly consumed culinary herbs and spices

Herbs and spices are added to foods to enhance the flavour and add variety to an otherwise bland dish. In many cases, it is not a single herb but a combination that gives the dish that subtle taste and aroma. Different nationalities characteristically have their own traditional dishes that vary considerably in the choice and number of spices added to different dishes whether they be savoury main courses on spicy deserts. Siong and co-workers who analysed eight food groups, some of which have already been mentioned above, has reported the iron concentrations of a number of spices and herbs (Siong, Khor Swan, and Siti Mizura 1989). Some examples will now be presented in the Table 13 below.

Spice or herb	mg Fe/100 g
Chilli, small	0.68
Nutmeg, fresh	0.22
Persimmon, dried	0.99
Chives, Chinese	0.62
Coriander, leaves	3.86
Garlic, bulbs	0.48
Garlic, plants	0.31
Parsley	9.90

Table 13. Iron concentration in mg Fe/100 g of a selection of spices and herbs

It can be seen in Table 13 that parsley is an excellent source of iron, 9.90 mg Fe/100 g, that is particularly important for those who eat little red meat or are vegetarians. Iron deficiency is a common problem particularly for women. Coriander leave being the next highest level of iron at 3.86 mg Fe/100 g. It is interesting to note that the iron in garlic bulbs is 0.48 mg Fe/100 g that is greater than in the plants at 0.31 mg Fe/100 g. Of this group nutmeg has the lowest iron level at 0.22 mg Fe/100 g.

10.3 Calcium levels in herbs and spices

In a related study that complements the above work on iron, (Siong, Khor Swan, and Siti Mizura 1989) analysed similar samples for the calcium concentrations. Again, some of these food groups have already been discussed in this review.

Spice	mg Ca/100gm
Anise seed, dried	950.6
Cardamon	1769.7
Cinnamon	600.9
Cumin seeds, black	816.8
Cumin seeds, white	1165.1
Curry powder	576.2
Fenugreek seeds	179.8
Pepper, powder, white	120.4

Table 14. Calcium content of selected spices in mgCa/100 g

It can be seen in Table 14 that cardamom has the greatest level of calcium at 1769.7 mg Ca/100 g followed by dried anise seed with a value of 950.6 mg Ca/100 g. While pepper, in contrast, has the lowest calcium content, namely, 120.4 mg Ca/100 g.

11. Conclusion

The chapter has taken examples of the literature on the mineral content, both Essential and Trace metals in meat, dairy products, fruit, vegetables and herbs and spices. The review is not exhaustive but a selection of examples of metals extracted from many authors with the data arranged in such a way as to highlight at a glance the concentrations of the metals in the above foods that are generally not processed. It is a collection of information in the one chapter assembled from published work, which allows a convenient comparison of the concentration of certain metals in the products discussed. Such information is of value to health care professionals, researchers and food manufacturers in preparing nutritious products. Levels of metals in some product may also be unexpected and hence informative and may lead on to further analyses and research.

12. References

Alfawaz, M. A. 2006. Chemical composition of hummayd (Rumex vesicarius) grown in Saudi Arabia. *Journal of Food Composition and Analysis: 19 (6-7, Biodiversity and nutrition: a common path) 552-555* 19 (6-7, Biodiversity and nutrition: a common path):552-555.

Alvarez, J., L. M. Marco, J. Arroyo, E. D. Greaves, and R. Rivas. 2003. Determination of calcium, potassium, manganese, iron, copper and zinc levels in representative samples of two onion cultivars using total reflection X-ray fluorescence and ultrasound extraction procedure. *Spectrochimica Acta Part B-Atomic Spectroscopy* 58 (12):2183-2189.

Brandao, G. C., G. D. Matos, and S. L. C. Ferreira. Slurry sampling and high-resolution continuum source flame atomic absorption spectrometry using secondary lines for the determination of Ca and Mg in dairy products. *Microchemical Journal: 98 (2) 231-233* 98 (2):231-233.

Cabrera, C., M. L. Lorenzo, and M. C. Lopez. 1995. Electrothermal atomic absorption spectrometric determination of cadmium, copper, iron, lead, and selenium in fruit slurry: analytical application to nutritional and toxicological quality control. *Journal of AOAC International: 78 (4) 1061-1067* 78 (4):1061-1067.

Carvalho, M. L., S. Santiago, and M. L. Nunes. 2005. Assessment of the essential element and heavy metal content of edible fish muscle. *Analytical and Bioanalytical Chemistry:* 382 (2, The European Conference on Analytical Chemistry XIII):426-432.

Dalton, E. F., and A. J. Malanoski. 1969. Atomic absorption analysis of copper and lead in meat and meat products. *Journal of the Association of Official Analytical Chemists:* 52 (5):1035-38.

Dogan, S., and W. Haerdi. 1980. Determination of total tin in environmental biological and water samples by atomic absorption spectrometry with graphite furnace. *International Journal of Environmental Analytical Chemistry: 8 (4) 249-257* 8 (4):249-257.

Dundas, M.J, and S Pawluk. 1977. Heavy metals in cultivated soils and in cereal crops in Alberta. *Canadian Journal of Soil Science* 7:329-339.

Foodstuffs. Determination of trace elements. Determination of tin by flame and graphite furnace atomic absorption spectrometry (FAAS and GFAAS) after pressure digestion. 2009. *European Standard: EN 15764, 16pp.*:16pp.

Formo, M. W., G. R. Honold, and D. B. MacLean. 1974. Determination of soy products in meat-soy blends. *Journal of the Association of Official Analytical Chemists: 57 (4) 841-846* 57 (4):841-846.

Germs, A. C., and H. Stennenberg. 1978. Estimating calcium in mechanically deboned poultry meat by oxidimetry and atomic absorption spectrophotometry. *Food Chemistry:* 3 (3):213-219.

Hannaford, P. 2002. Alan Walsh 1916-1998. *Historical Records of Australia Science* 13 (2):179-206.

Hoenig, M., and P. van Hoeyweghen. 1986. Determination of selenium and arsenic in animal tissues with platform furnace atomic absorption spectrometry and deuterium background correction. *International Journal of Environmental Analytical Chemistry:* 24 (3):193-202.

Howe, A., L. Fung, G. Lalor, R. Rattray, and M. Vutchkov. 2005. Elemental composition of Jamaican foods 1: A survey of five food crop categories. *Environmental Geochemistry and Health* 27 (1):19-30.

Huang, W. I., H. M. Wen, and J. C. Chern. 1987. [Selenium content of soils, crops and foods in Taiwan area.]. *Journal of the Chinese Agricultural Chemical Society:* 25 (2):150-158.

Hussein, L., and J. Bruggeman. 1999. Selenium analysis of selected Egyptian foods and estimated daily intakes among a population group. *Food Chemistry* 65 (4):527-532.

Jorhem, L., B. Sundstroem, C. Astrand, and G. Haegglund. 1989. The levels of zinc, copper, manganese, selenium, chromium, nickel, cobalt, and aluminium in the meat, liver and kidney of Swedish pigs and cattle. *Zeitschrift fuer Lebensmittel-Untersuchung und -Forschung:* 188 (1):39-44.

Kolasani, A, H Xu, and Millikan. 2011. Evaluation of mineral content of Chinese medicinal herb used to improve kidney funtion with chemometrics. *Food Chemistry* 127 (4):1467-1471.

Koops, J., and D. Westerbeek. 1993. Manganese in Dutch milk powder. Determination by Zeeman-corrected stabilized-temperature platform-furnace atomic-absorption spectroscopy. *Netherlands Milk and Dairy Journal: 47 (1) 1-13* 47 (1):1-13.

Larsen, E. H., and L. Rasmussen. 1991. Chromium, lead and cadmium in Danish milk products and cheese determined by Zeeman graphite furnace atomic absorption spectrometry after direct injection or pressurized ashing. *Zeitschrift fuer Lebensmittel-Untersuchung und -Forschung: 192 (2) 136-141* 192 (2):136-141.

Lopez, P., M. Castaneda, G. Lopez, E. Munoz, and J. L. Rosado. 1999. Iron, zinc and copper content of foods commonly consumed in Mexico. *Archivos Latinoamericanos De Nutricion* 49 (3):287-294.

Lovachev, L. N., I. F. Rodionova, E. V. Andreeva, and L. N. Fedorova. 1972. [Determination of copper and other elements in butter.]. *Voprosy Pitaniya: 35 (1) 68-71* 35 (1):68-71.

Lustre, A. O., and C. H. Lacebal. 1976. Zinc and magnesium in Philippine foods. I. Development of a method. *Philippine Journal of Nutrition:* 29 (1/2):25-32.

Maurer, J. 1977. [Extraction method for the simultaneous determination of sodium, potassium, calcium, magnesium, iron, copper, zinc and manganese in organic material using atomic absorption spectrophotometry.]. *Zeitschrift fuer Lebensmittel-Untersuchung und -Forschung* 165 (1):1-4.

McMahon, M., F. Regan, and H. Hughes. 2006. The determination of total germanium in real food samples including Chinese herbal remedies using graphite furnace atomic absorption spectroscopy. *Food Chemistry: 97 (3) 411-417* 97 (3):411-417.

Miquel, E., A. Alegria, R. Barbera, and R. Farre. 2005. Speciation analysis of calcium, iron, and zinc in casein phosphopeptide fractions from toddler milk-based formula by anion exchange and reversed-phase high-performance liquid chromatography-mass spectrometry/flame atomic-absorption spectroscopy. *Analytical and Bioanalytical Chemistry: 381 (5) 1082-1088* 381 (5):1082-1088.

Mojsiewicz-Pienkowska, K., and J. Lukasiak. 2003. Analytical fractionation of silicon compounds in foodstuffs. *Food Control: 14 (3) 153-162* 14 (3):153-162.

Nakamura, R. 1973. Estimation of water-extractable Ca in chicken breast muscle by atomic absorption. *Analytical Biochemistry:* 53 (2):531-537.

Regina de Amorim, F., M. Batista Franco, C. C. Nascentes, and J. Bento Borba da Silva. Direct determination of molybdenum in milk and infant food samples using slurry sampling and graphite furnace atomic absorption spectrometry. *Food Analytical Methods: 4 (1) 41-48* 4 (1):41-48.

Rigin, V. I. 1979. [Atomic absorption spectrometric determination of Sn in water and biological materials using electrolytic separation and atomization in gas phase.]. *Zhurnal Analiticheskoi Khimii: 34 (8) 1569-1573* 34 (8):1569-1573.

Ruiz, C., A. Alegria, R. Barbera, R. Farre, and M. J. Lagarda. 1995. Direct determination of calcium, magnesium, sodium, potassium and iron in infant formulas by atomic spectroscopy. Comparison with dry and wet digestions methods. *Nahrung:* 39 (5/6):497-504.

Schaefer, M. J., M. B. Kohrs, M. Howser, and S. Snider. 1979. The copper and zinc content of meat and meat analogs.

Siong, T. E., Choo Khor Swan, and Shahid Siti Mizura. 1989. Determination of calcium in foods by the atomic absorption spectrophotometric and titrimetric methods. *Pertanika:* 12 (3):303-311. 1989. Determination of iron in foods by the atomic absorption spectrophotometric and colorimetric methods. *Pertanika:* 12 (3):313-322.

Tanase, C. M., P. Griffin, K. G. Koski, M. J. Cooper, and K. A. Cockell. Sodium and potassium in composite food samples from the Canadian Total Diet Study. *Journal of Food Composition and Analysis* 24 (2):237-243.

Tinggi, U., C. Reilly, and C. Patterson. 1997. Determination of manganese and chromium in foods by atomic absorption spectrometry after wet digestion. *Food Chemistry: 60 (1) 123-128* 60 (1):123-128.

Turker, A. R., and E. Erol. 2009. Optimization of selenium determination in chicken's meat and eggs by the hydride-generation atomic absorption spectrometry method. *International Journal of Food Sciences and Nutrition: 60 (1) 40-50* 60 (1):40-50.

Uluozlu, O. D., M. Tuzen, D. Mendil, and M. Soylak. 2007. Trace metal content in nine species of fish from the Black and Aegean Seas, Turkey. *Food Chemistry:* 104 (2):835-840.

Vijan, P. N., and C. Y. Chan. 1976. Determination of tin by gas phase atomization and atomic absorption spectrometry. *Analytical Chemistry: 48 (12) 1788-1792* 48 (12):1788-1792.

Wehrer, C., J. Thiersault, and P. Laugel. 1976. [Determination of tin in canned fruits and vegetables by atomic absorption spectrometry.]. *Industries Alimentaires et Agricoles: 93 (12) 1439-1446* 93 (12):1439-1446.

Whitney, E.N, and S.R Rofles, eds. 2002. *Understanding Nutrition.* 9 ed: Thomas Learning, Inc.

Ybanez, N., R. Montoro, and A. Bueso. 1983. [Determination of cadmium, lead and copper in cooked meat products by flame atomic absorption spectroscopy.]. *Revista de Agroquimica y Tecnologia de Alimentos:* 23 (4):510-520.

Yongming, Liu, Gong Benling, Li Zhuanhe, Xu Yuli, and Lin Tiezheng. 1996. Direct determination of selenium in a wild fruit juice by electrothermal atomic absorption spectrometry. *Talanta: 43 (7) 985-989* 43 (7):985-989.

Zachariadis, G. A., J. A. Stratis, L. Kaniou, and G. Kalligas. 1995. Critical comparison of wet and dry digestion procedures for trace metal analysis of meat and fish tissues. *Mikrochimica Acta:* 119 (3/4):191-198.

Zuliani, T., B. L. Kralj, V. Stibilj, and R. Milacic. 2005. Minerals and trace elements in food commonly consumed in Slovenia. *Italian Journal of Food Science* 17 (2):155-166.

8

Application of Atomic Absorption for Determination of Metal Nanoparticles in Organic-Inorganic Nanocomposites

Roozbeh Javad Kalbasi[1,2] and Neda Mosaddegh[1,2]
[1]Department of Chemistry, Shahreza Branch,
Islamic Azad University, 311-86145 Shahreza, Isfahan,
[2]Razi Chemistry Research Center, Shahreza Branch,
Islamic Azad University, Shahreza, Isfahan
Iran

1. Introduction

The Pd-catalyzed Suzuki cross-coupling reaction has been shown as an efficient method for the construction of C-C bonds and plays an important role in pharmaceutical industry and organic synthesis (Makhubela et al., 2010; Venkatesan & Santhanalakshmi, 2010; Zhao et al., 2011). Numerous Pd complexes, such as palladacycles (Mu et al., 2011) and N-heterocyclic carbine (Chanjuan et al., 2008), have been developed for use in these reactions. However, these Pd(0) or Pd(II) complexes cause difficulties in the synthesis and purification of the final product. Another class of catalysts for these reactions, namely heterogeneous catalysts, is easy to prepare and readily separated from the products (Jana et al., 2009; Tamami & Ghasemi, 2010)

In catalytic applications, a uniform dispersion of nanoparticles and an effective control of particle size are usually expected. However, nanoparticles frequently aggregate to yield bulk-like materials, which greatly reduce the catalytic activity and selectivity. Therefore, they must be embedded in a matrix such as polymer or macromolecular organic ligands (Sanchez-Delgado et al., 2007; Luo & Sun, 2007). However, nanoparticle-polymer composites usually suffer from disadvantages such as absence of complete heterogeneity and high temperature annealing, which generally causes thermal degradation of organic polymers. In addition, to avoid the problems associated with metal nanoparticles such as homogeneity, recyclability and the separation of the catalyst from reaction system, some other works have focused on immobilizing metal nanoparticles on suitable support materials such as immobilization in pores of heterogeneous supports (Thomas et al., 2003; Jacquin et al., 2003), like ordered mesoporous silica. Although nanoparticle-mesoporous materials are completely heterogeneous, the hydrophilicity of these catalysts causes a reduction in the activity of such catalysts in organic reactions. Therefore, preparation of organic-inorganic hybrid catalysts with a hydrophobe-hydrophile nature is interesting.

The discovery of M41S-type ordered mesoporous materials opened a new class of periodic porous solids (Huo et al. 1996). Mesoporous silica structures have been regarded as ideal supports for heterogeneous catalysts due to their high surface area, tunable pore size, and alignment. In particular, the SBA-15 framework synthesized by Zhao et al. (Zhao et al., 1998) has a highly ordered hexagonal mesostructure with parallel channels and adjustable pore size in the range of 2-10 nm (Kalbasi et al, 2010). This size regime is relevant to catalysis, since the catalytically active components are metal particles in the 2-10 nm size range. SBA-15 is well suited as a structure that can contain individual metal particles within its mesopores, and the pores are wide enough to permit facile diffusion of reactants and products.

Hybrid organic–inorganic polymers have received increasing interest from research groups because of their unique properties (Mark, 2006; Zheng et al., 2008; Chung, 2002). Nevertheless, among the different researches on these materials, there are relatively a few reports on the application of organic-inorganic hybrid polymer as a heterogeneous catalyst (Morales et al., 2010; Ma et al., 2010; Alves et al., 2009). Recently, in our previous studies (Kalbasi et al., 2010; Kalbasi et al., 2010; Kalbasi et al., 2011; Kalbasi & Mosaddegh, 2011), Hybrid organic–inorganic polymers were used as catalysts. The hybrid materials could be obtained by combining organic polymers with inorganic materials (Run et al., 2007). These organic–inorganic hybrid materials could be prepared by various methods, depending on what kind of interaction is employed between organic polymers and inorganic elements, or on how organic moieties are introduced to inorganic phases. An in situ polymerization, which is the simultaneous polymerization of organic monomers in the presence of mesoporous materials, is an important method for the preparation of composite materials without chemical interaction.

There are a lot of methods for characterization of these nanoparticles supported on hybrid materials. However, one of the best methods for determination the amounts of the metal nanoparticles is atomic absorption spectrometry (AAS). For many years, atomic absorption spectrometry (AAS), both flame (FAAS) and electrothermal (ETAAS), has been widely used for the determination of metals in various materials (Ghanemi et al., 2011). At present, AAS is the most widely used method for determination of metal as long as the sample can be dissolved in acids. Separation, pre-concentration and dissolution of samples are the vital steps in many procedures, owing to the very low concentration of these metals in many samples and the complexity of the matrix. Among the AAS methods developed, the determinations for Pd and Rh are relatively sensitive (Scaccia & Goszczynska, 2004). Efforts have been made to make methods reliable and practical.

In this study, Pd nanoparticle-poly(N-vinyl-2-pyrrolidone)/SBA-15 (Pd-PVP/SBA-15) nanocomposite was prepared as a highly efficient catalyst by in situ polymerization method. The sample was studied by XRD, FT-IR, UV-Vis, TG, BET, TEM and AAS techniques. The main goal of this catalytic synthesis was to introduce a novel and efficient organic-inorganic composite to expand the use of these types of composites for carbon-carbon coupling reaction. This nanocomposite was tested for catalytic activity for Suzuki-Miyaura cross-coupling reactions between phenylboronic acid and several aryl halides. The reaction was carried out in the presence of water at room temperature. The amount of Pd nanoparticles has critical effect on the catalytic activity of the catalyst. So the effect of metal amounts for the synthesis of Pd-PVP/SBA-15 has been investigated. In addition, the effects of reaction temperature, the amount of catalyst, amount of support and solvent, were investigated as well as recyclability of the heterogeneous composite. The catalyst used for this synthetically useful transformation showed considerable level of reusability besides very good activity. Reusability of the catalyst is very important. So in this study we introduce a catalyst that can be reused up to 5 times without losing its catalytic efficiency. For determination of the amount of remained Pd after each cycle of the reaction on the catalyst structure, atomic absorption spectroscopy was used.

2. Experimental method

2.1 Catalyst characterization

The samples were analyzed using FT-IR spectroscopy (using a Perkin Elmer 65 in KBr matrix in the range of 4000–400 cm^{-1}). The BET specific surface areas and BJH pore size distribution of the samples were determined by adsorption–desorption of nitrogen at liquid nitrogen temperature, using a Series BEL SORP 18. The X-ray powder diffraction (XRD) of the catalyst was carried out on a Bruker D8Advance X-ray diffractometer using nickel filtered Cu Kα radiation at 40 kV and 20 mA. Moreover, scanning electron microscope (SEM) studies were performed on Philips, XL30, SE detector. The thermal gravimetric analysis (TGA) data were obtained by a Setaram Labsys TG (STA) in a temperature range of 30–650 °C and heating rate of 10 °C/min in N_2 atmosphere. Transmission electron microscope (TEM) observations were performed on a JEOL JEM.2011 electron microscope at an accelerating voltage of 200.00Kv using EX24093JGT detector in order to obtain information on the size of Pd nanoparticles. The DRS UV-Vis spectra were recorded with JASCO spectrometer, V-670 from 190 to 2700 nm. Moreover, Pd content of the catalyst was estimated using atomic absorption spectroscopy (AAS) (Perkin-Elmer, AAnalyst 300).

2.2 Catalyst preparation
2.2.1 Preparation of SBA-15

Mesoporous silica SBA-15 was prepared using Pluronic P123 ($EO_{20}PO_{70}EO_{20}$) template as a structure directing agent and tetraethylorthosilicate (TEOS) as the silica precursor. In a typical synthesis, Pluronic P123 (2 g) was dissolved at room temperature in H_3PO_4 (4.16 mL, 85%) and deionized water (75.37 mL), then TEOS (4.58 mL) was added to the solution and synthesis was carried out by stirring at 35 °C for 24 h in sealed Teflon breakers, and it was subsequently placed at 100 °C for 24 h. Then, the solution was filtered, washed with deionized water, and finally dried at 95 °C for 12 h in air. Template removal was performed by calcination in air using two successive steps; first heating at 250 °C for 3 h and then at 550 °C for 4 h.

2.2.2 Preparation of poly(N-vinyl-2-pyrrolidone)/SBA-15 (PVP/SBA-15)

N-vinyl-2-pyrrolidone (NVP) (0.5 mL, 4.6 mmol) and SBA-15 (0.5 g) in 7 mL tetrahydrofuran (THF) were placed in a round bottom flask. Benzoyl peroxide (3 % mol, 0.034 g) was added and the mixture was heated to 65-70°C for 5 h while being stirred under N_2 gas. The resulting white fine powder composite (PVP/SBA-15) was collected by filtration, washed several times with THF, and finally dried at 60 °C under reduced pressure.

2.2.3 Preparation of Pd nanoparticle-poly(N-vinyl-2-pyrrolidone)/SBA-15 (Pd-PVP/SBA-15)

Pd-PVP/SBA-15 nanocomposite was synthesized as follows (Scheme 1): PVP/SBA-15 (0.1 g) and 10 mL of an aqueous acidic solution (C_{HCl} = 0.09 M) of Pd(OAc)$_2$ (0.025 g, 0.111 mmol) were placed in a round bottom flask. The mixture was heated to 80 °C for 5 h while being stirred under N_2 gas. Then, 0.6 mL (9.89 mmol) aqueous solution of hydrazine hydrate (80 vol.%) was added to the mixture drop by drop in 15-20 minutes. After that, the solution was stirred at 60 °C for 1 h. Afterwards, the solution was filtered and precipitated, and was washed sequentially with chloroform and methanol to remove excess $N_2H_4.H_2O$ and was dried in room temperature to yield palladium nanoparticle-poly(N-vinyl-2-pyrrolidone)/SBA-15 composite (Pd-PVP/SBA-15).

In order to specify the amount of Pd, the composite should be decomposed by perchloric acid, nitric acid, fluoric acid and hydrochloric acid. For this purpose, Pd-PVP/SBA-15 nanocomposite (0.1 g), perchloric acid (0.5 mL), nitric acid (0.5 mL) and the known amount of water were placed in a round bottom flask. The mixture was heated for 1 h to evaporate the liquids. Then, to the precipitated, hydrofluoric acid (2.5 mL) and hydrochloric acid (2.5 mL) were added and the mixture was heated again for 2 h to evaporate the liquids. In the last step, hydrochloric acid (0.5 mL) and water (5 mL) were added to the precipitated and the mixture was heated for 30-60 minutes. Then, the Pd content of the catalyst was estimated by Atomic Absorption Spectroscopy (AAS). The Pd content of the catalyst estimated by AAS was 0.56 mmol g^{-1}.

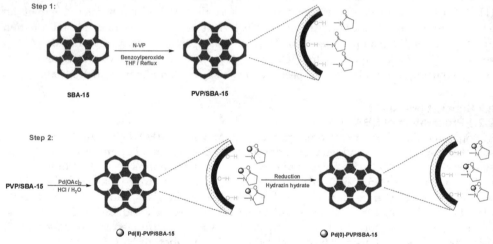

Scheme 1. Preparation of Pd-PVP-SBA-15

2.3 General procedure for Suzuki-Miyaura coupling reaction

The general procedure was as follows (Scheme 2): In the typical procedure for Suzuki-Miyaura coupling reaction, a mixture of iodobenzene (1 mmol), phenylboronic acid (1.5 mmol), K_2CO_3 (5 mmol), and catalyst (0.12 g, Pd-PVP/SBA-15) in H_2O (5 mL) was placed in a round bottom flask. The suspension was stirred at room temperature for 8 h. The progress of reaction was monitored by Thin Layer Chromatography (TLC) using n-hexane as eluent. After completion of the reaction (monitored by TLC), for the reaction work-up, the catalyst was removed from the reaction mixture by filtration, and then the reaction products were extracted with CH_2Cl_2 (3×5 mL). The solvent was removed under reduced pressure. The crude product was purified by flash column chromatography (hexane or hexane/ethyl acetate) to afford the desired coupling product (98% isolated yield).

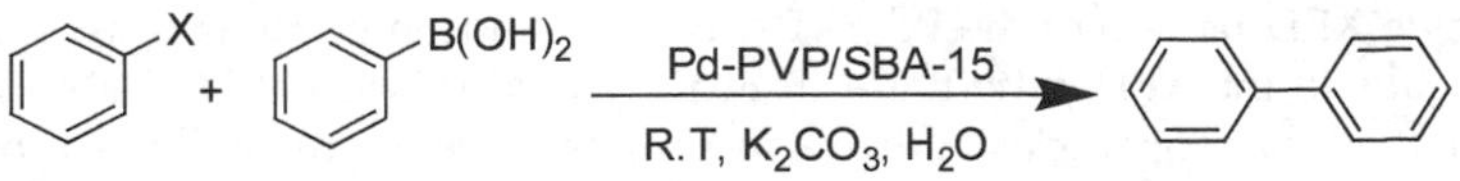

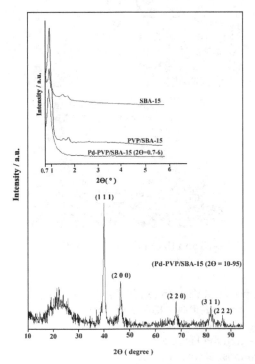

Scheme 2. Suzuki-Miyaura coupling reaction

3. Results and discussion

3.1 Catalyst characterization
3.1.1 XRD

The small angle XRD patterns of the SBA-15, PVP/SBA-15 and Pd-PVP/SBA-15 are shown in the Fig. 1. One intense peak at about 0.9–0.95° and two weak peaks at about 1.5–1.65° and 1.7–1.9° can be indexed as (100), (110), and (200) reflections associated with two-dimensional hexagonal symmetry ($P6mm$) (Kalbasi et al., 2010). The PVP/SBA-15 and Pd-PVP/SBA-15 ($2\Theta = 0.7$-6) samples show the same pattern indicating that the structure of the SBA-15 (100) is retained even after the support of the surface of the SBA-15 with PVP and Pd (Fig. 1). However, the intensity of the characteristic reflection peaks of the PVP/SBA-15 and Pd-PVP/SBA-15 ($2\Theta = 0.7$-6) samples are found to be reduced (Fig. 1). This may be attributed to the symmetry destroyed by the hybridization of SBA-15 which is also found in the ordered mesoporous silica loading with guest matter (Kalbasi et al., 2010). In addition, composites contain much less SBA-15 due to the dilution of the siliceous material by PVP and Pd; therefore, this dilution can also account for a decrease in the peak intensity.

Fig. 1. The powder XRD pattern of (a) mesoporous silica SBA-15, (b) PVP/SBA-15 and (c) Pd-PVP/SBA-15.

The wide angle XRD pattern of the Pd-PVP/SBA-15 nanocomposite is shown in Fig. 1. All diffraction peaks in the XRD pattern are from the *fcc* Pd (Harish et al., 2009). The average grain sizes of the Pd nanoparticles in the nanocomposite were estimated from the full width at half maximum of the diffraction peaks using Scherrer equation. The average grain size of the Pd nanoparticles is about 7 nm for Pd-PVP/SBA-15. The average grain size of the Pd nanoparticles is the same as the average particle size of the Pd nanoparticles determined by the TEM observations, which indicates that the Pd nanoparticles are single crystals.

3.1.2 TEM

The morphologies of SBA-15 (Fig. 2a), Pd-PVP/SBA-15 nanocomposite (Fig. 2b) and the distribution of Pd nanoparticles in the Pd-PVP/SBA-15 (Fig. 2c) were studied by the TEM observations. The typical TEM micrographs (Fig. 2) clearly show that the Pd-PVP/SBA-15 nanocomposite have a hexagonal pore array structure. In the side view micrographs of the nanocomposite (Fig. 2b), the dark spots are the Pd nanoparticles. It can be seen in Fig. 2 that most of the Pd nanoparticles are distributed within the channels of the mesoporous silica. The particle sizes of the Pd nanoparticles in Pd-PVP/SBA-15 are between 3 and 5 nm (Fig. 2c).

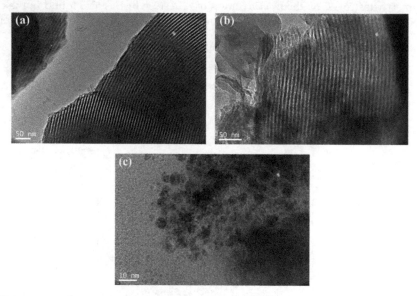

Fig. 2. TEM images for (a) SBA-15 and (b,c) Pd-PVP/SBA-15.

3.1.3 FT-IR

Fig. 3 shows the FT-IR spectra of SBA-15 (a), PVP/SBA-15 (b), and Pd-PVP/SBA-15 (c). The characteristic bands at around 1080, 810 and 480 cm^{-1} may be assigned to Si–O–Si asymmetrical stretching vibration, symmetrical stretching vibration and bending vibration, respectively, which is seen in the Fig 3. a,b,c. In addition, the band at around 950-965 cm^{-1} is related to Si–OH vibrations of the surface silanols (Fig. 3), which is characteristic of mesoporous silica. The existence of PVP in the PVP/SBA-15 composite is evidenced by the appearance of typical PVP vibration on the FT-IR spectrum (Fig. 3b). In the FT-IR spectrum of PVP/SBA-15 (Fig. 3b), the new band at 1666 cm^{-1} is corresponds to the carbonyl bond of PVP (Iwamoto et al., 2009). Moreover, the presence of peaks at around 2800-3000 cm^{-1} corresponds to the aliphatic C–H stretching in PVP/SBA-15 (Fig. 3b). The appearance of the above bands showed that PVP has been attached to the surface of SBA-15 and the PVP/SBA-15 has been obtained. As shown in Pd-PVP/SBA-15 spectrum (Fig. 3c), the band around 1666 cm^{-1} which corresponds to carbonyl bond of PVP, is shifted to lower wave numbers (1639 cm^{-1}) (red shift). Moreover, the peak intensity of the carbonyl bond in the spectrum of Pd-PVP/SBA-15 is lower than that of PVP/SBA-15. This may be due to the interaction between the Pd nanoparticles and C=O groups. This means that the double bond CO stretches become weak by coordinating to Pd nanoparticles. Thus, it is confirmed that PVP molecules exist on the surface of the Pd nanoparticles, and coordinate to the Pd nanoparticles (Metin et al., 2008; Hirai et al.,1985).

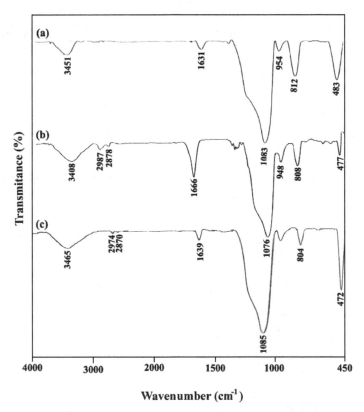

Fig. 3. FT-IR spectra of (a) mesoporous silica SBA-15, (b) PVP/SBA-15 and (c) Pd-PVP/SBA-15

3.1.4 BET

The representative N$_2$ adsorption–desorption isotherms of the SBA-15 (a), PVP/SBA-15 (b) and Pd-PVP/SBA-15 (c) are shown in Fig. 4. The mesoporous materials all exhibit clear H1-

type hysteresis loops indicative of mesoporous materials with one-dimensional cylindrical channels (Kalbasi et al., 2010) and narrow pore size distributions. The BET surface areas, the BJH pore diameters, and the pore volumes for the SBA-15, PVP-SBA-15 and Pd-PVP/SBA-15 nanocomposite are summarized in Table 1. A specific surface area of 1430 m²/g, a pore volume of 1.9 cm³/g, and a pore diameter of 9.9 nm are obtained from the isotherm of SBA-15. After hybridization with PVP through in situ polymerization, PVP/SBA-15 exhibits a smaller specific area, pore size and pore volume in comparison with those of pure SBA-15, which might be due to the presence of polymer on the surface of the SBA-15 (Table 1 and Fig. 4). However, there is a noticeable increase in pore diameter for Pd-PVP/SBA-15, and the pore volume of Pd-PVP/SBA-15 is smaller than that of PVP/SBA-15. It might be due to the incorporation of Pd nanoparticles into the pores of PVP/SBA-15 composite (Chytil et al., 2005).

According to the results, Pd-PVP/SBA-15 still has a mesoporous form with reasonable surface area and it is suitable to act as a catalyst.

Sample	BET surface area ($m^2 g^{-1}$)	V_P ($cm^3 g^{-1}$) [a]	BJH pore diameter (nm)
Mesoporous silica SBA-15	1430	1.09	9.90
PVP/SBA-15	465	0.93	4.03
Pd-PVP/SBA-15	102	0.13	4.61

[a] Total pore volume.

Table 1. Surface area analyses of mesoporous silica SBA-15, PVP/SBA-15 and Pd-PVP/SBA-15 samples obtained from N_2 adsorption.

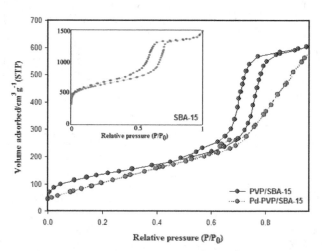

Fig. 4. N_2 adsorption-desorption isotherms of mesoporous silica SBA-15, PVP/SBA-15 and Pd-PVP/SBA-15.

3.1.5 TG-DTA

Fig. 5 presents the TGA curves of SBA-15 (a), PVP (b), PVP/SBA-15 (c) and Pd-PVP/SBA-15 (d) under N_2 atmosphere. The mass loss at temperature <100 °C (around 6%, w/w) is attributed to desorption of water present in the surfaces of the SBA-15 (Fig. 5a). The TGA curves of PVP show a small mass loss (around 7.5%, w/w) in the temperature range 50–150 °C, which is apparently associated with adsorbed water (Fig. 5b). At temperatures above 200 °C, PVP shows one main stage of degradation. The mass loss for PVP in the second step is equal to 80% (w/w) which corresponds to the effective degradation of the polymer (Fig. 5b). Thermo analysis of PVP/SBA-15 shows two steps of mass loss (Fig. 5c). The first step (around 3%, w/w) that occurs at temperature <150 °C is related to desorption of water. The second step (around 9%, w/w) which appeared at 220 °C is attributed to degradation of the polymer, and the degradation ended at 400 °C (Fig. 5c). By comparing the PVP and PVP/SBA-15 curves, one can find that PVP/SBA-15 has higher thermal stability and slower degradation rate than PVP (Fig. 5b,c). Therefore, after hybridization, the thermal stability is enhanced and this is very important for the catalyst application. However, for Pd-PVP/SBA-15 sample, two separate weight loss steps are seen (Fig. 5d). The first step (around 5%, w/w) appearing at temperature <100 °C corresponds to the loss of water. The second weight loss (about 200–500 °C) amounts around 6% (w/w) is related to the degradation of the polymer. Obviously, the hybrid Pd-PVP/SBA-15 shows higher thermal stability than PVP/SBA-15. It may be attributed to the presence of Pd nanoparticles in the composite structure. Therefore, it is very important for the catalyst application that the thermal stability was enhanced greatly after hybridization.

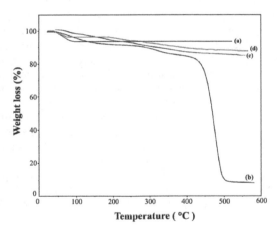

Fig. 5. TGA curves of (a) mesoporous silica SBA-15, (b) PVP, (c) PVP/SBA-15 and (d) Pd-PVP/SBA-15.

3.1.6 Uv-Vis

Fig. 6 displays the result of UV-Vis spectra of Pd-PVP/SBA-15. The UV-Vis spectra of Pd(OAc)$_2$ which reveal a peak at 400 nm refer to the existence of Pd(II) (Ahmadian et al.,2007). As mentioned in the experimental section, and according to scheme 1, Pd nanoparticle-PVP/SBA-15 was prepared by adding hydrazine hydrate to the Pd (II)-PVP/SBA-15. However, as can be seen in Fig. 6, there isn't any peak at 400 nm in the UV-Vis spectra of Pd-PVP/SBA-15, which indicates complete reduction of Pd(II) to Pd nanoparticles.

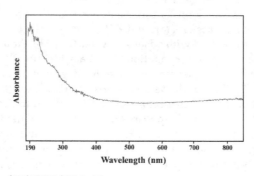

Fig. 6. UV-Vis spectra of Pd-PVP/SBA-15.

3.2 Catalytic activity

In this section, we firstly investigated the corresponding parameters for the Suzuki cross-coupling reaction. It includes different solvents and bases for the room-temperature Suzuki reaction, of iodobenzene, and phenylboronic acid in the presence of 0.12 g (Pd-PVP/SBA-15) catalyst. The molar ratio of iodobenzene to phenylboronic acid was set at 1:1.5 for the Suzuki cross-coupling reaction. Solvent plays a crucial role in the rate and the product distribution of Suzuki–Miyaura coupling reactions. Since water is known to increase the activity of the Suzuki–Miyaura catalyst (Bedford et al., 2003), two kinds of solvents were used: H_2O or $MeOH/H_2O$ (3:1 v/v). In recent years, a large number of studies have been devoted by academic and industrial research groups to the development of environmentally friendly processes. In this context, the use of water as a reaction medium in transition metal-catalyzed processes has merited increasing attention and is currently one of the most important targets of sustainable chemistry (Anastas et al., 2000). Water, an inexpensive, readily available, non-in flammable, non toxic solvent, provides remarkable advantages over common organic solvents both from an economic and an environmental point of view. The experimental results show that the time the reaction is completed is rarely shorter in the case of using $MeOH/H_2O$ (3:1 v/v) as a solvent, but neat H_2O was chosen because of the advantages of using water that were mentioned above.

We then examined the effect of bases for the Suzuki reaction. The inorganic bases including K_2CO_3 and Na_3PO_4 afforded high yields of 70–98%, as shown in Table 2. However, the organic base NEt_3 gave a lower yield of 30% as shown in Table 2, entry 3. Thus, K_2CO_3 was selected as the base and H_2O as the solvent.

Base	Yield (%) [b]
K_2CO_3	98
Na_3PO_4	70
Et_3N	30

[a] Reaction conditions: Pd-PVP/SBA-15 (0.12 g), iodobenzene (1 mmol), phenylboronic acid (1.5 mmol), base (5 eq), H_2O (5 mL), room temperature, 8 h.
[b] Isolated Yield.

Table 2. Effect of different bases on Suzuki-Miyaura reaction [a]

					Mp (°C)	
Entry	Substrate	Product	Yield (%)[b]	Time (h)	Found	Reported ref
1	(iodobenzene)	(biphenyl)	98	8 [c]	71-73	70-72 Xu et al.
2	(1-iodo-4-methoxybenzene, OCH₃)	(4-methoxybiphenyl, —OCH₃)	90	12 [c]	92-94	91-92 Xu et al.
3	(1-iodo-4-methylbenzene, CH₃)	(4-methylbiphenyl, —CH₃)	90	8 [d]	47-49	45-50 Xu et al.
4	(1-(4-bromophenyl)ethanone, Br)	(product)	60	12 [d]	121-123	120-122 Xu et al.
5	(1-chloro-2-iodobenzene, Cl, I)	(2-chlorobiphenyl, Cl)	96	4 [c]	31-35	32.5-33.5 Xu et al.
6	(bromobenzene, Br)	(biphenyl)	95	10 [c]	71-73	70-72 Xu et al.
7	(chlorobenzene, Cl)	(biphenyl)	90	14 [d]	71-73	70-72 Xu et al.

[a] Reaction conditions: Pd-PVP/SBA-15 (0.12 g), iodobenzene (1 mmol), phenylboronic acid (1.5 mmol), K_2CO_3 (5 eq), H_2O (5 mL).
[b] Isolated yield.
[c] R.T.
[d] 95 °C.

Table 3. Suzuki-Miyaura reaction of aromatic aryl halides and phenylboronic acid catalyzed by Pd-PVP/SBA-15 [a]

Using our optimized reaction conditions, we selected a series of aryl iodides, some aryl bromides and aryl chloride in the Suzuki reaction. All reactions were performed using 1:1.5 stoichiometric ratios of aryl halides and phenylboronic acid in air. The results are summarized in Table 3. Reactions were carried out in water at different times. In some cases, coupling reactions with aryl iodide (entry 3), aryl bromide (entry 4) and aryl chloride (entry 7) required higher temperature (95 °C) in order to obtain excellent yields. The results are listed in Table 3. It is well known that activation of C-Cl bond is much more difficult than C-Br and C-I bonds, and in general requires harsher reaction conditions in a heterogeneous catalysis system (Yin & Liebscher, 2007; Martin & Buchwald, 2008). Thus, the catalyst afforded average to excellent yields of biaryl products even at room temperature. In the literature, only a few catalysts are known for affecting the Suzuki-Miyaura cross-coupling reactions under mild conditions (Littke et al., 2000; Cuia et al., 2007; Marion et al., 2006, Navarro et al., 2006; Fairlamb et al., 2004).

The catalyst reuse and stability were checked using Suzuki reaction of iodobenzene with phenylboronic acid at room temperature in present water as solvent. The catalyst was separated from the reaction mixture after each experiment by simple filtration, washed with diethylether and acetone and dried carefully before using it in the subsequent run. The results showed that this catalyst could be reused without any modification, 5 times. It should be mentioned that there was rarely low catalyst leaching (4.5%) (Pd content of the catalyst was determined by AAS after each cycle) during the reaction and the catalyst exhibited high stability until 5 recycles (Table 4).

Cycle	Yield (%) [b]	Pd content of catalyst (mmol in 0.12 g catalyst) [c]
fresh	98	0.0672
1	98	0.0670
2	95	0.0661
3	93	0.0659
4	91	0.0651
5	87	0.0643

[a] Reaction conditions: Pd-PVP/SBA-15 (0.12 g), iodobenzene (1 mmol), phenylboronic acid (1.5 mmol), K_2CO_3
(5 eq), H_2O (5 mL), room temperature.
[b] Isolated yield.
[c] Estimated by AAS.

Table 4. The catalyst reusability for the Suzuki-Miyaura reaction [a]

4. Conclusion

In this chapter, we demonstrated a facile palladium nanoparticle-PVP preparation inside modified mesoporous silica and the utilization of this nanocomposite as a new heterogeneous organic hybrid catalyst system. The catalytic activity of this catalyst was excellent for Suzuki-Miyaura cross-coupling reaction of aryl chloride, bromide and iodides at room temperature under aerobic conditions. Further, easy catalyst recovery and reasonable recycling (at least 5 times) efficiency of the catalyst made it an ideal system for coupling reactions in the aqueous phase. For determination of the amount of remained Pd on the catalyst structure, atomic absorption spectroscopy was used.

There are various methods such as XRD, BET, TEM, FT-IR, UV-Vis and etc. for characterization of solid catalysts. Estimation of the stability of the catalysts is one the important factors for the performance of the catalysts. The stability of the catalysts is directly related to the amounts leaching of the active species of the catalysts. AAS is the best technique for determination of the metal leaching in the catalysts containing metal nanoparticles.

5. Acknowledgements

The support by Shahreza Branch, Islamic Azad University (IAUSH) Research Council and Center of Excellence in Chemistry is gratefully acknowledged.

6. References

Makhubela, B.C.E.; Jardine, A.; Smith, G.S. (2011). Pd nanosized particles supported on chitosan and 6-deoxy-6-amino chitosan as recyclable catalysts for Suzuki–Miyaura and Heck cross-coupling reactions. *Applied Catalysis A: General*, Vol.393, No.1-2, pp. 231-241

Venkatesan, P.; Santhanalakshmi, J. (2010). Synthesis, characterization and catalytic activity of trimetallic nanoparticles in the Suzuki C–C coupling reaction. *Journal of Molecular Catalysis A: Chemical*, Vol.326, No.1-2, pp. 99–106

Zhao, H.; Peng, J.; Xiao, R.; Cai, M. (2011). A simple, efficient and recyclable phosphine-free catalytic system for Suzuki–Miyaura reaction of aryl bromides. *Journal of Molecular Catalysis A: Chemical*, Vol.337, No.1-2, pp. 56–60

Mu, B.; Li, T.; Li, C.; Liu, P.; Shang, W.; Wu, Y. (2009). Langmuir–Blodgett films of cyclopalladated ferrocenylimine: preparation, characterization, and application in Suzuki coupling reaction. *Tetrahedron*, Vol.65, No.12, pp. 2599–2604

X. Chanjuan, W. Yongwei, Y. Xiaoyu; (2008). *Cis*-Fashioned palladium (II) complexes of 2-phenylbenzimidazole ligands: Synthesis, characterization, and catalytic behavior towards Suzuki–Miyaura reaction. *Journal of Organometallic Chemistry*, Vol.693, No.26, pp. 3842–3846

Jana, S.; Haldar, S.; Koner, S. (2009). Heterogeneous Suzuki and Stille coupling reactions using highly efficient palladium(0) immobilized MCM-41 catalyst. *Tetrahedron Letters*, Vol.50, No.34, pp. 4820–4823

Tamami, B.; Ghasemi, S. (2010). Palladium nanoparticles supported on modified crosslinked polyacrylamide containing phosphinite ligand: A novel and efficient heterogeneous catalyst for carbon–carbon cross-coupling reactions. *Journal of Molecular Catalysis A: Chemical*, Vol.322, No.1-2, pp. 98–105

Sanchez-Delgado, R.A.; Machalaba, N.; Ng-a-Qui, N. (2007). Hydrogenation of quinoline by ruthenium nanoparticles immobilized on poly(4-vinylpyridine) .*Catalysis Communication*. Vol.8, No.12, pp. 2115-2118

Luo, Y.; Sun, X. (2007). One-step preparation of poly(vinyl alcohol)-protected Pt nanoparticles through a heat-treatment method. *Materials Letters* Vol.61, No.10, pp. 2015-2017

Thomas, J.M.; Johnson, B.F.G.; Raja, R. (2003). *Accounts of Chemical Research*. Vol.26, No.4, pp. 20-30

Jacquin, M.; Jones, D.J.; Roziere, J. (2003). Novel supported Rh, Pt, Ir and Ru mesoporous aluminosilicates as catalysts for the hydrogenation of naphthalene. *Applied Catalysis A: General*. Vol.251, No.1, pp. 131-141.

Yuranov, I.; Moeckli, P.; Suvorova, E.; Buffat, P.; Kiwi-Minsker, L.; Renken, A. (2003). Pd/SiO_2 catalysts: synthesis of Pd nanoparticles with the controlled size in mesoporous silicas. *Journal of Molecular Catalysis A: Chemical*, Vol.192, No.1-2, pp. 239–251

Trzeciak, A.M.; Ziolkowski, J.J. (2007). Monomolecular, nanosized and heterogenized palladium catalysts for the Heck reaction. *Coordination Chemistry Reviews*. Vol.251, No.9-10, pp. 1281–1293

Huo, Q.; Margolese, D.I.; Stucky, G.D. (1996). Surfactant Control of Phases in the Synthesis of Mesoporous Silica-Based Materials. *chemistry materials*, Vol. 8, No. 5, pp. 1147-1160

Zhao, D.; Feng, J.; Huo, Q.; Melosh, N.; Fredrickson, G.H.; Chmelka, B.F.; Stucky, G.D. (1998). Triblock Copolymer Syntheses of Mesoporous Silica with Periodic 50 to 300 Angstrom Pores. *Science*, Vol.279, No.5350, pp. 548-552

Kalbasi, R.J.; Kolahdoozan, M.; Rezaei, M. (2010). Synthesis and characterization of PVAm/SBA-15 as a novel organic–inorganic hybrid basic catalyst. *Materials Chemistry and Physics*, Vol.125, No.3, pp. 784–790

Mark, J.E. (2006). *Accounts of Chemical Research*. Vol.39, No.12, pp. 881-888

Zheng, J.; Li, G.; Ma, X.; Wang, Y.; Wu, G.; Cheng, Y.(2008). Polyaniline–TiO$_2$ nano-composite-based trimethylamine QCM sensor and its thermal behavior studies. *Sensors and Actuators B*: Vol.133, No.2, pp. 374–380

Chung, C.M.; Lee, S.J.; Kim, J.G.; Jang, D.O. (2002). Organic–inorganic polymer hybrids based on unsaturated polyester. *Journal of Non-Crystalline Solid*, Vol.311, No.2, pp. 195–198

Morales, G.; Grieken, R.V.; Martin, A.; Martinez, F. (2010). Sulfonated polystyrene-modified mesoporous organosilicas for acid-catalyzed processes. *Chemical Engineering Journal*, Vol.161, No.3, pp. 388-396

Ma, Z.H.; Han, H.B.; Zhoua, Z.B.; Nie, J. (2009). SBA-15-supported poly(4-styrenesulfonyl(perfluorobutylsulfonyl)imide) as heterogeneous Brønsted acid catalyst for synthesis of diindolylmethane derivatives. *Journal of Molecular Catalysis A: Chemical*, Vol.311, No.1-2, pp. 46–53

Alves, M.H.; Riondel, A.; Paul, J.M.; Birot, M.; Deleuze, H.(2010). Polymer-supported titanate as catalyst for the transesterification of acrylic monomers. *Comptes Rendus Chimie*, Vol.13, No.10, pp. 1301-1307

Kalbasi, R.J.; Kolahdoozan, M.; Massah, A.R.; Shahabian, K. (2010). Synthesis, Characterization and Application of Poly(4-Methyl Vinylpyridinium Hydroxide)/SBA-15 Composite as a Highly Active Heterogeneous Basic Catalyst for the Knoevenagel Reaction. *Bulletin of the Korean Chemical Society*, Vol.31, No.9, pp. 2618-2626

Kalbasi, R.J.; Nourbakhsh, A.A.; Babaknezhad, F. (2011). Synthesis and characterization of Ni nanoparticles-polyvinylamine/SBA-15 catalyst for simple reduction of aromatic nitro compounds. *Catalysis Communications*, Vol.12, No.11, pp. 955-960

Kalbasi, R.J.; Kolahdoozan, M.; Shahabian, K.; Zamani, F. (2010). Catal. Commun. Vol.11, No.9, pp. 1109–1115

Kalbasi, R.J.; Mosaddegh, N. (2011). Synthesis and characterization of poly(4-vinylpyridine)/MCM-48 catalyst for one-pot synthesis of substituted 4H-chromenes. *Catalysis Communications*, Vol.12, No.13 , pp. 1231-1237

Run, M.T.; Wu, S.Z.; Zhang, D.Y.; Wu, G. (2007). A polymer/mesoporous molecular sieve composite: Preparation, structure and properties. *Materials Chemistry and Physics*, Vol.105, No.2-3, pp. 341–347

Ghanemi, K.; Nikpour, Y.; Omidvar, O.; Maryamabadi, A. (2011). Sulfur-nanoparticle-based method for separation and preconcentration of some heavy metals in marine samples prior to flame atomic absorption spectrometry determination. *Talanta*, Vol.85, No.1, pp. 763–769

Scaccia, S.; Goszczynska, B. (2004). Sequential determination of platinum, ruthenium, and molybdenum in carbon-supported Pt, PtRu, and PtMo catalysts by atomic absorption spectrometry. *Talanta* Vol.63, No.3, pp. 791–796

Harish, S.; Mathiyarasu, J.; Phani, K.L.N.; Yegnaraman, V. (2009). Synthesis of Conducting Polymer Supported Pd Nanoparticles in Aqueous Medium and Catalytic Activity Towards 4-Nitrophenol Reduction. *Catalysis Letters*, Vol.128, No.1, pp. 197–202

Iwamoto, T.; Matsumoto, K.; Matsushita, T.; Inokuchi, M.; Toshima, N. (2009). Direct synthesis and characterizations of fct-structured FePt nanoparticles using poly(N-vinyl-2-pyrrolidone) as a protecting agent. *Journal of Colloid and Interface Science*, Vol.336, No.2, pp. 879–888

Metin, O.; Ozkar, S. (2008). *Journal of Molecular Catalysis A: Chemistry*, Vol.295, No.8, pp. 39–46

Hirai, H.; Chawanya, H.; Toshima, N. (1985). Colloidal palladium protected with poly(N-vinyl-2-pyrrolidone) for selective hydrogenation of cyclopentadiene. *Reactive Polymers*, Vol.3, No.2, pp. 127–141

Chytil, S.; Glomm, W.R.; Vollebekk, E.; Bergem, H.; Walmsley, J.; Sjoblom, J.; Blekkan, E.A. (2005). Platinum nanoparticles encapsulated in mesoporous silica: Preparation, characterisation and catalytic activity in toluene hydrogenation. *Microporous and Mesoporous Materials*, Vol.86, No.1-3, pp. 198-206

Ahmadian Namini, P.; Babaluo, A.A.; Bayati, B. (2007). *International Journal of Biomedical Nanoscience and Nanotechnology*, Vol.3, pp. 37-43

Bedford, R. B.; Cazin, C. S. J.; Coles, S. J.; Gelbrich, T.; Horton, P. N.; Hursthouse, M. E.; Light, M. E. (2003). *Organometallics*, Vol.22, pp. 987–999

Anastas, P.; Heine, L. G.; Williamson, T. C., (2000). Green Chemical Syntheses and Processes, *Journal of American Chemical Society*, Washington DC

Li, C.-J. (2005). For a recent review on C-C bond forming reactions in aqueous media. *Chemical Reviews*, Vol.105, No.8, pp. 3095-3165

Yin, L.; Liebscher, J. (2007). *Chemical Reviews*, Vol.107, pp. 133

Martin, R.; Buchwald, S. L. (2008). *Acc. Chem. Res.* Vol.41, pp. 1461

Littke, A. F.; Dai, C.; Fu, G. C. (2000). *Journal of American Chemical Society*, Vol.122, pp. 4020-4028

Cuia, X.; Zhoua, Y.; Wanga, N.; Liu, L.; Guo, Q. (2007). N-Phenylurea as an inexpensive and efficient ligand for Pd-catalyzed Heck and room-temperature Suzuki reactions. *Tetrahedron Letters*. Vol.48, No.1, pp. 163-167

Marion, N.; Navarro, O.; Mei, J.; Stevens, E. D.; Scott, N. M.; Nolan, S. P. (2006). *Journal of American Chemical Society*, Vol.128, pp. 4101- 4111

Navarro, O.; Marion, N.; Mei, J.; Nolan, S. P. (2006). Chemistry - A European Journal Vol.12, pp. 5142-5148

Fairlamb, I. J. S.; Kapdi, A. R.; Lee, A. F. (2004). Ç²-dba complexes of palladium (0): electron rich dba ligands enhance the reactivity of PdLn catalyst species in Suzuki-Miyaura coupling. *Organic Letters*. Vol.6, No.24, pp. 4435-4438

Xu, Q.; Duan, W-L.; Lei, Z-Y.; Zhu, Z-B.; Shi, M. (2005). A novel cis-chelated Pd(II)–NHC complex for catalyzing Suzuki and Heck-type cross-coupling reactions. *Tetrahedron*, Vol.61, No.47, pp. 11225–11229

9

Mineral Content and Physicochemical Properties in Female Rats Bone During Growing Stage

Margarita Hernández-Urbiola[1,2] et al.*
*1Posgrado en Ciencias Biomédicas, Instituto de Neurobiología,
Universidad Nacional Autónoma de México, Campus Juriquilla, Querétaro, Qro.
2División de Ciencias de la Salud, Universidad del Valle de México,
Campus Querétaro, Querétaro, Qro.
México*

1. Introduction

The atomic absorption spectrophotometry (AAS) is a technique extensively used for trace and ultra trace analysis of organic and inorganic materials. AAS is a novel method that determines in liquid samples the presence of metals such as: Ca, Fe, Cu, Al, Pb, Zn, and Cd from different sources. The determination of minerals is important in environmental and biological studies as well as in the clinical practice. The determination of mineral content is a key to understand changes in some metabolism that conduct to diseases as result of the increase or decrease of mineral components in diet and consequently, to develop new models in the field of animal and human nutrition.

In the case of biological samples including plant leaves, fruits, vegetables, organic vegetables i.e. nopal (*Opuntia ficus indica*), Rodriguez-Garcia et al., (2007) reported that the major mineral components in this cactus was as follows: Calcium, Magnesium, Sodium, Manganese, Iron, Zinc, Potassium, and minor mineral components were: Lithium, Vanadium, Cobalt, Arsenic, Selenium, Cadmium, Thallium (Hernández-Urbiola et al., 2010, 2011). AAS has been used also to analyze muscle tissue, blood, urine, hair, bones, among others (Martinez-Flores et al., 2002; Christian, 1972). In most cases complex nature of biological materials requires dry ashing followed by wet digestion with oxidizing acids, i.e. HNO_3 and HCl_4.

* Astrid L. Giraldo-Betancur[3], Daniel Jimenez-Mendoza[4], Esther Pérez-Torrero[2,4], Isela Rojas–Molina[5], María de los Angeles Aguilera-Barreiro[5], Carolina Muñoz-Torres[6] and Mario E. Rodríguez-García[7,*]
3Centro de Investigación y de Estudios Avanzados del IPN, Sede Querétaro, Querétaro, Qro, México
4Division de Estudios de Posgrado, Facultad de Ingeniería, Universidad Autónoma de Querétaro, Querétaro, Qro, México
5Facultad de Ciencias Naturales, Licenciatura en Nutrición, Universidad Autónoma de Querétaro, Avenida de las Ciencias, s/n, Juriquilla, Querétaro, México
6Centro de Geociencias, Universidad Nacional Autónoma de México, Campus Juriquilla, Querétaro, Qro, México
7Departamento de Nanotecnología, Centro de Física Aplicada y Tecnología Avanzada, Universidad Nacional Autónoma de México, Campus Juriquilla, Querétaro, Qro., México

The use of AAS spectroscopy to the study of nutrimental components in food specially Ca has been reported by Rojas-Molina et al., (2009), Palacios-Fonseca et al., (2009). Cornejo-Villegas et al., (2010), analyzed calcium content in instant corn flours obtained during traditional thermo-alkaline-treatment of corn kernels (nixtamalization), as well as, commercial corn flours and corn flours supplemented with a vegetable i.e. Nopal. AAS technique is an excellent tool to determine anti-nutrimental compounds that compromise the bioavailability of some important minerals such Ca in diet. Contreras-Padilla et al., (2011) use the AAS to study anti-nutrimental components in fresh nopal and nopal powders as a function of the maturity stage. They found that calcium oxalates decreases as the maturity stage increases, while the total Ca increases. This fact could be an indicative of a better bio-availability or bio-accessibility of Ca in diet. The determination of minerals as nutrimental factors in food have also been an important issue in the development of new products and to establish physicochemical criteria for food processing, i.e, Rojas-Molina et al., (2007) and Gutiérrez-Cortez et al., (2010) reported that Ca content in nixtamalized corn kernels is associated with gelatinization of the starch granules. Additionally, Galvan-Ruiz et al. (2007) used AAS to study the incorporation of Ca as a food additive in food.

In Biological studies   AAS has been used to determine changes in minerals content, Brem et al., (2004) studied mineral changes in blood and ash bone, as well as the bone mineral density in three different regions of the body, and the minerals related to the osseous tissue activity. This study was carried out in order to evaluate the osteoporosis type I, that is associated with vertebral compression and hip fracture on ovaricectomized rats in contrast with normal rats. They analyzed different ways to ameliorate the health alterations. Findings showed that the bones of ovaricectomized rats had significant decreases in Ca, P, and Mn respect to the control group, but no differences were found in the case of Cu and Mg between these groups. They also reported significant differences in bone mineral density in ovariectomized rats compared to the control group, after 6 and 9 months of treatment.

On the other hand, for inorganic materials as is the case of bones; the AAS has been used to determine their mineral contents. This technique have been used in order to analyze the most important bone minerals as Ca, Mg and Zn in human studies, and Ca and Mg contents in animals (Ma et al., 2000). In the field of nutrition it is well known the importance of the Ca/P ratio for the bone formation. Ca is determined using AAS, where P has been determined colorimetrically (Nowicka et al., 2006). The human skeleton develops through infancy, childhood and adolescence and the peak bone mass is achieved during maturity. High bone mass density (BMD) at skeletal maturity is the best protection against age-related bone loss and fractures, the loss of BMD conduce to develop osteoporosis (Heaney, 1993). This degenerative disease is associated with a decrease of minerals in bone. Fractures occur when skeleton is exposed to minimal or moderate trauma (Matkovic, 1991).

Several researches have demonstrated that Ca and P play an important role in skeletal mineralization and also on peak bone mass (Masuyama et al., 2003; Matkovic & Ilich, 1993). Both minerals are present in hydroxyapatite (HAp), this component $[Ca_{10}(PO_4)_6(OH)_2]$ constitutes seventy percent of bone. Collagen and the inorganic mineral HAp, combine, providing hardness and mechanical strength to the bone (Smith et al., 1983). From a nutrimental point of view, an adequate intake of Ca in the human diet during growth is a critical factor for acquiring and maintaining an adequate bone mass, the regulation of body weight; protect organism against high blood pressure and certain types of cancer (Rodríguez-Rodríguez et al., 2010). Ščančar et al., (2000), studied the mineral content in human bone from the iliac crest, using AAS. The concentration of Al, Cu, Fe, and Ca was reported for normal individuals.   They found that microwave method for digestion procedure was appropriate technique for dissolution of bone biopsy samples and will be an excellent method for evaluated trace elements in bone samples from dialysis patients.

Martinez–Flores et al., (2002) found that the femurs of rats fed with diets based on corn (tortilla), and with a Ca/P ratio at 1.1, were higher in length, and they were heavier, thicker that bones of rats fed with a Ca/P less than 1. Mexican diets are high in P with Ca/P ratio higher than recommend for an optimal bone development. Bean and tortillas are the primary source of energy and minerals in the typical Mexican diet. Wyatt et al., (2000), studied the effect of different Ca/P ratios in Mexican diets on the growth and composition of rat femurs; they found that levels of dietary Ca affect bone mass and Ca bone deposition to a much greater extent than dietary P levels.

Environmental applications of AAS are also important because these are close related to the health in individual and community. The determination of contaminants in water, air, and soil are fundamental in order to improve treatments or to resolve problems that affect the human and animal health.   On the other hand, in the case of cultivars the soil mineral content is the most important condition to ensure a well grown and development of plants.

2. Bone mineral content

The amount of minerals in a human bone is an important parameter which is of interest for many reasons. The measurement of this parameter could be used, for example, to diagnose and follow the treatment of various demineralizing diseases, such as osteoporosis and osteomalacia. The skeleton of the human body is a complex matrix that suffers changes along life; these changes are related to the incorporation of several minerals into the matrix and this is a kinetic process that is affected by factors such as hormones, lifestyle, nutrition amount others. Peak bone mass in humans is achieved after sexual maturity and is then maintained for two decades. Thereafter, the mass of virtually all bones declines until death. Thus, it has been established that Ca deposition in bone in the growing stage contributes to the prevention of generated bone diseases (Takahara et al., 2000). The bone mineral composition consists of Ca and P in amorphous and crystalline fractions as a major components, as well as Ma, Na, Zn, K, as minor elements. According to Blokhuis et al., (2000), mineral components are composed of approximately 30% amorphous calcium phosphates and around 70% fine crystalline HAp [$Ca_{10}(PO_4)_6(OH)_2$].

The bone is a matrix constituted by minerals immerse in a collagen matrix, both determine its physical and functional properties. There are several methods to evaluate the mineral content in a bone sample. Direct evaluation through AAS and inductively couple plasma (ICP), and an indirect manner by using single or dual X-ray diffraction and computer tomography, to obtain the bone mineral density and the bone mineral content that involves in both cases the use a calibration sample or a matrix. In the second case, it is possible to obtain an average information about the density and mineral content by in a semi quantitatively manner (Bonnick, 2010). By using direct methods, it is possible to obtain quantitative information regarding to each one of the bone minerals components, while using indirect methods the information is close related the optical absorption of the bone tissues. The measurement of BMD has been performed during the last decade, and different technologies have been used for this purpose. These technologies include: computer tomography (CT) scans (Langton et al., (2009), ultrasound (Higuti & Adamowski (2002), and dual energy x-ray absorptiometry (Norcross & Van Loan, 2004).

Bone loss could induce osteopenia or osteoporosis; this loss may be due to increased bone resorption and decreased bone formation. Osteoporosis with a decrease in bone mass is widely recognized as a major public health problem. Pharmacological and nutritional factors may have the potential to prevent bone loss with increasing age. Nutritional factors may be especially important in the prevention of osteoporosis, although this is poorly understood (Ma et al., 2000). Several studies have been done to establish the role of minerals in bone. Bone stores 99% of the body's Ca, and calcium salts are responsible for the hardness of bones. Mg is not only a critical ion in mammals as a cofactor for many enzymes of the energy extraction system and protein synthesis pathways, but it is also necessary for bone formation. Thus, Mg might also play an important role in bone structure or the hardness of bone (Takahara et al., 2000). Zn, an essential trace element, has been demonstrated to have a potent stimulatory effect on bone formation and an inhibitory effect on bone resorption. Zn can stimulate protein synthesis in osteoblastic cells in vitro by activating aminoacylt RNA synthetase. The oral administration of a Zn compound can prevent bone loss in ovariectomized rats, an animal model of osteoporosis. (Ma et al., 2000).

HAp is a crystalline compound constituted of three molecules of calcium phosphate and one molecule of calcium hydroxide. HAp concentrate is the matrix of Ca from bone protein found in the unprocessed bone. This natural substance contains about 14% collagen protein and 4% other proteins and small amino acids (especially hydroxyproline, glycine and glutamic acid). Ca comprises between 24-30% of the hydroxyapatite matrix and together with several minerals (Zn, K, silicon manganese, and Fe) is a group active bio-available Ca (Gómez et al., 2004). The HAp and dibasic calcium phosphate are the only calcium phosphate phases that are chemically stable to temperature and pH of the human body (37 °C and about 7, respectively), The mineral part of bone that support nearly all the organic and the mechanical load (collagen) acts as a bonding material, which also absorbs shock, providing flexibility to the bones; Additionally to mechanical functions, bones have an essential role in metabolic activity as mineral reservoirs that is able to absorb and release ions.

Skeletal allow mobility of the body and protect internal organs. While the properties of bone tissue and the proportions of the constituent minerals vary with the different parts of the skeleton. Bone is specialized tissue that changes along the life. The bones, teeth, cells, fat, and natural polymers such as: polysaccharides, collagen, and polyphosphates are composed by minerals. It can be considered that bones contain about two thirds of inorganic material and a third of portion of organic matter. Bone tissue is composed of a mineral phase of 69% of its total weight, 9% water and 22% of an organic matrix, which in turn is composed mainly of collagen (90-96%). In terms of structure, bones can be considered as a dispersion of mineral particles (bio-minerals) embedded in an organic matrix, which forms the contiguous phase. Bioapatite $[Ca_{10}(PO_4)_6(OH)]$ is considered the major component in the mineralized part of mammalian bone, is a calcium-phosphate biomineral with a structure that closely resembles HAp (Meneghini, 2003).

Bone mineral composition depends of several factors such as: age, nutritional condition, diseases and habits. These factors make difficult to establish the composition and structure of HAp (Meneghini, 2003). HAp is able to store and release minerals as calcium, phosphorous and several other ions such Na. K, Mg, F, CO_3 and OH. For this reason HAp is an important mineral reservoir for the metabolic activity of the organism. Atomic absorption technique combined with other techniques such as ICP and UV-VIS are adequate methods to determine mineral contents in bones. Ca is the most important mineral in HAp because > 90% of the this element in the human body is deposited as calcium-phosphate within the skeleton and teeth; the apatite crystals are one of the major constituents in bone and other mineralized vertebrate tissues; their presence, which accounts for about 65 % of weight bone, provides most of the stiffness and strength of bone. With the normal increase of apatite deposition during tissue aging and maturation, bone mechanical properties increase greatly. The physical character of bone, including the morphology, dimensions, and distribution of its composite apatite crystals, affect its mechanical properties (Su et al., 2003).

3. Biological studies

The rat constitutes a well established model for study the bone health research and the preclinical evaluation of agents used in the prevention or treatment of osteoporosis (Frost & Jee, 1992; Ke et al., 2001). Consequently, Ca and P content as majority minerals in rat bones constitutes and adequate tool to evaluate bone formation and development during growing stage and to correlate mineral content with peak bone mass.

3.1 Animal model

Fifty four female Wistar rats, 21 days old, and 50 ± 4.3 g of weight, from the bioterium of the Neurobiology Institute, UNAM, Queretaro, Qro., Mexico were fed with a control diet (AIN 93G, Reeves et al., 1993) and water, and fed *ad libitum* during 8 weeks. The rats were individually placed in stainless steel cages under controlled temperature (25°C ± 1) and light conditions (12:12 h light-dark cycle), and the body weigh was weekly verified. The rats were cared for in compliance with the "principles of Laboratory Animal Care"formulated by the National Institutes of Health (National Institutes of Health publication no. 96-23, revised 2003) and the protocol was approved by the Ethics committee of the Universidad Nacional Autónoma of México.

3.2 Experimental protocol

Six female rats were sacrificed each week by carbon dioxide aspiration. The left and right femoral bones were dissected and introducing in paraformaldehyde solution (4%) in order to eliminate the adhering bone tissue; after that, they were introducing in a buffer solution to eliminate paraformaldehyde excess and finally in distilled water in order to eliminate the residual buffer and soft tissues were removed. Following, bones were dried at 90°C for 12 hours (Farris & Griffith, 1949). The length and weight of the dry femurs were measured according to Gómez-Aldapa et al., (1999) methodology. Subsequent bones were charred in order to obtain a sample without organic component eliminated at 550°C for 4 h, immediately powders were passed for a sieve US 60 (250 μm) to carry out ray diffraction experiments and also to AAS measure.

3.3 Physical development of rats

The rats in this study were maintained under controlled environmental conditions: temperature, light cycles, water consumption. The physical development of rats is related to the changes in the skeletal and muscle, which depends drastically on four factors: genetic, feeding conditions, mechanical (living conditions as physical activity), and hormonal.

(Fernández-Tresguerres et al., 2010). However, changes in one of these parameters have important effects in the development and maintenance of the bone.

Figure 1 shows the increases of the weight for female rats from 3 to 11 weeks that corresponds to the called self- accelerating phase. According to the data showed in this figure the changes in body weight of the rats in this period is around 600%. This curve exhibits two different behaviors: for the first 4 weeks, the growth rate is bigger than for the final 4 weeks. After the week 11 according to Tamaki et al., (1995) the rats are in the adult phase in which their organs and functions reach the state call adult phase. This curve represents only the self-accelerating phase and is important because this can be considered as a normal reference curve for future clinical or biological studies. This accelerating phase is the period where individuals are more sensitive to the adverse effects that influence their future development for long-time. Although, the individual survives, some physiological adaptations have to appear in the adult phase, and one of them could be directly related to the bone formation and mineralization.

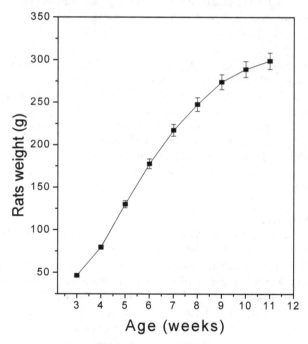

Fig. 1. Body weight for female rats as a function of the age.

A very important aspect related to the rats development, are the hormonal changes where thyroid hormone and parathormone, calcitonin, growth hormone, estrogen, progesterone, and calcitriol are indispensable for the bone development and osseous remodeling (Bogden et al 2008). In order to study the changes in the inorganic-organic bone matrix, the weight and the length of the femur (right-left) were obtained. As was mentioned, bone is a matrix composed by fat, protein, and minerals. It means that changes in these constituents affect the final physicochemical composition of the bones.

Figure 2.a and b shows the femur weight and length of female rats during the growth stage. As can be seen, these physical characteristics have the same trend found in the case of total weight. It is important to remark, that a change in the slope of this curve occurs between the 7th and 8th week. The environmental conditions, as well as feeding and exercise or physical activity, and hormonal changes are fundamental to explain this behavior. At the begging of the growing, the rats exhibit a mayor physical activity due to the dimension of the cage (Martin, 2000). On the other hand, the hormonal activity in the case of male and female rats plays an important role in their physicochemical development that includes muscle and bones. In female rats, estrogen is one of the major regulating factors responsible for bone formation and maintenance and the lost of this hormone after menopause often leads to bone loss in the case of human (McMillan, 2007).

3.4 Evaluation of mineral content in femurs

Ca, Na, K, and Mg content in the femurs rats were determined according to the dry-ashing procedure A.O.A.C. 968.08 (1998), using lanthanum chloride. The concentration was measured with a double beam atomic absorption spectrometer (Analyst 300 Perkin Elmer, USA) equipped with a deuterium lamp, background corrector and a hollow cathode lamp. The equipment was operated with 12 psi of dry air, 70 psi of acetylene, a 422.7 nm flame, a 10 mA. lamp current, and a 0.7 nm slit width.

In order to have a precise determination of minerals content in the rats bones as a function of the age, a certified minerals bone ash was used as calibration standard. The standard of bone ash used in this experiment was certified by the National Institute of Standards & Technology (NIST), with Standard reference Material 1400. This standard reference material (SRM) is intended for use in evaluating analytical methods for the determination of selected major, minor, and trace elements in bone and in material of a similar matrix.

The certified concentrations of the constituent elements are shown in table 1. These concentrations are based on the results of a definitive analytical method or the agreement of results by at least two dependent analytical methods. In this table, it was included the mineral content obtained for the same sample using AAS in our laboratory. Our results are in good agreement with those reported by NIST.

ELEMENT	CONCENTRATION (wt. Percent)	AAS CONCENTRATION (wt. Percent)
Calcium	38.18 ± 0.13	37.70± 0.11
Phosphorous	17.91 ± 0.19	18.02 ± 1.29*
Magnesium	0.684 ± 0.013	0.638 ±0.021
Iron	0.066 ± 0.003	N/A
Potassium	0.0186 ± 0.001	0.0195 ±0.002
Strontium	0.0249 ± 0.001	N/A

Table 1. Certified concentrations of constituent elements for NIST bone ash sample, and experimental values obtained using AAS. *Phosphorous value was obtained by UV-VIS. These concentrations are based on the results of a definitive analytical method or the agreement of results by at least two dependent analytical methods. Additionally Phosphorous in femurs was determined according to the A.O.A.C official methods 985.35 (2000).

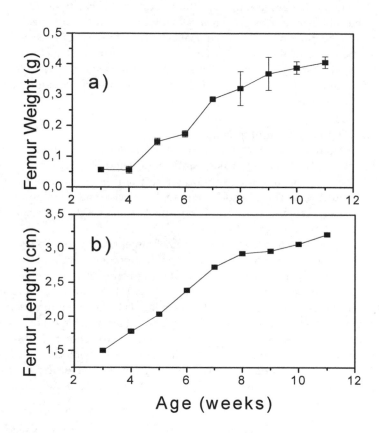

Fig. 2. a. Femur and b. Femur length for female rats as a function of the age.

Bone is a composite formed by inorganic and organic compounds. The organic material can be eliminated during the calcination process and the ash is composed only by minerals such as Ca, P, Mg, Na, P, among others. Figure 3 shows the ash content in the femurs as a function of the age. The increase in the ash content along the growth period is due to the decreases of the organic content (proteins) and the incorporation of new minerals into the bone matrix. Tamaki & Uchiyama (1995) found that the total weight of the rats has an abrupt increase until the 10th week of age. It is important to mention, that the growth of the living body depends on the development of the skeletal system, tissues, and organs.

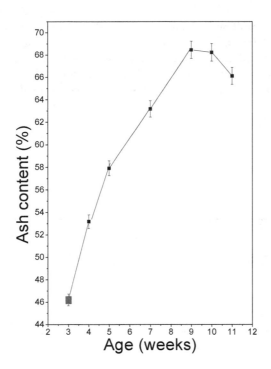

Fig. 3. Ash content in femurs of female rats as a function of the age.

The mineral content of femurs was obtained for rats between 3 and 11 weeks, but is important to know that the first data, for 3 weeks, corresponds to 21 day of life (red square in Figure 3) and at this time, rats were at weaning. Figure 4.a shows the Ca content in ash as a function of the age. During the period from 4th to 7th week, an increment in the Ca content is evident, while for the 7th to 11th week it has small changes.

Figure 4.b shows the P content as a function of the age during the acceleration phase. After weaning a reduction of P is evident, due to changes in the diet, and increases during the 7th and 10th weeks is present. This result is related to the formation of amorphous and crystalline compound of the bone as will be explained in the X-ray section as well as hormonal fluctuations. The bone mineral phase is composed mainly of microscopic crystals of calcium phosphates, in which the HAp [$Ca_{10} (PO_4)_6 (OH)_2$], is the main component. Other mineral phases that are present in bone are dicalcium phosphate ($Ca_2P_2O_7$), dibasic calcium phosphate ($CaHPO_4$), and some amorphous phases of calcium phosphate. There are also ions such as: citrate ($C_6H_5O_7^{4-}$), carbonate (CO_3^{2-}), fluoride (F^-) and hydroxyl ions (OH^-), which can lead to subtle differences in bone microstructure. There are also some impurities such as Mg and Na with Cl and Fe traces.

Several researches employing biological models describe that, even if the corporal Ca concentration is adequate and the daily Ca intake covers recommendations, bones demineralization takes place if the Ca^{2+}/P ratio in diet is not adequate (Anderson & Draper, 1972). Nnakwe & Kies (1985) reported that rats fed with rich P diets showed high risk of bone fracture. Similarly, Spencer et al., (1988) showed that high P content in diet has a negative effect in Ca deposition in bones. Calvo & Park (1996) observed that P modifies significantly the Ca metabolism. These authors detected that the Ca^{2+}/P ratio has a greater effect on bone mass formation and Ca deposition than the Ca content itself. According to Figure 5, the Ca/P ratio increases from 1.75 to 2.2; it is important to remember that there are amorphous compounds and ectopic Ca in the bone. The optimal Ca/P ratio for HAp formation, about 1.664, was reported by Benhayoune et al., (2001).

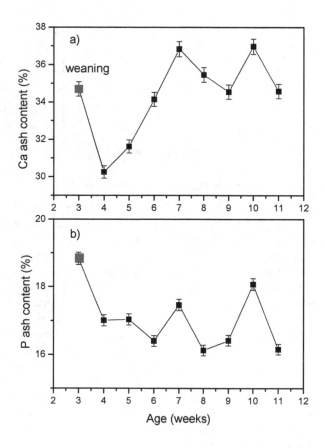

Fig. 4.a. Calcium and b. Phosphorous content in ash as a function of the age.

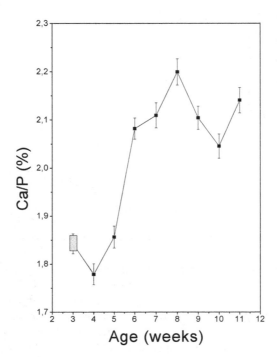

Fig. 5. Ca/P ratio for femur ash of female rats as a function of the age.

Figure 6 shows the changes in the ash mineral content as a function of the age decreases in the mineral content from the 3 to 4 weeks corresponds to the change in the diet. It is important to remark that the 10th week shows strong changes in K, Na and Mg. This increase is close related to those observed in the ash content, but as can be seen, when these minerals increase, a reduction in Ca and P is observed. At this point Boskey (2005) reported that there are minerals substitutions in HAp structure.

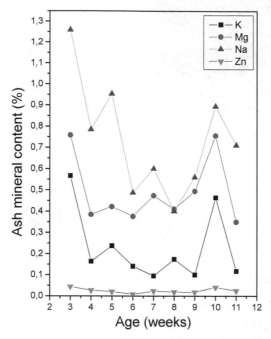

Fig. 6. Mineral content in femur of female rats as a function of the age.

3.5 Evaluation of the structural composition in bones

The study of the crystalline structures present in ash bones was obtained using the method described by Rodríguez et al., (1995), using X-ray diffraction. The ash bone was ground to a fine powder passing through a 150-µm mesh screen. The powder samples were densely packed into an aluminum frame. X-ray diffraction patters of the samples were carried out by using a diffractometer (Siemens D5000) operating at 35 Kv, and 15 mA, Cu K_α radiation. Data was collected from 4 to 30° on 2Θ scale with a step size of 0.05°.

Changes in the amorphous and crystalline structures of the bone tissues are attributed to the bone forming process and mineral changes during the growth stage. At this point Bandyopadhyay-Ghosh (2008) reported that the osteoblast release Ca, Mg and Phosphate ions, which chemical combine and harden within the matrix (collagen) into several compounds mainly HAp and phosphates.

The configurational crystalline order, gives important information regarding the synthesis of bone formation which is closely related to the diet and especially to the Ca/P ratio. Many of the biological and mechanical functions of the bone are determined by the composition and crystal structure of the bone. It is widely known that the bone mineral crystals (compositionally and structurally similar to the synthetic mineral calcium hydroxyapatite) are nano-crystalline, highly disordered, compositionally non stoichiometry and labile, and exists as a composite mineral embedded in the collagen organic matrix. Using the data obtained in Figures 4 and 5, it is clear that at the beginning of the development the femur rats are deficient in Ca, while in the case of P no significant changes were found. This means that for the structural formation of crystalline phase the amount of minerals available plays the most important role.

Figure 7 shows the X-ray diffraction patterns of the bone at the beginning and at the end of the growing stage, in this figure it was included bone ash that was used in this experiments as reference values for minerals, According to Figure 7, there are structural and compositional changes in the inorganic matrix as a function of the age, that are evident in bone ash from rats to 3th and 11th weeks. The bone ash used as standard reveals the existence of one phase identified as Hydroxyapatite. On the other hand, the X-ray diffraction patterns obtained from femurs at several ages, exhibit two different crystalline   structures: Calcium phosphate hydrated at the earlier growing stage and Hydroxyapatite at the end of the growth stage. These crystalline phases were identified using Powder Diffraction Files PDF# 09-0432 and PDF# 41-0483. It means that the formation of bio-hydroxyapatite as the main inorganic component of bone in rats is formed after the 7th week.

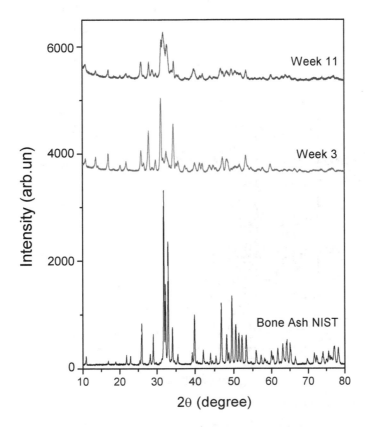

Fig. 7. X-ray diffraction patterns of bone ash for 3 and 11 weeks for female rats, as well as the bone ash used as standard (NIST).

In figure 8 the continuous lines correspond to the complete identification of Calcium phosphate hydrated (PDF# 41-0483), as can be seen, the transformation of this structure to bio HAp can be detected by studying the decrease of its peak intensity as the rat age increase, an important point is that until the end of the growing process this crystalline structure does not disappear. It means that this is a phase present during the growing stage. A careful examination of Figure 8 shows that Hydroxyapatite phase increases as age increases. The structural development is close related to the hormonal changes. An abrupt change in the crystalline structures between these two phases is evident at the 5th week, in which thin peaks are observed in comparison to the other X-ray diffraction patterns. By direct inspection of Figure 5 for the Ca/P ratio, is clear that at this age the amount of Ca and P are forming mainly crystalline structures as was mentioned before, but after the 6th weeks increase in the Ca is evident. The remaining elements and the excess of Ca and P allow the formation of amorphous structures.

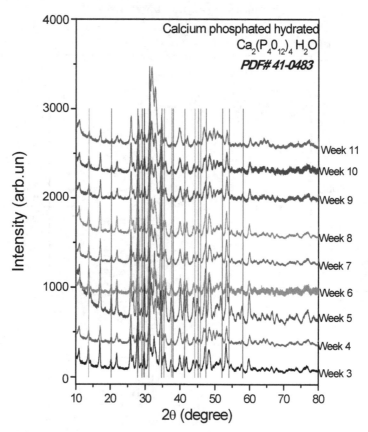

Fig. 8. X-ray diffraction patterns of bone ash from 3th to 11th weeks.

Taking into account the results of Figure 6 and the HAp showed in Figure 7, we could conclude that the main structure of the bone corresponds to bio-Hydroxyapatite, where some ions substitutions are present. It means that Na, K; Mg and other minerals were incorporated into the HAp structure and also into de amorphous materials. Additionally, by inspection of the figure 7, the quality of the bio-hydroxyapatite from rats ash bones is smaller than bone ash from NIST, likely due to the minerals present in the organic part that were not eliminated in the calcined process. Possible, the NIST sample is calcinated after removing the fat and protein matrix that contain minerals. The structural conformation of the bone has strong implication on the mechanical and functional properties of it. The structural conformation of the bone is an amazing kinetic process that takes place at body temperature. It is well known that for the crystallinization of HAp is required a temperature close to 700°C and the formation of the bone is still an unresolved and open problem for future researches.

Figure 9, shows the X-ray diffraction patterns of bone ash corresponding to different ages of the rats. In this case, the continuous lines correspond to the identification of HAp (PDF 09-0432). X-ray patterns confirm that if the rat age increases, the minerals content into the bone begin to form HAp. This fact is important; because the new bone formation influences the mechanical properties of the bone that can be reflected in more breaking force, which is an excellent indicative to prevent bone fractures.

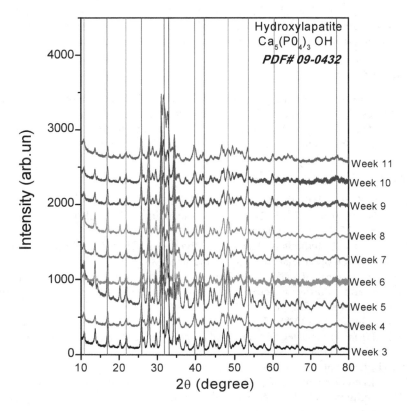

Fig. 9. X-ray diffraction patterns of bone ash for female rats.

4. Conclusion

In the case of biological studies, AAS is an excellent tool that allows a quantitative determination of mineral content in bones and other materials. The physical development depends on four factors: genetic, feeding conditions, mechanical, and hormonal. These factors are responsible and determine the performance of all structures of the rat in adult phase. During acceleration phase the alteration of any of the aforementioned factors have strong effects on the rat develop. Nevertheless, during this phase these factors can be modified according to the experimental model proposed for any study.

The ash content shows an increase as a function of the age, during the acceleration phase, being this behavior the normal develop of the rats. This current data indicate that the organic material into the bone matrix decreases in the same manner, and it is confirmed by the determination of Ca, P, Mg, and K.   The Ca/P ratio in bone is bigger than 1.67 in all cases, but this increases with the age, indicating that part of these minerals are conforming the bio-Hydroxyapatite and the excess of minerals are amorphous compound based in these minerals. The analysis of X-ray diffraction patterns of bone ashes showed that during this period, two crystalline phases are present: Calcium phosphate hydrated at the earlier growing stage and bio-Hydroxyapatite during the whole stage.

These results are very important for the biological models, because these constitute the normal curves for growing, weight, femur length in female rats. Additionally, for the first time the normal curves for ashes, Calcium, Phosphorus, Sodium, Zinc, Potassium, and Magnesium for female rats bones during growing were obtained.

5. Acknowledgment

This work was supported by project Conacyt-Fondos Mixtos-Gobierno del Estado de Querétaro, Project: QRO-2009-C01-118072, México, Esther Pérez-Torrero wants to thanks Conacyt-México for the grant # No. 21211. The authors want to thank Dra. Beatriz Millan and M. en C. Rosa Maria Lima for their technical support of this work.

6. References

A.O.A.C. (1998). Official Methods of Analysis. 16th edition, Official Method 968.08, Official Methods 925.10, 965.17, 974.24, 992.16. Association of Official Analytical Chemists, Maryland, U.S.A. AOAC international.

A.O.A.C. (2000). Official Methods of Analysis. 17th edition. Official Methods 935.35. Ed. by the Association of Official Analytical Chemists, Maryland, U.S.A.

Anderson, G.H. & Draper, H.H. (1972). Effect of dietary phosphorus on calcium metabolism in intact and parathyroidectomized adult rats. *The Journal of Nutrition*, Vol.102, No.9, (September 1972), pp. 1123-1132, ISSN 1541-6100

Bandyopadhyay-Ghosh, S. (2008). Bone as a collagen–hydroxiapatite composite and its repair. *Trends in Biomaterials & artificial organs*, Vol.22, No.2, (August 2008), pp. 116-124, ISSN 0971-1198

Benhayoune, H.; Charlier, D.; Jallot, E.; Laquerriere, P.; Balossier, G. & Bonhomme, P. (2001). Evaluation of the Ca/P concentration ratio in hydroxyapatite by STEM-EDXS: influence of the electron irradiation dose and temperature processing.

Journal of Physics D: Applied Physics, Vol.34, No.1, (January 2001), pp. 141–147, ISSN 1361-6463

Blokhuis, T.J.; Termaat, M.F.; den Boer, F.C.; Patka, P.; Bakker F.C. & Haarman, H.J. (2000). Properties of calcium phosphate ceramics in relation to their in vivo behavior. *The journal of Trauma: Injury, Infection, and Critical Care*, Vol.48, No.1, (January 2000), pp. 179–86, ISSN 1529-8809

Bogden, J.D.; Kemp, F.W.; Huang, A.E.; Shapses, S.A.; Ambia-Sobhan, H.; Jagpal, S.; Brown, I.L. & Birkett, A.M. (2008). Bone mineral density and content during weight cycling in female rats: effects of dietary amylase-resistant starch. *Nutrition & Metabolism*, Vol.5, No.34, (November 2008), pp. 1-12, ISSN 1743-7075

Bonnick, S.L. (2010). *Bone densitometry in clinical practice: Application and interpretation* (3), Humana Press, ISBN 978-1-60327-498-2, Denton, Texas, USA

Brem, J.J.; Trulls, H.E.; Lanari Zubiaur, A.E.; Ortíz de Rus, L.M.; Pochón, D.O.; Picot, J.A. & Brem, J.C. (2004). Osteoporosis experimental en ratas ovariectomizadas: densitometría ósea y concentración de minerales en sangre y cenizas de hueso. *Revista Veterinaria*, Vol.15, No.2, (November 2004), pp. 56-61, ISSN 1669-6840

Boskey, A.L.; (2005). The organic and inorganic matrices, In: *Bone tissue Engineering*, Hollinger, J.O; Eirnhorn, T.A.; Doll, B.A. & Sfeir, C., pp. 91-123, CRC press, Available from http://es.scribd.com/doc/51217314/Bone-Tissue-Engineering

Calvo, M. S., & Park, Y. K. (1996). Changing phosphorus content of the U.S. diet: Potential for adverse effects on bone. *Journal of Nutrition*, Vol.126, No.4, (April 2006), pp. 1168S-1180S, ISSN 1541-6100

Christian, G.D. (1972). Atomic absorption spectroscopy for the determination of elements in medical biological samples, In: *Inorganic and Analytical Chemistry*, pp. 77-112, Springer Berlin/Heidelberg, Available from http://dx.doi.org/10.1007/BFb0051600

Contreras-Padilla, M.; Pérez-Torrero, E.; Hernández-Urbiola, M.I.; Hernández-Quevedo, G.; del Real, A.; Rivera-Muñoz. E.M. & Rodríguez-García, M.E. (2011). Evaluation of oxalates and calcium in nopal pads (*Opuntia ficus-indica var. redonda*) at different maturity stages. *Journal of Food Composition and Analysis*, Vol.24, No.1, (September 2010), pp. 38-43, ISSN 0889-1575

Cornejo-Villegas, M.A.; Acosta-Osorio, A.A.; Rojas Molina, I.; Gutiérrez-Cortéz, E.; Quiroga, M.A.; Gaytán, M.; Herrera, G. & Rodríguez-García, M.E. (2010). Study of the physicochemical and pasting properties of instant corn flour added with calcium and fibers from nopal powder. *Journal of Food Engineering*, Vol.96, No.3, (February 2010), pp. 401-409, ISSN 0260-8774

Farris, E.J. & Griffith, J.Q. (1949). *The rat in laboratory investigation* (2), J.B. Lippincott Company, ISBN 978-0028445502, England. http://www.amazon.com/Rat-Laboratory-Investigation-Edmond-Farris/dp/0028445503

Fernández-Tresguerres, J.A.; Ariznabarreta-Ruiz, C.; Cachofeiro, V.; Cardinali, D.; Escrich-Escriche, E.; Gil-Loyzaga, P.E.; Lahera-Juliá, V.; Mora-Teruel, F.; Romano-Pardo, M.; Tamargo-Menéndez, J. (2010). *Fisiología Humana* (4), Mc Graw Hill, ISBN 978-607-15-0349-7. México D.F., México.

Frost, H.M. & Jee, W.S. (1992). On the rat model of human osteopenias and osteoporoses. *Bone Mineral*, Vol.18, No.3, (September 1992), pp.227-236, ISSN 8756-3282

Galvan-Ruiz, M.; Baños, L. & Rodriguez-Garcia, M.E. (2007). Lime characterization as a food additive. *Sensing and Instrumentation for Food Quality and Safety.* Vol.1, No.4, (August 2007), pp. 169–175, ISSN 1932-9954

Gómez-Aldapa, C.A.; Martínez-Bustos, F.; Figueroa, C.J.D.; Ondorica F.C.A. (1999). A comparison of the quality of whole corn tortillas made from instant corn flours by traditional or extrusion processing. *International Journal of Food Science & Technology,* Vol.34, No.4, (August 1999), pp. 391–399, ISSN 1365-2621

Gómez del Río, J.A.; Morando, P.J. & Cicerone, D.S. (2004). Natural materials for treatment of industrial effluents: comparative study of the retention of Cd, Zn and Co by calcite and hydroxyapatite. Part I: batch experiments. *Journal of Environmental Management,* Vol.71, No.2, (June 2004), pp. 169–177, ISSN 0301-4797

Gutiérrez-Cortez, E.; Rojas-Molina, I.; Rojas, A.; Arjona, J.L.; Cornejo-Villegas, M.A.; Zepeda-Benítez, Y.; Velázquez-Hernández, R.; Ibarra-Alvarado, C. & Rodríguez-García, M. E. (2010). Microstructural changes in the maize kernel pericarp during cooking stage in nixtamalization process. *Journal of Cereal Science,* Vol.51 No.1, (January 2010). Pp. 81-88, ISSN 0733-5210

Hernández-Urbiola, M.I.; Contreras-Padilla, M.; Pérez-Torrero, E.; Hernández-Quevedo, G.; Rojas-Molina, J.I.; Cortes, M.E. & Rodríguez-García. M.E. (2010). Study of Nutritional Composition of Nopal (Opuntia ficus indica cv. Redonda) at Different Maturity Stages. *The Open Nutrition Journal,* Vol.4, (2010), pp. 11-16, ISSN 1874-2882.

Hernández-Urbiola, M.I.; Pérez-Torrero, E. & Rodríguez-García, M.E.; (2011). Chemical Analysis of Nutritional Content of organic Prickly Pads (Opuntia ficus indica) at Varied Ages in an Organic Harvest. *International Journal of Environmental Research and Public Health,* Vol.8, No.5, (April 2011), pp. 1287-1295, ISSN 1660-4601

Heaney, R.P. (1993). Nutritional factors in osteoporosis. *Annual Review of Nutrition,* Vol.13, (July 1993), pp. 287-316, ISSN 978-0-8243-2813-9

Higuti R.T. & Adamowski, J.C. (2002). Ultrasonic densitometer using a multiple reflection technique. *IEEE transactions on ultrasonics, ferroelectrics, and frequency control,* Vol. 49, No.9, (September 2002), pp. 1260–1268, ISSN 0885-3010

Ke, H.Z.; Crawford, D.T.; Qi, H.; Chidsey-Frink, K.L.; Simmons, H.A.; Li, M.; Jee, W.S. & Thompson, D.D. (2001). Long-term effects of aging and orchidectomy on bone and body composition in rapidly growing male rats. *Journal of Musculoskel Neuron Interaction,* Vol.1, No.3, (March 2001), pp. 215-224, ISSN 1108-7161

Langton, C.M.; Pisharody, S.; Mathers, C. & Keyak, J.H. (2009). Generation of a 3D proximal femur shape from a single projection 2D radiographic image. *Osteoporosis International,* Vol.20, No.3, (June 2008), pp. 455-461, ISSN 0937-941X

Ma, Z.J.; Igarashi, A.; Inagaki, M.; Mitsugi, F. & Yamaguchi, M. (2000). Supplemental Intake of Isoflavones and Zinc- Containing Mineral Mixture Enhances Bone Components in the Femoral Tissue of Rats with Increasing Age. *Journal of health Science,* Vol.46, No.5 (July 2000), pp. 363-369, ISSN 1344-9702

McMillan, J.; Kinney, R.C.; Ranly, D.M.; Fatehi-Sedeh, S.; Schwartz, Z. & Boyan, B.D. (2007). Osteoinductivity of demineralized bone matrix in immunocompromised mice and rats is decreased by ovariectomy and restored by estrogen replacement. *Bone,* Vol.40, No.1, (January 2007), pp. 111-121, ISSN 8756-3282

Martin, R.B. (2000). Toward a unifying theory of bone remodeling. *Bone,* Vol.26, No.1, (January 2000), pp. 1-6, ISSN 8756-3282

Martinez-Flores, H.E.; Figueroa, J.D.C.; Martinez-Bustos, F.; González-Hernández, J.; Rodríguez-García, M. E, Baños- López, L. & Garnica-Romo M.G. (2002). Physical properties and composition of femurs of rat fed with diets based on corn tortillas made from different processes. *International Journal of Food Science and Nutrition*, Vol.53, No.2 (2002), pp. 155–162, ISSN 1465-3478

Masuyama, R.; Nakaya, Y.; Katsumata, S.; Kajita, Y.; Uehara, M.; Tanaka, S.; Sakai, A.; Kato, S.; Nakamura, T. & Suzuki, K. (2003). Dietary calcium and phosphorus ratio regulates bone mineralization and turnover in vitamin D receptor knockout mice by affecting intestinal calcium and phosphorus absorption. *Journal of bone mineral research*, Vol.18, No.7, (July 2003), pp. 1217-1226, ISSN 1523-4681

Matkovic, V. (1991). Calcium metabolism and calcium requirements during skeletal modeling and consolidation of bone mass. *American Journal of Clinical Nutrition*, Vol.54, No.1, (July 1991), pp. 245S-260S, ISSN 1938-3207

Matkovic, V. & Ilich, J.Z. (1993). Calcium requirements for growth: Are current recommendations adequate?. *Nutrition Reviews*, Vol.51, No.6, (June 1993), pp. 171-180, ISSN 1753-4887

Meneghini, C.; Chiara-Dalconi, M.; Nuzzo, S.; Mobilio, S. & Wenk R.H. (2003). Rietveld Refinement on X-Ray Diffraction Patterns of Bioapatite in Human Fetal Bones. *Biophysical Journal*, Vol.84, No.3, (March 2003), pp. 2021–2029, ISSN 1542-0086

National Research Council. (2003). Guidelines for the care and use of mammals in neuroscience and behavioral research. National Academic Press. ISBN-10: 0-309-08903-4, ISBN-13: 978-0-309-08903-6

Nnakwe, N. & Kies, C. (1985). Mouse bone composition and breaking strength, In: *Nutritional bioavailability of Calcium*, pp. 89-104, ACS publications, ISBN 9780841209077, Washington, DC.

Norcross, J. & Van Loan, M.D. (2004), Validation of fan beam dual energy x ray absorptiometry for body composition assessment in adults age 18-45 years. *British Journal of Sport Medicine*, Vol.38, No.4, (August 2004), pp. 472-476, ISSN 1473-0480

Nowicka, W.; Machoy, Z.; Gutowska, I.; Noceń, I.; Piotrowska, S. & Chlubek, D. (2006). Contents of Calcium, Magnesium, and Phosphorus in Antlers and Cranial Bones of the European Red Deer (*Cervus Elaphus*) from Different Regions in Western Poland. *Polish, Journal of Environmental Studies*, Vol. 15, No. 2, (2006), pp. 297-301, ISSN 1230-1485

Palacios-Fonseca, A.J.; Vazquez-Ramos, C. &. Rodríguez-García, M.E. (2009). Physicochemical Characterizing of Industrial and Traditional Nixtamalized Corn Flours. *Journal of Food Engineering*, Vol.93, No.1, (July 2009), pp. 45-51, ISSN 0260-8774

Reeves, P.G.; Nielsen, F.H. & Fahey, G.C. (1993). AIN-93 purified diets for laboratory rodents: final report of the American Institute of Nutrition ad hoc writing committee on the reformulation of the AIN-76A rodent diet. *The Journal of Nutrition*, Vol.123, No.11, (November 1993), pp. 1939-1951, ISSN 1541-6100

Rodríguez-Garcia, M.E.; de Lira, C.; Hernández-Becerra, E.; Cornejo-Villegas, M.A.; Palacios-Fonseca, A.J.; Rojas-Molina, I.; Reynoso, R.; Quintero, L.C.; del Real, A.; Zepeda, T.A. & Muñoz-Torres, C. (2007). Physicochemical characterization of nopal pads (*Opuntia ficus indica*) and dry vacuum nopal powders as a function of the maturation. *Plant Foods for Human Nutrition,* Vol.62, No.3, (August 2007), pp.107-112, ISSN 0921-9668

Rodríguez, M.E.; Yañez, J.M.; Cruz-Orea, A.; Alvarado-Gil, J.J.; Zelaya-Angel, O.; Sánchez-Sinencio, F.; Vargas, H.; Figueroa, J.D.; Martínez-Bustos, F. & Martínez-Montes, J.L. (1995). The influence of slaked lime content on the processing conditions of cooked maize tortillas: changes of thermal, structural and rheological properties. *Zeitschrift für Lebensmitteluntersuchung und -Forschung A,* Vol.201, No.3, (May 1995), pp. 233-240, ISSN 1431-4630

Rodríguez-Rodríguez, E.; Navia-Lombán, B.; López-Sobaler, A.M. & Ortega-Anta, R.M. Review and future perspectives on recommended calcium intake. *Nutrición Hospitalaria,* Vol.25, No.3, (May-June 2010), pp.366-374, ISSN 0212-1611

Rojas-Molina, I.; Gutiérrez-Cortez, E.; Palacios-Fonseca, A.; Baños, L., Pons-Hernández, J.L.; Guzmán, S.H.; Pineda-Gómez, P. & Rodríguez, M.E. (2007). Study of structural and thermal changes in endosperm of quality protein maize during traditional nixtamalization process. *Cereal Chemistry,* Vol.84, No. 4, (July/August 2007), pp. 304-312, ISSN 0009-0352

Rojas-Molina, I.; Gutiérrez, E.; Rojas, A.; Cortés-Álvarez, M.; Campos Solís, L.; Hernández-Urbiola, M.; Arjona, J.L.; Cornejo, A. & Rodríguez-García. M.E. (2009). Effect of Temperature and Steeping Time on Calcium and Phosphorus Content in Nixtamalized Corn Flours Obtained by the Traditional Nixtamalization Process. *Cereal Chemistry,* Vol.86, No.5, (September/October 2009), pp.516-521, ISSN 0009-0352

Ščančar, J. Milačič, R.; Benedik, M. & Bukovecc, P. (2000). Determination of trace elements and calcium in bone of the human iliac crest by atomic absorption spectrometry. *Clinica Chimica Acta,* Vol.293, No.1-2, (March 2000), pp. 187-197, ISSN 0009-8981

Smith, E.L., Hill, R.L., Lehman, I.R., Lefkowitz, R.J., Handler, P., White, A. (1983). Principles of Biochemistry. Mammalian Biochemistry. Seventh edition. N.Y. McGraw-Hill Book Co.

Spencer, H.; Kramer, L. & Osis, D. (1988). Do protein and phosphorus cause calcium loss? *The Journal of Nutrition,* Vol.118, No. 6, (June 1988), pp. 657-660, ISSN 1541-6100

Su, X.; Sun, K.; Cui, F.Z. & Landis, W.J. (2003). Organization of apatite crystals in human woven bone. *Bone,* Vol.32, No.2, (February 2003), pp. 150-162, ISSN 8756-3282

Takahara, S.; Morohashi, T.; Sano, T.; Ohta, A.; Yamada, S. & Sasa, R. (2000). Fructooligosaccharide Consumption Enhances Femoral Bone Volume and Mineral Concentrations in Rats. *The Journal of Nutrition, Nutrient Metabolism-Research communication,* Vol.130, No. 7 (July 2000), pp. 1792-1795, ISSN 1541-6100

Tamaki, T. & Uchiyama, S. (1995). Absolute and relative growth of rat skeletal muscle. *Physiology & behavior,* Vol.57, No. 5, (May 1995), pp. 913-919, ISSN 003-9384

Wyatt, C.J.; Hernández-Lozano, M.E.; Méndez, R.O. & Valencia, E. (2000). Effect of different calcium and phosphorus content in Mexican diets on rat femur bone growth and composition. *Nutrition Research,* Vol.20, No.3, (March 2000), pp. 427-437, ISSN 0271-5317

10

Comparative Assessment of the Mineral Content of a Latin American Raw Sausage Made by Traditional or Non-Traditional Processes

Roberto González-Tenorio[1], Ana Fernández-Diez[2],
Irma Caro[2] and Javier Mateo[2]
[1]University Autónoma del Estado de Hidalgo Mexico
[2]University of León
[1]Mexico
[2]Spain

1. Introduction

Mineral content of food in general, and of meat and meat products in particular, has been widely studied mainly due to its essential role in human nutrition and implication on safeness issues, i.e., mineral toxicity and deficiencies. Furthermore, since a few decades ago, instrumental analytical techniques based on atomic absorption or emission spectrometry applied to the determination of the mineral content coupled to multivariate statistical analysis have been proved to produce suitable methods to characterise food products, discriminate between food quality categories and control food authenticity, i.e., determination of the geographical origin of food, discrimination between cultivation methods (e.g. organic vs convenience crops), varieties of fruits and vegetables, or food processing practices (Grembecka et al., 2007; Kelly & Bateman, 2010; Luykx & van Ruth, 2008; Sun et al., 2011).

Inductively coupled plasma-atomic emission spectroscopy (ICP-AES) is a prevailing instrumental technique utilised for the simultaneous determination of a considerably high number of metals and some non-metals in biological samples at ppm or ppb levels. As well as in AES conventional techniques, in ICP-AES technique, analyte atoms in solution are aspirated into the excitation region where they are desolvated, vaporised, and atomised, and finally the optical emission measurement from excited atoms is used to determine analyte concentration. However, ICP-AES provides higher reproducibility and quantitative linear range compared to conventional AES, and reduces molecular interferences due to a higher temperature (7000–8000 K) in the excitation source (plasma). On the other hand, ICP-AES is more expensive than conventional AES, and in complex samples emission patters can be of difficult interpretation (Ibáñez & Cifuentes, 2001; Luykx & van Ruth, 2008).

There is a wide variety of meat products all over the world and, among them, fresh sausages represent an important part. Two types of fresh sausages can be found in shops or markets: emulsion-type and minced comminuted meat products. Fresh sausages are sold without having suffered any heat treatment, and they are generally stored and commercialised chilled or frozen (Feiner, 2006). However, many times, in Latin America fresh sausages are slightly dried at room temperature during some hours as part of the making-process and afterwards the sausages are stored and commercialised for short periods (few days) at room temperature, without the use of the cool chain. During those periods additional drying and lactic acid fermentation can take place (Kuri et al., 1995). In this case, sausages can be considered to be raw-semidry sausages instead of fresh sausages.

Meat products are generally made from various meat and non-meat matters (from different origins and suppliers), which are combined at the formulation stage in obedience to criteria of composition, technological factors, sensory characteristics, legal regulations, functionality and also production cost (Jiménez-Colmenero et al., 2010). In view of that, fresh sausages (and eventually the raw-semidry sausages derived from them) are made from different types of meat such as beef, pork, mutton, chicken, turkey, etc. and usually pork fat or fatty tissues. Moreover, numerous non-meat ingredients (salt, herbs, spices, juices, vinegar, etc.) and additives (nitrites, phosphates, sorbates, etc.) can be added depending on each sausage type, geographical traditions or manufacturing practices (Feiner, 2006).

Current fresh sausage making process includes both traditional and non-traditional methods. On the one hand, there is an increasing attention in natural products, sometimes proceeding from a particular region, elaborated according to traditional methods. On the other hand, the demand for non-traditional commercial food products is steadily increasing due to different factors, namely, cheapness, extended-shelf life or convenience. Variations in meat and non-meat ingredients and the processing conditions, with respect to traditional methods, can result in changes in nutritional composition. In order to avoid drastic and negative impact on consumers, those changes are controlled or limited by regulations. As common examples, the fat (maximum) and lean meat (minimum) content is generally legally established in most countries. Furthermore, usually it is required a minimun protein content; however, proteins could originate from meat or cheaper sources such as wheat gluten or soy protein (Feiner, 2006). In this way, low cost sausages are largely sold in many countries.

Chorizos are sausages extremely popular in Latin American countries. These sausages can be considered as variations of those that were brought to America by colonizers, usually Iberians (Mateo et al., 2008). In fact, at present there are sausages in Spain (chorizo) or Portugal (Chouriço) that resemble those from Latin America and vice versa. Chorizos are usually made from pork meat and fat. Grounded meat and fat are normally added and mixed with salt, herbs and spices and, optionally, vinegar or lemon juice, sugar and suitable additives (curing agents, phosphates, etc.). Finally, the mixture is stuffed into casings, commonly natural casings. Chorizos are made with different herbs and spices in different Latin American regions, with garlic, clove, cumin and dry *Capsicum spp.* fruits (named as chile) being typically used in Mexican chorizo. Once stuffed, these sausages are dried for a short period of time (from fours to few days) at room temperature before selling or consuming.

The mineral content of fresh or raw-semidry chorizo is the result of the sum of the contributions from all the ingredients, i.e., lean meat, fatty tissues, common salt, dry *Capsicum* spp. fruits, textured soy flour, etc., and additives included in their formulations. In order to better ascertain the eventual contribution of ingredients to the mineral content of chorizo, the mineral composition of the main ingredients used in chorizo making process is shown in Table 1. Moreover, mineral content can be further increased due to the solute concentration by evaporation which takes place during an eventual drying.

Lean meat is the main ingredient of a sausage and therefore the mineral composition of a
raw sausage it is closely related to the meat content. Meat is considered as a significant
dietary source of several minerals, specifically highly-available Fe and Zn (Hazell, 1982;
McAfee et al., 2010). The mineral content of meat is rather stable under conditions of normal
supply (average mineral composition of raw lean pork is shown in Table 1); however, this
depends on pre-mortem factors such as animal genetic, physiological and environmental,
e.g., age and feeding (Jiménez-Colmenero et al., 2006), and post-mortem factors such as
meat cuts, meat processing and preparation (Jiménez-Colmenero et al., 2010; Lombardi-
Boccia et al., 2005). Fat is the second ingredient in importance of sausages. Conversely,
compared to meat, the contribution of fat to the mineral content of a sausage is poor due to
the relatively low mineral content in fat or fatty tissues (Table 1).
Regarding the non-meat ingredients, common salt is the major source of Na in sausages.
Apart from Na, common salt contains other minerals but at trace levels (Table 1).
Nevertheless, common salt can be a notorious source of Mg or Ca because these mineral
salts are on occasion added to common salt as anti-caking agents at the food industry level.
Dry *Capsicum* spp. fruits (in paste or powdered form) are extensively used in Latin
American and also Iberian raw sausages. The amount usually added in chorizos is
approximately 2-3% of the weight of the initial mixture of sausage. An addition of dry
capsicum fruits at that amount represents an appreciable source of some mineral elements,
i.e., Mg, Ca, Fe and Mn, in sausage mixtures (Aguirrezábal et al., 1998; Fonseca et al., 2011).
Textured soy flour is used as filler or meat extender in a large variety of commercially
prepared meat products. The use of soy-derived ingredients results in increased yields and
lower cost of production (Hoogenkamp, 2008). The amount of soy flour added to those meat
products range from 2 to 10% (Rocha McGuire, 2008). In Mexico, the use of textured soy
flour in raw sausages is common, particularly in low-cost sausages commercialized in
wholesale markets (Personal communication: technical personnel, Fabpsa Corporation,
www.fabpsa.com.mx). In accordance with the mineral content of textured soy flour (Table 1)
and the amount used, the contribution of this ingredient to the levels of some minerals, i.e.,
Ca, Mg, Cu and Mn, in raw sausages appears to be relevant. Finally, the additives
commonly used for commercial fresh sausages (Feiner, 2006), such as antioxidants
(erythorbate or ascorbate), preservatives (nitrite, sorbate) or others (phosphates, etc.)
provide Na, P and K to the raw sausages.

	Pork Lean	Pork Backfat	Salt	Paprika, Dry *Capsicum annuum*	Defatted soy flour
Sodium	59	11	38758	68	20
Potassium	364	65	8	2280	2384
Phosphorus	217	38	0	314	674
Calcium	13	2	24	229	241
Magnesium	24	2	1	178	290
Zinc	2.2	0.4	0.1	4.3	2.5
Iron	0.8	0.2	0.3	21.1	9.2
Copper	0.1	0.0	0.0	0.7	4.0
Manganese	0.0	0.0	0.1	1.6	3.0

Table 1. Mineral composition of main ingredients used in Mexican chorizo (mg/100 g;
Source: USDA, 2010)

As far as the authors are aware, there is great interest about the usefulness of mineral content analysis in meat industry. Apart from nutritive and safety issues, to assessing the differences in minerals between sausages from different qualities could reflect fraudulent manufacturing practices. The aim of the present study is therefore to determine the mineral composition of a popular Latin American raw sausage (Mexican chorizo) using ICP-AES, and to compare the mineral contents between sausages made using traditional and those made using non-traditional making processes in order to explore differences between both groups regarding nutritive value and discrimination purposes.

2. Material and methods

2.1 Samples and sampling procedure

Forty sausages belonging to two different groups (twenty per group) were used in the present study. One group consisted of Mexican chorizos elaborated in a traditional manner by small-sized producers (traditional sausages) and the other was comprised of mass-produced commercial Mexican chorizos (non-traditional sausages).

Ten samples of sausages from the first group were purchased from randomly-chosen butcheries in the Hidalgo State largest cities: Tulancingo and Pachuca (approximately 150 and 300 thousands inhabitants, respectively). The sausages from this type were locally produced by the butchers. The other ten traditional sausages sampled were purchased from rural street markets in ten randomly-chosen small villages (5 to 20 thousands inhabitants) in Hidalgo State (one sample per village). These were homemade sausages which were produced by local small-sized farmers.

On the other hand, ten out of the twenty samples of the non-traditional sausages were purchased from supermarkets at Tulancingo and Pachuca cities. These sausages were manufactured by nationwide companies and were characterised by their long shelf life. The other ten samples were purchased from wholesale large-city markets in Tulancingo, Pachuca and Mexico City. The sausages were manufactured by ten different medium-sized companies to be sold at low price. All sausage samples weighted about 1 kg each. Once taken, samples were immediately transported to the laboratory in Tulancingo, where they were first blended in a domestic food processor and then frozen and stored at -18 °C until further analysis.

2.2 Chemical analysis

Determinations of moisture, fat, protein and ash contents in the sausage samples were performed in duplicate according to methods recommended by the AOAC International {AOAC} (AOAC, 1999) – Official Methods nos. 950.46, 991.36, 981.10 and 920.153, respectively.

The analysis of mineral composition of sausages was performed by ICP-AES on wet digested samples. Duplicate aliquots of approximately 1 g (±0.01) of the previously homogenised samples were digested with 10 ml of concentrated HNO_3 in tightly closed screw cap glass tubes for 16 h at room temperature, and then for a further 4 h at 90 °C. For the analysis of sodium, potassium and phosphorus, 1 ml of the mineralised solution was added with 8 ml of deionised water and 1 ml of scandium solution as internal standard. In order to determine the levels of calcium, copper, iron, magnesium, and zinc, 3 ml of the digested solution was added with 6 ml of deionised water and 1 ml of Sc solution.

The instrumental analysis was performed with a Optima 2000 DV ICP optical emission spectrometer (PerkinElmer, Waltham, MA, USA). Instrument operating conditions were: radiofrequency power, 1400 W; plasma gas flow, 15.0 l/min; auxiliary gas flow, 0.2 l/min; nebulizer gas flow 0.75 l/min, crossed flow; standard axial torch with 2.0 mm i.d. injector of silica; peristaltic pump flow, 1 ml/min; no. of replicates, 2. The spectrometer was calibrated for Cu, Mn, Zn, Fe, Ca and Mg determinations (at 224.7, 257.61, 213.9, 238.2, 393.4 and 279.6 nm, respectively) with nitric acid/water (1:1, v/v) standard solutions of 2, 5 and 10 ppm of each element, and for Na, P and K (at 589.6, 213.6 and 766.5, respectively) with nitric acid/water (1:9, v/v) standard solutions of 30, 50 and 100 ppm, respectively.

2.3 Statistical analysis

A one-way analysis of variance was performed; the values obtained for each variable (moisture, fat, protein, ash or mineral elements) on all fresh weight (FW), dry matter (DM) and nonfat dry matter (NFDM) basis were compared between groups (traditional and non-traditional) and types (as function of the place where the sausages were purchased from) of sausages. Moreover, linear correlation coefficients were calculated to describe the degree of correlation between each variable (only values expressed on DM basis were considered). A principal component (PC) analysis, unrotated method, was also performed to obtain a better perception of differences in mineral composition between the groups and types of sausages. In the PC-analysis model the contents of the mineral elements determined, except for Cu, were included. The software STATISTICA for Windows (StatSoft Inc., 2001) was used for the statistical treatment of data.

3. Results and discussion

3.1 Proximate composition

The proximate composition of the sausages studied on all FW, DM, and NFDM basis is shown in Table 2. Mean values are given for the traditional (Trad) and non-traditional (Non-Trad) sausages and for the individual types of sausages samples: Btch, traditional sausages from butcheries; RuSM, traditional sausages from rural street markets; Smkt, non-traditional sausages from supermarkets; WhCM, non-traditional sausages from wholesale city markets. The Table also shows the level of significance (P-level: NS, not significant; *, $P<0.05$; **, $P<0.01$ and ***, $P<0.001$) of the analysis of variance, and the standard error of the mean (SEM). Significant differences between types of sausages (Duncan-test, $P<0.05$) are marked with different superscripts. Carbohydrate content was calculated by difference.

Most of the sausages can be assigned to the category of semidry sausages according to the scheme proposed by Adams (1986), because their humidity content was inside or near the range of 40 to 50%. According to that author, weight (drying) losses for this type of sausages are estimated to be between 10 and 20%.

The variability observed within each group of sausage (as assessed by standard deviations) was high. This could be mainly attributed to large variations among samples in the amount of fat used in the sausage making process and the degree of sausage drying. In spite of the variability, statistical differences ($P<0.05$) in the proximate composition between sausage groups were observed. Traditional sausages had lower ash and carbohydrate (calculated by difference) contents than the non-traditional ones. Furthermore, protein content on DM and NFDM basis were higher in traditional than in non-traditional sausages.

	Trad n=20	Non-Trad n=20	P-level	Btch n=10	RuSM n=10	Smkt n=10	WhCM n=10	SEM n=40
Fresh weight								
Moisture	42.8±11.4	37.6±7.6	NS	42.9	42.7	36.5	38.7	1.6
Fat	33.4±12.0	36.7±9.0	NS	34.9	31.9	33.3	40.2	1.7
Protein	18.2±3.8	16.3±4.1	NS	16.3[b]	20.2[a]	19.5[a]	13.1[c]	0.6
Ash	2.9±0.8	3.9±1.1	**	2.8[b]	2.9[b]	4.6[a]	3.2[b]	0.2
Carbohydrate	2.7±1.8	5.4±2.6	***	3.2[bc]	2.3[b]	6.1[a]	4.7[c]	0.4
Dry matter								
Fat	57.6±11.5	58.5±10.4	NS	60.1[ab]	53.2[b]	52.2[b]	64.8[a]	1.7
Protein	33.7±12.4	26.4±6.6	*	29.6[ab]	37.8[a]	30.9[ab]	21.9[b]	1.7
Ash	5.1±1.4	6.3±1.7	*	5.0[b]	5.2[b]	7.3[a]	5.3[b]	0.3
Carbohydrate	4.5±3.1	8.8±4.4	**	5.3[bc]	3.7[b]	9.6[a]	7.9[ab]	0.7
Nonfat dry matter								
Protein	76.2±9.4	63.9±7.4	***	73.0[ab]	79.5[a]	65.6[bc]	62.3[c]	1.7
Ash	12.1±2.6	15.3±1.9	***	12.7[b]	11.5[b]	15.4[a]	15.3[a]	0.4
Carbohydrate	11.6±8.3	20.8±7.5	***	14.3[bc]	9.0[b]	19.0[ab]	22.4[a]	1.4

Table 2. Proximate composition (g/100 g) of traditional and non-traditional sausages, and of types of sausages as a function of the places where they were purchased from

Differences in ash content, assuming that common salt is the principal source of ash in sausages, are indicative of higher amounts of common salt being used in non-traditional than in traditional sausages. The amount of carbohydrate (which include lactate formed by lactic acid fermentation) found in traditional sausages (around 2-3% on FW basis) was the expected for a conventional fermented sausage, i.e., Polish sausage or summer sausage (USDA, 2010). However, the amount found in non-traditional sausages (6% on FW basis) was higher. This increase must be the result of the inclusion of hydrocolloids (gums, starches, dextrins) or other fillers in the formulations of this group of sausages (Personal communication: technical personnel, Fabpsa Corporation). Finally, the higher proportion of protein found in traditional sausages on both DM and NFDM basis is related to a higher proportion of lean meat used in this group compared to the non-traditional sausages. Lean meat is the main ingredient in sausage formulations and has a protein content relatively high, of approximately 65% of DM (USDA, 2010).

3.2 Mineral composition
Mineral composition, expressed as mg per 100 g on all FW, DM and NFDM basis is reported in Table 3. This Table shows the results found for the two groups of sausages: Traditional (Trad) and non-traditional (Non-Trad); and for the four types of sausages: Btch, traditional sausages from butcheries; RuSM, traditional sausages from rural street markets; Smkt, non-traditional sausages from supermarkets; WhCM, non-traditional sausages from wholesale city markets. The Table also shows the level of significance (P-level: NS, not significant; *, $P<0.05$; **, $P<0.01$ and ***, $P<0.001$) of the analysis of variance, and the standard error of the mean (SEM). Significant differences between types of sausages (Duncan-test, $P<0.05$) are marked with different superscripts.

There were statistically significant differences between both groups of sausages for practically all the minerals studied (all except for K). These differences could be attributed to variations in the ingredients and additives used, and in the dryness level of sausage samples. From the nutritional value point of view – minerals are essential nutrients –, FW-basis data provide useful information to what consumers are eating. On the other hand, DM- and NFDM-basis results could be related to type and amount of meat and non-meat ingredients and additives used in the making process and hence could be useful for sausage characterization and discrimination purposes.

	Trad n=20	Non-Trad n=20	P-level	Btch n=10	RuSM n=10	Smkt n=10	WhCM n=10	SEM n=40
Fresh weight								
Sodium	856±279	1287±453	***	845[b]	867[b]	1528[a]	1047[b]	68
Potassium	395±117	440±119	NS	376[b]	414[b]	519[a]	361[b]	19
Phosphorus	188±64	268±110	**	185[b]	191[b]	319[a]	217[b]	15
Calcium	36.8±22.2	60.5±34.8	*	40.5[ab]	33.1[b]	62.5[a]	58.6[ab]	4.9
Magnesium	32.2±12.9	43.0±11.4	**	28.0[b]	36.7[ab]	40.9[a]	45.1[a]	2.1
Zinc	3.32±0.81	2.64±0.71	**	3.19[ab]	5.41[a]	2.95[ab]	2.33[b]	0.83
Iron	3.05±1.14	2.47±0.45	*	2.55[b]	3.55[a]	2.34[b]	2.60[b]	0.14
Copper	-	-	-	-	-	-	0.10	-
Dry matter								
Sodium	1482±396	2063±669	**	1448[b]	1476[b]	2399[a]	1727[a]	97.6
Potassium	720±291	710±181	NS	685	755	817	602	37.8
Phosphorus	344±148	432±175	NS	336[b]	352[ab]	501[a]	362[ab]	26.3
Calcium	63.2±34.3	100±66.6	*	71.4	55.0	103.0	97.0	8.8
Magnesium	57.1±20.6	69.7±21.1	NS	49.4[b]	64.9[ab]	64.4[ab]	75.0[a]	3.4
Zinc	6.05±2.06	4.27±1.17	**	5.83[a]	6.26[a]	4.65[b]	3.88[b]	0.30
Iron	5.50±2.29	4.01±0.91	*	4.52[a]	6.47[a]	3.70[b]	4.32[a]	0.30
Copper	-	-	-	-	-	-	0.17	-
Nonfat dry matter								
Sodium	3632±1166	4997±1069	***	3839[b]	3424[b]	5017[a]	4976[a]	206
Potassium	1666±460	1717±188	NS	1668	1663	1716	1718	55
Phosphorus	787±234	1026±312	**	815[ab]	759[b]	1051[a]	1003[ab]	47
Calcium	159±101	243±131	*	180[ab]	138[b]	208[ab]	278[a]	19
Magnesium	137±54	175±55	*	126[b]	148[b]	136[b]	214[a]	9
Zinc	14.0±3.0	10.6±2.4	***	14.3[a]	13.7[a]	10.1[b]	11.0[b]	0.5
Iron	12.7±3.6	10.1±2.8	*	11.5[a]	13.9[a]	7.9[b]	12.4[a]	0.5
Copper	-	-	-	-	-	-	0.42	-

Table 3. Mineral content (mg/100 g) of traditional (Trad) and non-traditional (Non-Trad) sausages, and of types of sausages as a function of the places where they were purchased from (-: value under the detection limit; 0.1 mg/100 g)

In the sausages studied, considering the results expressed on FW basis, traditional sausages contained higher amounts of Fe and Zn, which are the mineral micronutrients most significantly related to the benefits to health of meat consumption (McAfee et al., 2010). On the contrary, non-traditional sausages presented higher Na, P, Ca and Mg levels than traditional ones. The increased levels of Na and P in the non-traditional sausages are not considered as advantageous. Excessive intake of Na derived from meat products consumption, and of P from the consumption of phosphate-added meat products have been linked to health risks: hypertension-derived pathologies and anomalous bone metabolism, respectively (Calvo & Park, 1996; Ruusunen and Poulanne, 2005).

Having into account the results expressed on DM basis, the concentrations of the mineral elements detected in the Mexican sausages (Mexican chorizos) were generally into de range of values reported for other chorizos from Latin America and Spain (Table 4): Traditional Peruvian fresh chorizo (Reyes-García et al., 2009), Spanish fresh and dry-ripened chorizo (Jiménez-Colmenero et al., 2010), and traditional Spanish dry-ripened chorizo (Fonseca et al., 2011) – all of them including dry capsicum fruits as characteristic ingredient. Furthermore, in general, chorizos had notable higher levels of Na, K, Mg and Fe than those reported for a mixture of ground pork, lean and fat (USDA, 2010; Table 4), at a similar percentages than that used in sausage making (72% to 28%, respectively). This could be explained by the presence, apart from lean and fat, of non-meat ingredients, i.e., salt and spices, and additives, in the sausage mixtures. Moreover, comparing the mineral content of ground pork, lean and fat (Table 4), with that of the traditional and non-traditional Mexican sausages (DM basis; Table 3), it can be noted that the composition of non-traditional sausages was more distant from ground pork than that of traditional sausages. This leads to the suggestion that the first type of chorizo would include larger proportions of non-meat ingredients and additives than the second group.

	Traditional Peruvian fresh chorizo	Spanish fresh chorizo	Spanish dry-ripened chorizo	Traditional Spanish dry-ripened chorizo	Ground pork, raw (72% lean/28% fat)
Sodium	1401	2364	1675	2504	157
Potassium	Nd	842	593	1106	404
Phosphorus	311	Nd	Nd	431	300
Calcium	117.4	19.7	22.7	27.4	36.4
Magnesium	Nd	50.6	32.3	58.2	29.5
Zinc	Nd	5.5	3.0	7.4	4.3
Iron	8.4	2.5	2.1	3.5	2.0

Table 4. Mineral composition (mg/100 g of dry matter) of several Latino American and Iberian sausages (chorizos) which include dry capsicum fruits as characteristic ingredient, and of ground pork (Nd: Not determined)

As shown in Table 3, non-traditional sausages had higher levels (on DM basis) of Na and Ca and lower of Zn and Fe than the traditional counterparts. When results were expressed on a NFDM basis, the differences between both groups were more pronounced than when results were expressed on DM. This is because there was considerable variation in the fat content of sausages, and mathematically removing the fat from the dry sausage eliminates the fat dilution effect on mineral contents – fat is the ingredient with lower mineral content. Thus, differences between traditional and non-traditional sausages when results being expressed as NFDM, were detected for all the mineral elements except for K. Non-traditional sausages

had higher amounts of Na, P, Ca and Mg and lower of Zn and Fe than traditional sausages –
differences in Na and Zn being the more statistically significant ($P<0.001$; Table 3).

Differences between the two groups of sausages regarding Na content can be explained
considering that higher salt amounts are usually used in non-traditional making processes.
A similar trend to that of Na was observed for P concentrations. Physiological P of animal
tissues is a constituent of protein structures, and its concentration in meat products may be
estimated from the concentration of protein. P to protein content ratio has been estimated as
approximately 0.01 for pork (Tyszkiewicz et al., 2001). Consequently, the ratio found in
traditional sausages was 0.010 (0.011 in sausages purchased from butcheries and 0.009 in
those purchased from rural street markets; data not shown in Tables). However, the mean
ratio in non-traditional sausages was 0.016 in the sausages from supermarkets and 0.017 in
those from wholesale city markets. The addition of phosphates and textured soy flour (a
non-meat proteinaceous ingredient with high P content), that are frequently used in
commercial sausage making (Feiner, 2006; Personal communication: technical personnel,
Fabpsa Corporation), could be responsible for the higher P concentrations in non-traditional
sausages. The phosphate addition to meat products is regulated by national and
international standards, for example, USDA-FSIS (2007), EC (1995), or the Codex
Alimentarius (2009). As an illustration, the maximum level of addition of phosphates, as
established by the European Union (EC, 1995), is 5 g/kg (expressed as P_2O_5). With respect to
the textured soy flour, its P to protein ratio is higher than that of meat and has been
calculated to be 0.015 (USDA, 2010).

The concentrations of Ca and Mg were also higher in non-traditional sausages and among
them those from wholesale city markets (sausages commercialized at the lowest cost)
contained the higher amounts. Those increased concentrations could be related to the use of
the following ingredients: mechanically recovered meat (residual meat recovered using
mechanical equipment from animal bones), texture soy flour and/or mineral fillers.
Mechanical recovered meat is cheaper than convectional meat and, thus, its use in sausages
is aimed to decrease production costs. Mechanically recovered meat has higher Ca content
than the conventional manually deboned meat, from 40 to 500 mg depending on the raw
matter and equipment used (Newman, 1981; USDA, 2010). Texture soy flour (compared to
meat) can be considered as a relatively Ca- and Mg-rich ingredient (Table 3). Finally,
mineral fillers such as Hubersorb (Akrochem, Akron, OH, USA) are included in some of the
formulations commercialised for sausages (Personal communication: technical personnel,
Fabpsa Corporation). That filler (Hubersorb) contains Ca silicate at a large proportion and it
is used in sausages for its high absorbency properties.

On the contrary, levels of Zn and Fe were higher in traditional sausages than in non-
traditional ones. This could be the result of a higher proportion of lean used and
additionally the use of meat from older animals (with higher Fe content) for the first group
of sausages. Fe content was especially high in sausages from rural street markets. This
finding can be related not only to the age of animals but also to the feasible migration of Fe
ions to meat and sausage mixture from the surfaces of cast iron equipment, i.e., pans,
mincers (Quintaes et al., 2004), which are frequently present at small homemade sausage
producing facilities in small villages. In spite of the lower amounts of Fe found in non-
traditional sausages, the Fe concentration in sausages from wholesale city markets was
comparable to that of traditional sausages. This could be indicative of the use of
mechanically recovered meat, which contains higher amounts of Fe, approximately twice
higher, than manually deboned meat (Newman, 1981).

3.3 Correlations and principal component analysis

Table 5 shows the correlation coefficients between the composition variables studied considering the values expressed on DM basis, for traditional and non-traditional sausages separately. Regardless the sausage group, significant correlations ($P<0.05$) detected for the

	Protein	Ash	Na	K	P	Ca	Mg	Zn	Fe
Fat									
Traditional	-0.96*	-0.57*	0.01	-0.70*	-0.70*	0.14	-0.43	-0.68*	-0.61*
Non-traditional	-0.88*	-0.90*	-0.75*	-0.92*	-0.70*	-0.38	-0.30	-0.39	-0.24
Protein									
Traditional	1.00	0,42	-0.19	0.73*	0.64*	-0.13	0.46*	0.68*	0.59*
Non-traditional	1.00	0.75*	0.58*	0.83*	0.56*	0.13	0.19	0.47*	0.12
Ash									
Traditional		1.00	0.61*	0.22	0.52*	-0.09	-0.04	0.52*	0.17
Non-traditional		1.00	0.92*	0.87*	0.74*	0.45*	0.29	0.19	0.22
Na									
Traditional			1.00	-0.13	0.20	0.03	-0.11	0.17	-0.07
Non-traditional			1.00	0.76*	0.75*	0.29	0.20	0.05	0.12
K									
Traditional				1.00	0.69*	0.48*	0.74*	0.70*	0.48*
Non-traditional				1.00	0.64*	0.39	0.42	0.28	0.26
P									
Traditional					1.00	0.17	0.54*	0.43	0.41
Non-traditional					1.00	0.40	0.16	0.41	0.20
Ca									
Traditional						1.00	0.48*	0.25	-0.14
Non-traditional						1.00	0.48*	0.23	0.51
Mg									
Traditional							1.00	0.25	0.42
Non-traditional							1.00	0.11	0.87*
Zn									
Traditional								1.00	0.49*
Non-traditional								1.00	0.30

Table 5. Simple linear correlation coefficients between the composition variables studied considering the values expressed in dry matter basis for the two groups sausages (n=20 for each group; *: $P<0.05$)

two groups of sausages were as follows: fat content was inversely correlated with protein and ash contents and with some of the mineral elements studied, namely K and P. As said before, fat has relatively low mineral content, therefore, the more the fat proportion is, the lower the mineral content will be. Conversely, protein was positively correlated with ash and with K, P and Zn; and ash was positively correlated with several minerals but mainly with Na. Furthermore, K was positively correlated with P, and Ca with Mg.

The comparison of coefficients between each group of sausages can be useful for a better understanding of the group-related differences found as described above. The correlation coefficient between protein and fat in non-traditional sausages was lower than that in traditional sausages. This denotes that in some of non-traditional sausages carbohydrate sources or mineral fillers were used. Moreover, lower correlation coefficients were detected between protein and both Zn and Fe (which appeared to be mineral elements mainly provided by meat to sausages) and between K and all P, Ca, Mg, Zn and Fe in non-traditional sausages, with respect to the traditional sausages. This leads to the suggestion that proteinaceous non-meat ingredients were used in a number of non-traditional sausages. The really high coefficient (0.92) between Na and ash observed in non-traditional sausages indicates that Na exerted a heavy significant weight on ash content for sausages of this group, and that ash explained a great part of ash-content variance. Finally, the higher correlations found between Na and P in non-traditional sausages could be partially explained by the use of phosphates as additives, which contain it their molecules not only of P but also of Na (Fonseca et al., 2011).

Results of PC analysis are shown in Table 6 and Figs. 1 to 3. PC analysis was carried out for all FW-, DM- and NFDM-basis data. Table 6 shows the factor loadings and cumulative variance explained by PC1 and PC2. The first PC accounted for a variance from 36 to 43% and the sum of both PCs from 61 to 66%. In absolute values, the elements with higher loading scores (>0.7) for PC1 were Na, K, P and Mg using data on FW basis; K, P, Mg using data on DM basis; and Ca and Mg using data on NFDM. For PC2, Zn and Fe were the elements showing higher loading scores using all FW, DM and NFDM basis data.

	Fresh weight		Dry matter		Nonfat dry matter	
	PC1	PC2	PC1	PC2	PC1	PC2
Sodium	0.792	-0.187	-0.480	0.654	-0.601	0.472
Potassium	0.845	0.263	-0.867	-0.221	-0.681	-0.382
Phosphorus	0.805	-0.031	-0.791	0.246	-0.551	0.442
Calcium	0.679	-0.257	-0.598	0.419	-0.818	-0.092
Magnesium	0.750	-0.036	-0.721	0.098	-0.827	-0.294
Zinc	0.050	0.834	-0.435	-0.727	0.128	-0.723
Iron	0.051	0.835	-0.445	-0.722	-0.052	-0.780
Cumulative variance	43.37	64.88	41.04	66.38	35.75	61.30

Table 6. Loading scores and cumulative variance explained for the two principal components (PC1 and PC2)

Figs. 1, 2 and 3, obtained from FW, DM and NFDM results, respectively, show that samples from each group of sausage are located in two defined sets of results. Among them, the neatest separation between traditional and non-traditional samples can be observed in the plot of NFDM results (Fig. 3), where moisture and fat effect in mineral concentrations were avoided by mathematically removing those components. Thus, following the NFDM results, PC1 seemed to relate to the use of ingredients comparatively rich (with respect to meat) in

Ca and Mg such as textured soy, mechanically recovered meat and PC2 with the proportion of lean meat, which is relatively rich in Zn and Fe.

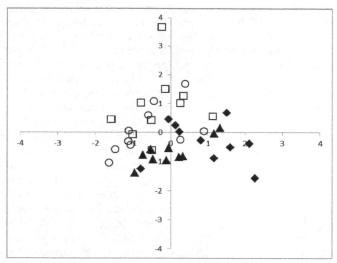

Fig. 1. Principal component (PC) score plot (PC 1 and PC 2, horizontal and vertical axis, respectively), considering mineral composition on fresh weight basis, and showing samples according to sausage types: (○) Traditional, from butcheries; (□) traditional, from rural street markets; (♦) non-traditional, from supermarkets; (▲) non-traditional, from wholesale city markets

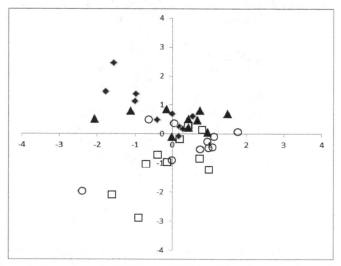

Fig. 2. Principal component score plot (PC 1 and PC 2, horizontal and vertical axis, respectively) considering mineral composition on dry matter basis, and showing samples according to sausage types: (○) Traditional, from butcheries; (□) traditional, from rural street markets; (♦) non-traditional, from supermarkets; (▲) non-traditional, from wholesale city markets

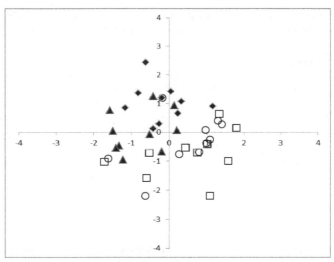

Fig. 3. Principal component score plot (PC 1 and PC 2, horizontal and vertical axis)
considering mineral composition on nonfat dry matter basis, and showing samples
according to sausage types: (○) Traditional, from butcheries; (□) traditional, from rural street
markets; (♦) non-traditional, from supermarkets; (▲) non-traditional, from wholesale city
markets

4. Conclusion

The nutritional composition (the proximate composition and mineral content, in the case of
this study) of mass-produced commercial raw sausages (non-traditional) undertakes
changes by reason of variations in meat and non-meat ingredients and additives used with
respect to the traditional production manners. Differences in ash content and in the
concentrations of most of the mineral elements studied were found between traditional and
non-traditional sausages. Among those, Na and Zn had the highest statistical significance.
Those changes in mineral composition can represent a disadvantage regarding nutritive
value of sausages associated with lower Zn and Fe contents and higher Na and P higher in
non-traditional than in mass-produced commercial sausages. Differences between
traditional and non-traditional sausages are presumably due to the use of textured soy flour,
mechanically recovered meat, and higher amounts of common salt in non-traditional
sausages, and, conversely, higher proportion of lean meat in traditional sausages.
Determination of mineral content of raw-semidry or fresh sausages could be useful for
sausage characterization purposes. Moreover, mineral content analysis performed by
inductively coupled plasma atomic emission spectroscopy together with multivariate
statistical analysis (chemometrics) seems to be a useful tool for discrimination purposes
regarding the control of sausage authenticity (traditional vs non-traditional sausages) – with
the term traditional being used for sausages made with lean and fat, salt and spices and
eventually sodium nitrite, if appropriate. It seems more appreciate to conduct the
multivariate analysis with the data expressed on nonfat dry matter.
In order to complete this study, further research would be needed to a) determine the
mineral composition of raw-semidry or fresh sausages from other regions and countries and

b) to determine mineral elements not included in this study such as manganese, selenium, sulphur, iodine, chlorine and heavy metals (arsenic, lead, mercury, and cadmium).

5. Acknowledgment

This study was supported by a pre-doctoral grant (Programa de Mejoramiento del Profesorado, PROMEP) from the Subsecretaría de Educación Superior, Secretaría de Educación Pública (SEP). México, 2008, Folio No. UAEH-112. Moreover, the authors would like to thanks the University of Leon (Spain) and the University Autónoma del Estado de Hidalgo (Mexico) for funding the analysis on which this study is based. We also thank Mrs. Tania Trinidad Enrique for her technical laboratory support and the Instrumental Techniques Laboratory of the University of León where the mineral analysis was carried out.

6. References

Adams, M.R. (Ed.) (1986). *Progress in Industrial Microbiology, Vol.23 : Microorganisms in the Production of Food*, Elsevier, ISBN 0-444-42661-2, Amsterdam, Netherlands

Aguirrezábal, M., Mateo, J., Domínguez, C., & Zumalacárregui, J.M. (1998). Spanish paprika and garlic as sources of compounds of technological interest for the production of dry fermented sausages. *Sciences des Aliments*, Vol.18, No.4, pp. 409-414, ISSN 1684-5615

AOAC (1999). Official Methods 920.153 ash, 950.46 moisture, 981.10 crude protein, and 991.36 fat (crude) contents in meat and meat products. In: *Official methods of nalysis of the AOAC International* P.A. Cunniff (Ed.), Vol. II, 16th ed., 1–15, AOAC International, ISBN 0-935-58454-4, Gaithersburg, MD, United States of America

Calvo, M.S., & Park, Y.K. (1996). Changing phosphorus content of the U.S. diet: potential for adverse effects on bone. *The Journal of Nutrition*, Vol.126, No.4 Sup., pp. 1168S-1180S, ISSN 0022-3166

Codex Alimentarius (2009). *General Standard for Food Additives.* Food Agriculture Organization and World Health Organization, Retrieved from http://www.codexalimentarius.net/gsfaonline/foods/index.html?id=127

EC (1995). *Council Directive 95/2/EC, February 20th 1995, on Food Additives Other than Colours and Sweeteners.* Official Journal of the European Union, No L 61, pp. 1–56, Amended by Directive 96/85/EC, Directive 98/72/EC, and corrected by C1 Corrigendum, No. L 248, pp. 60, European Union, Luxembourg

Feiner G. (2006). *Meat Products Handbook: Practical Science and Technology.* Woodhead Publishing Ltd., ISBN 1-845-69050-8, Cambridge, United Kingdom

Fonseca, B., Kuri, V., Zumalacárregui, J.M., Fernández-Diez, A., Salvá, B.K., Caro, I., Osorio, M.T., Mateo, J. (2011). Effect of the use of a commercial phosphate mixture on selected quality characteristics of 2 Spanish-style dry-ripened sausages. *Journal of Food Science*, Vol.76, No.5, pp. S300-S305, ISSN 0022-1147

Grembecka, M. Malinowska, E., & Szefer, P. (2007). Differentiation of market coffee and its infusions in view of their mineral composition. *Science of the Total Environment*, Vol.383, No.1-3, pp. 59–69, ISSN 0048-9697

Hazell, T. (1982). Iron and zinc compounds in the muscle meats of beef, lamb, pork and chicken. *Journal of the Science and Food Agriculture*, Vol.33, No.10, pp. 1049–1056, ISSN 1097-0010

Hoogenkamp, H.W. (2005). *Soy Protein and Formulated Meat Products*. CABI Publishing, ISBN 0-851-99864-X, Oxfordshire, United Kingdom

Ibáñez, E., & Cifuentes, A. (2001). New analytical techniques in food science. *Critical Reviews in Food Science and Nutrition*, Vol.41, No.6, pp. 413–450, ISSN 1040-8398

Jiménez-Colmenero, F., Reig, M., & Toldrá, F. (2006). New Approaches for the Development of Functional Meat Products. In: *Advanced Technologies for Meat Processing*, L.M.L. Nollet & F. Toldrá (Eds.), pp. 275-308, CRC Press-Taylor & Francis Group, ISBN 1-574-44587-1, Boca Raton, FL, United States of America

Jiménez-Colmenero, F., Pintado, T., Cofrades, S., Ruíz-Capillas, C., & Bastida, S. (2010). Production variations of nutritional composition of commercial meat products. *Food Research International*, Vol.43, No.10, pp. 2378-2384, ISSN 0963-9969

Kelly S. D., & Bateman, A. S. (2010). Comparison of mineral concentrations in commercially grown organic and conventional crops – Tomatoes (*Lycopersicon esculentum*) and lettuces (*Lactuca sativa*). *Food Chemistry* Vol.119, No.2, pp. 738–745, ISSN 0308-8146

Kuri, V., Madden, R.H., & Collins, M.A. (1995). Hygienic quality of raw pork and chorizo (raw pork sausage) on retail sale in Mexico city. *Journal of Food Protection*, Vol.59, No.2, pp. 141-145, ISSN 0362-028X

Lombardi-Boccia, G., Lanzi, S., & Aguzzi, A. (2005). Aspects of meat quality: trace elements and B vitamins in raw and cooked meats. *Journal of Food Composition and Analysis*, Vol.18, No.1, pp. 39–46, ISSN 0889-1575

Luykx, D.M.A.M., & van Ruth, S.M. (2008). An overview of analytical methods for determining the geographical origin of food products. *Food Chemistry*, Vol.107, No.2, pp. 897-911, ISSN 0308-8146

Mateo, J., Caro, I., Figueira, A.C., Ramos, D., & Zumalacárregui, J.M. (2008). Meat Processing in Ibero-American Countries: a Historical View. In: *Traditional food production and rural sustainable. A European challenge*, V.T. Noronha, P. Nijkamp, & J.L. Rastoin (Eds.), 121-134, Ashgate Publishing Ltd., ISBN 0-754-67462-2, Surrey, Reino Unido

McAfee, A.J., McSorley, E.M., Cuskelly, G.J., Moss, B.W., Wallace, J.M.W., Bonham, M.P., & Fearon, A.M. (2010). Red meat consumption: An overview of the risks and benefits. *Meat Science*, Vol.84, No.1, pp. 1-13, ISSN 0309-1740

Newman, P.B. (1981). The separation of meat from bone – a review of the mechanics and the problems. *Meat Science*, Vol.5, No.3, pp. 171-200, ISSN 0309-1740

Quitaes, K.D., Amaya-Farfan, J., Tomazini, F.M., Morgano, M.A., & Mantovani, D.M.B. (2004). Mineral migration from stainless steel, cast iron and soapstone pans (steatite) onto food simulants. *Ciência e Tecnologia de Alimentos*, Vol.24, No.3, pp. 397-402, ISSN 0101-2061

Reyes García, M., Gómez-Sánchez Prieto, I., Espinoza Barrientos, C., Bravo Rebatta, F., & Ganoza Morón, L. (2009). *Tablas Peruanas de Composición de Alimentos*, Instituto Nacional de Salud, Retrieved from http://www.ins.gob.pe/insvirtual/images/otrpubs/pdf/Tabla%20de%20Aliment os.pdf

Rocha McGuire, A.E. (2008). El uso de soya texturizada como extensor de productos cárnicos. Carnetec, artículos técnicos, Retrieved from http://www.carnetec.com

Ruusunen, M., & Poulanne, E. (2005). Reducing sodium intake from meat products. *Meat Science*, Vol.70, No.3, pp. 531-541, ISSN 0309-1740

StatSoft Inc. (2001). *STATISTICA* (data analysis software system), version 6. www.statsoft.com

Sun, S., Guo, B., Wei, Y., & Fan, M. (2011). Multi-element analysis for determining the geographical origin of mutton from different regions of China. *Food Chemistry*, Vol.124, No.3, pp. 1151-1156, ISSN 0308-8146

Tyszkiewicz, S., Wawrzyniewicz, M., & Borys, A. (2011). Factors influencing physiological phosphorus content in pork meat. *Acta Agrophysica*, Vol.17, No.2, pp. 387-393, ISSN 1234-4125

USDA (2010). *USDA National Nutrient Database for Standard Reference, release 23*, U.S. Department of Agriculture, Agricultural Research Service, Retrieved from http://www.nal.usda.gov/fnic/foodcomp/search/

USDA-FSIS (2007). *Safe and Suitable Ingredients Used in the Production of Meat and Poultry Products*. Food Safety and Inspection Service Directive 7120.1, amendment 10, attachment 1, January 18th 2007. U. S. Department of Agriculture, Washington D.C., United States of America

11

Analysis of High Solid Content in Biological Samples by Flame Atomic Absorption Spectrometry

Lué-Merú Marcó Parra

Universidad Centro-Occidental Lisandro Alvarado, Decanato de Agronomía,
Dpto. Química y Suelos, Núcleo Tarabana, Cabudare, Edo. Lara
Venezuela

1. Introduction

The analysis of organic samples by flame atomic absorption spectrometry (FAAS) involves the difficulties of the digestion step. This fact was partially overcome by the use of the microwave assisted digestion technique (Skip, 1998). The digestion of the samples has the analytical advantage of an appropriated presentation for the analysis by different techniques. Nevertheless it has some disadvantages as could be analyte losses, risk of contamination, higher cost, longer analysis time and obligatory dilution of analyte, in some cases to undetectable levels (Bugallo *et al*, 2007; Marcó and Hernández, 2004). It is desirable the development of methods those avoid the sample digestion step. The technique of flow injection atomic absorption spectrometry is well suited for this purpose (Trojanowicz, 2000), in the determination of elemental levels in slurry samples and high solid content samples. It allows to the analysis of solid samples in a simple manner. (Arroyo *et al*, 2002; Koleva and Ivanova, 2008).

The analysis of slurry samples gives the advantage of a liquid while allowing the introduction of a solid. The slurry method is reported for the analysis of prior dried samples (Januzzi *et al*, 1997; Da Silva *et al*, 2006; Mokgalaka *et al*, 2008), precalcined (Andrade *et al*, 2008). The analysis by FAAS of solid samples or high solid content samples, as could be crude clinical samples is not frequently found in the literature. It is reported the use of slurries from crude tissues combining the FAAS technique with nebulization with a Babington type nebulizer for the introduction of high solids content samples (Mohamed and Fry, 1981; Fry and Denton, 1977).

Brandao *et al*, 2011 reported a simple and fast procedure for the sequential multi-element determination of Ca and Mg in dairy products employing slurry sampling and high resolution-continuum source flame atomic absorption spectrometry (HR-CS FAAS). The main experimental conditions optimized were 2.0 mol L^{-1} hydrochloric acid, sonication time of 20 min and sample mass of 1.0 g for a slurry volume of 25 mL. The elements were determined using aqueous standards for the external calibration with limits of quantification of 0.038 and 0.016 mg g^{-1}, respectively. The precision expressed as relative standard deviation varied from 2.7 to 2.9% for a yogurt sample containing Ca and Mg concentrations of 1.40 and 0.13 mg g^{-1}, respectively.

Erik G.P. da Silva *et al*, 2008 evaluated the slurry sampling flame atomic absorption spectrometric method   for the determination of copper, manganese and iron in oysters (*Crassostrea rhizophora*), clams (*Anomalocardia brasiliana*) and mussels (*Mytella guiyanensis; Perna perna*). They optimized the variables nature and concentration of the acid solution for slurry preparation, sonication time and sample mass. The optimized conditions were 80 mg of sample grounded in a cryogenic mill, dilution using 1.0 mol L^{-1} nitric /hydrochloric acid solution, sonication time of 30 min and a slurry volume of 10 mL. The calibration curves were prepared matching the acid concentration. This method allowed the determination of copper, manganese and iron by FAAS, with detection limits of 0.17, 0.09 and 0.46 µg g^{-1}, respectively. The precision, expressed as relative standard deviation (RSD), was 3.0%, 2.9% and 3.8% (n= 10), for concentrations of copper, manganese and iron of 17, 22 and 719 µgg^{-1}, respectively. The accuracy of the method was confirmed by analysis of the certified oyster tissue (NIST 1566b).   The results showed no significant differences using the proposed method respect to those obtained after complete digestion and determination by inductively coupled plasma-optical emission spectroscopy (ICP-OES).

The   ultrasonic extraction is an interesting aid to the slurry sampling but its use is not widely spread as the digestion procedures (Taylor *et al*, 2002; Taylor *et al*, 2006).   An ultrasound-assisted solid–liquid extraction procedure by using diluted mixed acid solution was developed by Manutsewee, *et al*, 2007 for determination of cadmium, copper and zinc in fish and mussel samples. They evaluated the effects of several parameters such as nitric acid concentration, hydrochloric acid concentration, hydrogen peroxide concentration, leaching solution volume, temperature and sonication time. After the optimization of these parameters the elements cadmium and copper were determined by graphite furnace atomic absorption spectrometry, and zinc was determined by flame atomic absorption spectrometry.   The results were compared to those obtained by microwave-assisted digestion. The recoveries (%) of metal amount obtained by leaching technique to the amount obtained by digestion technique for cadmium, copper and zinc ranged from 92% to 114% for fish and from 88% to 103% for mussel samples. The accuracy of the developed method was verified with the dogfish muscle certified reference material (DORM-2). The Recoveries were in the order of 80.9 ± 0.3 and 87.2 ± 0.6%.

Bugallo *et al*, 2007 compared the method of slurry sampling to the microwave assisted digestion for the determination of calcium, copper, iron, magnesium and zinc in fish tissue samples by flame atomic absorption spectrometry. They found that in comparison to microwave-assisted digestion, the analysis of slurries is simple, requires short time and overcome the difficulty of the total sample dissolution before analysis. It is necessary the addition of acid for some analites as iron, to enhance the recovery. For Ca and Cu the quantification must be performed using standard addition. Both methods were accurate and the standard deviations obtained using slurry sampling method and microwave-assisted digestion were not significantly different and the mean relative standard deviation of the slurry sampling method for different concentration levels was below 12%.

In this chapter will be discussed the analysis by FAAS of two kinds of biological samples: brain and onion bulb tissues. The preparation procedures were simplified in order to perform the analysis of the crude samples with the aid of ultrasonic acid extraction for brain slurries and crude onion leachates. The introduction using a Flow Injection System avoids transport effects. The results by FAAS and flow injection atomic absorption spectrometry (FIAAS) were evaluated using the independent technique of TXRF.

1.1 Determination of Zn and Cu in crude human brain slurry samples by flow injection flame atomic absorption spectrometry

The determination of metals in brain samples is required for the study of brain physiology, biochemistry and neurochemistry, or clinical purposes to find correlations between metal levels and some pathologies (Schizophrenia, Wilson disease, Alzheimer, etc.) (Andrási et al, 1995; Religa et al, 2006). In connection to neurochemistry, it is necessary to determine the elements Zn and Cu for the evaluation of neurotransmission processes. (Horning et al, 2000).

The brain matrix is mainly formed by high molecular weight carboxilic acids or greases of long chains, a complex organic matrix. The digestion of such a matrix prior to the analysis is reported in the literature, for different techniques (Andrade et al, 2008; Taylor et al, 2002; Taylor et al, 2006; Lech and Lachowicz, 2009). The main objective of this work was the determination of Cu and Zn in crude brain slurry samples by the method of flow injection atomic absorption spectrometry and the development of precise, accurate, efficient and cheap method of determination of metals in human brain samples. The results were compared to those obtained after microwave aided digestion of the samples. The independent technique of total reflection X-ray fluorescence (TXRF) was used for accuracy evaluation.

Crude brain dissected samples were homogenized with a high speed homogenator to obtain slurries. These slurry samples were properly diluted in 5% V/V HNO_3 to aid the analyte extraction to the aqueous phase, to carry out the determination of copper and zinc by flame atomic absorption spectrometry, following a procedure reported by Marco et al, 2003.

1.2 Determination of calcium, potassium, manganese, iron, copper and zinc levels in representative samples of two onion cultivars by flame atomic absorption spectrometry using ultrasonic assisted leaching

The onion is one of the most important cultivars in the world. The determination of major, minor and trace element levels is an important tool for the enhancement of production efficiency in the field of agriculture, provenance, and contamination risk evaluation (Ariyama et al, 2007; Abdullahi et al, 2008; Abdullahi et al, 2009). There is a necessity for new analysis methods and simple sample preparation procedures. The chemical characterization of the cultivar samples, becomes important due to the fact that chemical composition is closed related to the quality of the products. (Akan et al, 2010; Hashmi et al, 2007). Alvarez et al, 2003, reported a preparation procedure for the elemental characterization, involving the acid extraction of the analytes from crude samples by means of an ultrasonic bath, avoiding the required digestion of samples in vegetable tissue analysis. The technique of total reflection X-ray fluorescence (TXRF) was successfully applied for the simultaneous determination of the elements Ca, K, Mn, Fe, Cu and Zn. The procedure was compared with the wet ash and dry ash procedures for all the elements using multivariate analysis and the Scheffe test. The technique of flame atomic absorption spectrometry (FAAS) was employed for comparison purposes and accuracy evaluation of the proposed analysis method. A good agreement between the two techniques was found when using the dry ash and ultrasound leaching procedures. The levels of each element found for representative samples of two onion cultivars (Yellow Granex PRR 502 and 438 Granex) were also compared by the same method.

In this work a sample preparation procedure for onion bulb elemental characterization by FAAS is proposed, involving the acid extraction of the analytes by means of an ultrasonic bath. The procedure was compared with the wet ash and dry ash procedures for all the elements using multivariate analysis and the Scheffé Test. The accuracy was also verified with the TXRF technique. The onion samples were grounded and homogenized with deionized water (1:1) in a domestic homogenator. In a second step, nitric acid was added at different concentrations for the optimization of acid levels. The samples were placed for 30 minutes in the ultrasonic bath. It was found an optimal concentration of 5% V/V of nitric acid and 10% of wet sample mass. Then samples were filtrated. The filtrates were also analyzed by flow injection flame atomic absorption spectrometry (FI-AAS) for all the elements, using deionized water as carrier, an optimized injection volume of 150 μL and an optimized flow of 3.5 ml/min. The results obtained were compared to those obtained using the methods of wet ash and dry ash sample digestion and flame atomic absorption analysis after the humidity correction (dry base). For two kind of cultivars (Yellow Granex PRR 502 and 438 Granex). it was found that the dry ash method was statistical equal to the method of ultrasonic extraction-FIAAS. No significant differences were found between the results obtained by FAAS and TXRF. The precision was always below 5% of relative standard deviation in all the cases. It was concluded that the proposed method is the most reliable in the basis of its simplicity, shorter analysis time and minor use of reagents and glassware.

2. Experimental

2.1 Analysis of brain samples
2.1.1 Samples
Brain samples from healthy, male individuals, who suffered accidental and/or instantaneous death were taken at Morgue of the Central Hospital of Barquisimeto, Edo. Lara, Venezuela. Brains were dissected, not more than 24 hours after death, and kept at -50 °C until sample preparation.

2.1.2 Sample treatment
The brain sections, such as cerebellum, hypothalamus, frontal cortex and encephalic trunque, were weighed and homogenised with deionized water with a high speed homogeneiser at 23000 rpm (Ultra-Turrax P25 Janke and Kumkel, IKA registered mark-LABORTECHNIK). Homogenates with a 50-60% w/V (wet weight) of brain tissue were obtained and kept at -20 °C until analysis. Some samples were lyophilized after homogenization at –50 °C in a digital LABCONCO lyophiliser, LYPH-LOCK.
The digestion of the homogenates and lyophilized samples for comparison purposes was carried out in a Domestic microwave oven using closed teflon vessels, in two steps: 15 minutes at medium power and 10 minutes at maximum power. An ultrasonic bath Cole Palmer was employed for slurry treatment and for homogenization. The whole sample treatment is detailed by Marcó and Hernández, 2004.

2.1.3 Slurries
Slurries were prepared taking aliquots of the 50-60% w/V crude brain homogenates with volumetric pipettes and transferring to calibrated flasks, following strictly the next procedure:
1.   An appropriate aliquot of homogenate is taken with the glass volumetric pipette, depending on the desired concentration of the slurry .

2. The homogenate portion is then transferred with the pipette to the volumetric flask. In this step, the pipette is rinsed with deionised water helping the remaining homogenate in the inner wall of the pipette falling to the volumetric flask.
3. Deionized water is added to the sample into the flask to fill approximately half of its volumetric capacity and flask was slightly agitated for few seconds to form an homogeneous slurry.
4. In the case of acid addition, the appropriate amount of the nitric acid is added to the flask, depending of the desired acid concentration.
5. The calibrated flask with the slurry is finally filled to the labelled volume, with deionised water.
6. The sample slurries in the flask are treated for 10 minutes in ultrasonic bath at 25 °C.

It is important to remark that this procedure must be followed in all the cases, in order to get stable, and homogeneous slurry. When the step order is not followed, it is frequently to observe the instantaneous denaturalization of the homogenate and the formation of particles of non desirable size.

Additional reagents for assurance of the slurry stability, such as Viscalex, Triton among others were not necessary. Problems with foaming were not found.

2.1.4 Digested samples

In a similar way as described in steps 1 and 2 of the slurry preparation procedure, a sample aliquot of 5 ml of homogenate was transferred to the teflon vessels, instead of the volumetric flasks. Then 5 ml of ultrapure concentrated nitric acid and drops of hydrogen peroxide were added to the teflon vessels. Vessels were closed for digestion in the microwave oven.

About 0.35 g of Lyophilised brain samples were weighed and digested in closed teflon vessels with 5 ml. of concentrated nitric acid and drops of hydrogen peroxide using the same microwave assisted digestion followed with the brain homogenates.

2.1.5 Standards and reagents

Aqueous calibration standards of Zinc and Copper were prepared by serial dilution of the stock solution (1000 μg mL^{-1}), Titrisol, Merk. Zinc determination was carried out using always aqueous calibration curves. Copper determination was carried out using the aqueous calibration curve and also the standard addition method. Suprapur, 65% v/v HNO_3 (MERCK, Germany) and 30% v/v H_2O_2 (Riedel de Haen, Germany) were employed for leaching and digestion purposes. Nitric acid (Riedel de Haen, Germany) was used for cleaning quartz reflectors for the TXRF analysis. Distilled, deionized water (16 MVcm) was employed for rinsing and dilution purposes and also as FIAAS carrier.

2.1.6 FAAS analysis

Measurements were performed in a 2100 Perkin Elmer flame atomic absorption spectrometer. FIAAS set up without peristaltic pump, using the nebulizer aspiration flow for sample and carrier propulsion as shown in figure 1 was used to avoid the clogging of the system. This system was made of Teflon pieces from a chromatographic column kit. In one position of the valves the loop is charged while the carrier is passing direct to the nebulizer. In the other position of the valves, the sample in the loop is aspirated and the carrier immediately passes through the loop to the nebulizer. As the system has the possibility of two loops, when one of the loops is being charged the carrier is passing through the other loop and viceversa. The carrier container is an extraction funnel fixed 30 cm above the entry of nebulizer. The volume of the sample loop was approximately 100 microliters. The length of the tube from injection port to nebuliser was 35 cm and the internal diameter 0.8 mm.

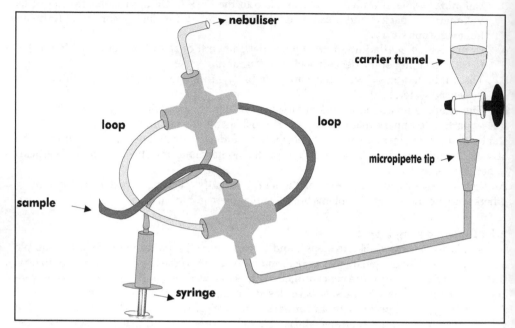

Fig. 1. Sample introduction system

TXRF measurements

A CANBERRA Energy Dispersive X-Ray Spectrometer, with Si-Li detector and set up for TXRF, Excitation with Mo tube, 17.4 Kev line was used for X-Ray analysis. TXRF set up with TLAP crystal as monochromator. Details of the TXRF measurements are given in the works of Marcó *et al*, 1999 and Marcó and Hernández, 2004.

2.2 Analysis of onion samples. (According to Álvarez *et al*, 2003).
2.2.1 Reagents and standards
Titrisol, 1000 standard solutions (MERCK, Germany) were employed for preparation of calibration curves. Suprapur, 65% v/v HNO_3 (MERCK, Germany) and 30% v/v H_2O_2 (Riedel de Haen, Germany) were employed for leaching and digestion purposes. Nitric acid (Riedel de Haen, Germany) was used for cleaning quartz reflectors for the TXRF analysis. Distilled, deionized water (16 MVcm) was employed for rinsing and dilution purposes and also as FIAAS carrier.

The standards for FAAS and FIAAS analysis were prepared by serial dilution of titrisol standards, according to the linear range of each of the analyzed elements, as reported by the manufacturer.

For the TXRF analysis aqueous, multielement (K, Ca, V, Mn, Cu, Se and Sr) standards were prepared by mixing and dilution of the corresponding stock solutions with distilled, deionized water to yield final concentrations of 5, 10 and 20 mgL^{-1}. The element Co was used as internal standard and added to standards and samples. The further quantification is well explained by Alvarez *et al*, 2003.

2.2.2 Sampling procedure

Two kinds of onion cultivars were collected, Yellow Granex PRR of Sumblex and Texas Grano 438, Asgrow, at the main market of Barquisimeto following the next procedure: ten bags of 60 Kg from each cultivar were random selected. Then from each bag were taken 10 onions. A total amount of 100 onions for each cultivar were collected.

Sample preparation

The onion samples were grounded and homogenized with deionized water (1:1) in a domestic homogenizator in a previous step (30 random selected onions for each preparation procedure from each cultivar) . An amount of 5 g of wet weight or 10 g of the homogenate was digested by a wet ash procedure and prepared by ultrasonic leaching, as explained bellow. In similar way an amount of dry onion corresponding to 5 g of wet weight (calculated on the basis of the dry matter content) was used for the dry ash procedure. The humidity correction and the determination of dry masses were performed separately by drying in oven at a fixed temperature of 60 ºC. The values of dry matter percentage were 8% for Yellow Granex and 9% for 438 Granex cultivar. The results are expressed in dry basis. In all the cases four independent replicas were prepared.

2.2.3 Ultrasonic extraction

An ultrasonic bath, Cole Palmer (USA) with temperature control was used. Temperature was fixed to 70 ºC. Time of sonication was 30 min. Ten grams of the homogenate (5 g of wet sample) were placed in flasks and mixed with different nitric acid concentrations (0, 5, 10 and 15% v/v) for the optimization of acid levels. The samples were placed for 30 min in the ultrasonic bath with a fixed temperature of 70 ºC. Then samples were filtrated with Whatman filters by gravity and the supernatants were quantitatively transferred to 50 mL volumetric flasks.

2.2.4 Wet ash

The wet digestion was performed weighing 5 g of the wet sample and adding 15mL of concentrated HNO_3 and drops of H_2O_2. The digestion was performed in a hot plate. After the digestion procedure the sample were aphorized to a final volume of 50 mL.

2.2.5 Dry ash

The dry digestion was performed weighing 0.4 g of dry sample (approx. 5 g of wet sample) and calcining at 700 ºC for 2 h. Then, the ashes were dissolved with nitric acid and samples were quantitatively transferred to 50 mL volumetric flasks.

2.2.6 FAAS analysis

The samples were analyzed in a Perkin Elmer (USA) 3110 Atomic Absorption Spectrometer under conditions suggested by the manufacturer.

The FIAAS manifold was designed for low dispersion and two channels (carrier and sample), using a An ISMATEC peristaltic pump IPC model for sample and carrier introduction controlled by a Temporizer GrabLab model 900 and a control valve Cole-Palmer, model 625E Bunker CT. See Figure 2.

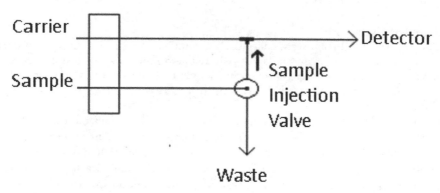

Fig. 2. FIAAS manifold

2.2.7 TXRF analysis

The TXRF analysis was carried out in a Canberra XRF Spectrometer with a modified TXRF module designed at the Atominstitut der Osterreichschen Universitaeten, Vienna. The excitation was performed with the Ka (17.4 keV) line of a molybdenum anode X-ray tube, operated at 40 kV and 20 mA, as detailed by Alvarez *et al*, 2003. A carbon -molybdenum multilayer crystal was used for monochromation of the incident beam and a Si(Li) detector (Resolution 180 eV at Mn line, 5.8 keV) was used for the detection of the fluorescence signal. The spectra were collected in a PC based multichannel analyzer (Canberra S100), with live collection time of 200 s. The spectral data analysis was conducted with the AXIL fitting program and QXAS package supplied by the International Atomic Energy Agency.

3. Results

3.1 Analysis of brain samples

The feasibility of the crude brain slurry direct introduction was tested at a first stage using different concentration (w/V) of brain tissue in water with and without acid. The introduction of the samples was performed by the use of a simple flow injection system described in the figure 1 due to frequent obstruction of the valve of the FIAAS manifold with peristaltic pump. Optimal slurry concentration was in the range of 2.3% w/V to 12.5% w/V for zinc determination while for copper the slurry concentration must be higher than 20% w/V for an appropriate detection and less than 24% w/V to avoid matrix interferences. The results obtained by the slurry method were compared to those obtained after sample digestion and also to the independent technique total reflection X-ray fluorescence. A good agreement between results confirmed the accuracy of the proposed sample preparation procedure. The mean precision for the zinc and copper determination was less than 5% for most of the samples.

3.1.1 Optimization of experimental conditions

Experimental conditions for the FAAS method were fixed following the routine recommended by the spectrometer manual. The parameters as slit, lamp current and gas flow were automatically selected. For the FIAAS method, variations of the gas flow were performed. Gas flow was changed from 5 L/min to 8 L/min. A slight increment in sensitivity was observed when using 5L/min of gas flow, nevertheless the obstruction of the capillary tube was frequently observed under this condition due to the high solid content of the brain slurry samples. High gas flow was necessary to avoid the capillary tube obstruction.

Sample volume: three sample loop sizes were tested: 100, 150 and 200 microliters. The 100 microliters loop was selected as optimal volume, since no significant changes are observed in peak height respect to higher volumes, dispersion is low and risk of memory effect are minimized. The dispersion was 1.3 for Cu and Zn in the optimized set up.

slurry concentration and stability: Slurries with concentration w/V of 2.3%, 12.5% and 23 % were tested. Simultaneously, the effect of nitric acid 5% w/V was evaluated. The slurry 2.3% corresponded to the minimal concentration that allowed to the measurement of Zinc signal in the lower value of the working calibration range. The 12.5% is a concentration in the range recommended by Mohamed and Fry, 1981 for homogeneized tissues and 23.5% corresponds to the critical matrix due to the high solid content. The stability of the absorbance signal does not depend on the acid at concentration levels lower than 12.5%, evidencing that the analyte is mostly in the aqueous phase. It was deduced that for Zn a slurry concentration less than 12.5 % w/V and 5% HNO_3 V/V is adequate for the analysis. Concentrations higher than 13% are over the linear calibration range (0.1-1 μgml^{-1}). In the case of copper, slurry concentration must be higher than 12.5% for an appropriate detection since linear range lies at higher concentration values, between 1-10 μgml^{-1}. The critical higher concentration was 24%, due to matrix effects. See figure 3.

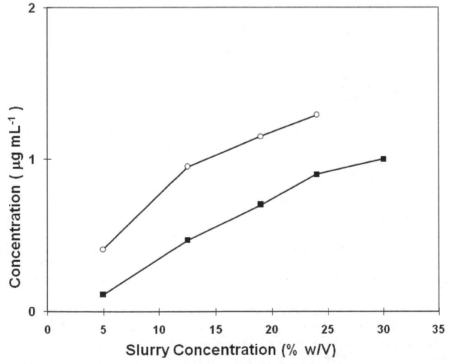

Fig. 3. Effect of the Slurry Concentration on Zn (■) and Cu(○) Signals.

Effect of nitric acid: the effect of nitric acid was evaluated at concentrations of 5% w/V, 10% w/V and 20% w/V. An increment of the absorbance signal is observed when the acid is added. The quality of the slurries in terms of particle size is improved. The clogging of the capillary was observed for samples without acid. Acid concentrations above 5% v/V do not affect significantly the absorbance signal. The acid blank was tested and its influence discarded on the measurement.

3.1.2 Determination of Zn

The determination of Zn can be carry out by slurry sampling at slurry concentration levels (%w/V) between 2.3 and 13% if nitric acid is added. Nevertheless, for concentration levels near the lower limit (2.3%) a high relative standard deviation is obtained. See table 1.

SAMPLE	CONCENTRATION	RSD (n=5)
Ce-2	8.8 (12%)	3%
	8.5 (4.8%)	4%
Te-2	7.7 (12%)	2%
Te-1	5.6 (4.8%)	3%
Cf-1	9.6 (4.8%)	2%
Ce-1	11.0 (4.8 %)	2%
Hp-1	10.9 (4.8%)	3%
Te-3	14.2 (2.3%)	9%
Hp-3	10.0 (2.7%)	10%
Cf-3	5.6 (4.7%)	5%

Table 1. Concentration ($\mu g \ g^{-1}$) of Zn in Crude Brain Slurry Samples. The values in parenthesis indicate the slurry concentration. RDS: Relative Standard Deviation. Ce: Cerebellum; Te: Encephalic trunque; Hp: Hypotalamus; Cf: Frontal Cortex.

A good concordance between results obtained by slurry sampling and those obtained after microwave digestion of crude brain samples was observed, as shown in table 2. If the results for crude samples are compared to the corresponding liophylized and digested samples, the analyte concentration is in concordance to the dry weight correction (about 20% in brain samples (Andrási *et al*, 1995). The test of the correlation factor F at the 0.05 level of confidence was applied (F = 120.7 p= 0.002 and critical F 10.1) demonstrating the agreement between results. See table 2.

The precision ranges between 2% of relative standard deviation (RSD) and 10%. This parameter is independent of the sample preparation procedure (slurry or digested) or the brain section. The precision depends on the analyte concentration. For slurry concentration above 4.8% w/V the value is under 5% of RSD. The precision when comparing independent replicates was similar.

The accuracy was evaluated by comparison to the technique of TXRF for the slurry and digested samples (see table 3). The t-Student at 95% of confidence level was applied (t= 0 and p=1).

Sample	Crude	Digested	Lyophilised and Digested
CE-2	8.8 (3%)	9.0 (2%)	
Et-1	5.6 (3%)	5.1 (4%)	5x5.4
			27 (3%)
CE-1	11.0 (2%)	10.5 (5%)	5x13.2
			66 (2%)
Hp-1	10.9 (3%)	9.9 (2%)	
Et-3	14.2 (9%)	12.9 (3%)	

Table 2. Comparative Results of Zn Concentration (μgg^{-1}) in Crude Brain Slurries and Digested Brain Samples. The values in parenthesis correspond to relative standard deviation.

Sample	TXRF	FIAAS
Cf1	2.3 +/-0.2	
	**2.1 +/-0.2	2.3 *(3%)
Ce1	2.3 +/-0.2	2.6 (2%)
	**2.4 +/-0.2	**2.5 (5%)
Cf2	2.2 +/-0.3	2.1 (1%)
		**2.2 (2%)
Ce2	2.2+/-0.3	2,0 (2%)

Table 3. Comparative Results of Zn Concentration (μgg^{-1}) Obtained by TXRF of Slurries, TXRF of Digested Samples , FIAAS of Slurries and FAAS of Digested Samples. The values in parenthesis indicate the relative standard deviation. ** The value corresponds to digested sample. Cf: Frontal Cortex; Ce: Cerebellum

3.1.3 Determination of Copper

The slurries for Copper determination should have a concentration higher than 20% w/V, due to the low concentration of the analyte in the samples. Slurries with concentration less than 20% (w/V) have an analyte concentration bellow the lowest point of the working calibration range (1 μg mL^{-1}). Values of slurry concentration higher than 24% (w/V) have the lack of matrix effects, transport effects and capillary tube obstruction among others. The nitric acid should be added to the slurries in order to extract efficiently the analyte to the aqueous phase and to enhance the nebulisation and atomization processes in the spray chamber and into the flame. See table 4.

The copper determination was performed by direct FIAAS analysis and by the standard addition method (see reference 78), due to the high concentration (w/V) of the slurries. As shown in table 4 no significant differences were found between results using standard addition method and the direct method. The correlation coefficient between results was 0.994 the slope 0.99 and intercept 0.017. The t-Student test at 95% confidence level was t=

0.11, p= 0.92. These results demonstrate that even for slurries at 24% w/V no matrix effects are observed. This fact is a consequence of the addition of the nitric acid and the analyte extraction to the aqueous phase.

The precision of the results for the FIAAS-Slurry method was between 3% of RSD and 11% of RSD. Nevertheless precision values higher than 5% RSD were not observed for digested samples. Then the parameter is affected by the matrix.

The accuracy was evaluated by comparison to the technique of TXRF, as shown in table 5. When FIAAS method is compared to TXRF using the t-Student test at 95% of confidence level (t=0; p=1), demonstrating the good agreement between the results obtained by both techniques.

Sample	Cu Concentration μgg^{-1} (%RSD)	
	Standard Addition	Calibration Curve
Ce1	0.93*(10%)	0.92(4%)
Ce3	0.68(6%)	0.71(4%)
Te5	0.52(6%)	0.52(7%)
Cf1	0.78(3%)	0.80(11%)
Ce4	0.98(5%)	0.94(10%)
Ce2	0.94(3%)	0.92(4%)
Cf4	0.72(2%)	0.71(3%)
Cf3	0.72(5%)	0.62(3%)
P1	1.55(3%)	1.50(3%)
P2	1.58(2%)	1.59(1%)

Table 4. Concentration of Cu (μgg^{-1}) in Crude Brain Slurry Samples Determined Using Standard Calibration Curve vs. Standard Addition Method of Determination. The values in parenthesis indicate relative standard deviation.

Concentration of Cu (μgg^{-1})		
Sample	TXRF	FIAAS
Cf1	0.8 +/-0,1	0.77*(3%)
Cf4	0.8 +/-0,1	0.72 (2%)
Ce3	0.9 +/-0,1	0.68 (6%)
Cf2	0.6 +/-0,1	0.69 (9%)
Ce1	1.1 +/-0,1	0.93*(10%)
Te3	0.5 +/-0,1	0.52 (6%)
Ce2	1.0 +/-0,1	0.93 (3%)
Te2	0,6 +/-0,1	0.56 (6%)

Table 5. Copper Levels in Crude Brain Slurries by Standard Addition-FIAAS Method vs. TXRF. The values in parenthesis indicate relative standard deviation.

3.2 Analysis of onion samples
3.2.1 Optimization of experimental conditions

Recovery efficiency in acid medium: It is expected that the partial extraction of the analite to the supernatant occurs when acid is added to a slurry. The atomization efficiency is enhanced, in addition due to the fact that the acid helps to decrease the particle size.

Taking this fact into account, diluted nitric acid was employed at concentration levels of 0, 5, 10 y 15% v/v, in water. The signal increases due to the HNO_3 effect, being the optimal value 5%, with 100% of recovery for the determined elements. A slight suppression of the signal was observed for HNO_3 (10 y 15%) and in consequence lower % recovery for Mn, Zn and Fe. The addition of HNO_3, induces a predigestion of the solid phase, decrement of particle size and almost the total extraction of the analites into the aqueous phase. (See table 6). The elements Ca and K are extracted to the aqueous phase with deionized water.

Element	% of recovery			
	0 % V/V HNO_3	5 % V/V HNO_3	10 % V/V HNO_3	15 % V/V HNO_3
Fe	30	101	103	104
Mn	34	100	98	96
Zn	33	102	80	85
Cu	25	98	94	93
K	100			
Ca	100			

Table 6. Percent of recovery in onion bulb samples after ultrasonic extraction procedure as function of the nitric acid concentration (% V/V).

Optimization of FIAAS parameters: the optimized FIAAS parameters were pump flow rate (3.5 mL/min), suction flow rate set 0.2 units bellow pump flow rate (3.3 mL/min). As it is deduced from table 7 there are not significant differences in the recovery, but the highest rate tested ensures a minor residence time of the sample in the nebuliser and chamber, and in consequence a lesser memory effect. The optimal sample volume was fixed at 300 µl. It was found that sample volumes higher than 350 µl did not allow to a significant enhancement. In this case there is not compensation of the matrix effect by dispersion and the signal was similar to that of the classical FAAS analysis. See Figures 4 and 5.

Element	% of recovery			
	2 mL/min	2.5 mL/min	3 mL/min	3.5 mL/min
Fe	86	89	88	89
Mn	92	93	93	93
Cu	97	98	97	99
Zn	100	101	101	103
Ca	100			
K	100			

Table 7. Percent of recovery for Fe, Mn, Cu, Zn, Ca and K as function of the pump flow rate in the analysis of onion bulb leachates by FIAAS. Nitric acid (5% v/V) and sample loop volume 350 µL.

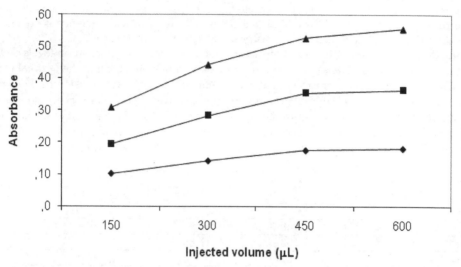

Fig. 4. Relative Absorbance signal as function of the injected sample volume and nitric acid concentration. 5% V/V (♦), 10 % V/V (■) and 15% v/v (▲).

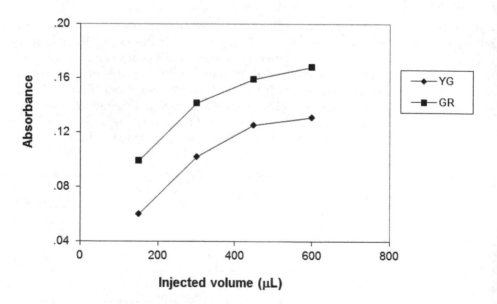

Fig. 5. Relative absorbance as function of the injected volume for leachates of the onion varieties Yellow Granex (YG) and Texas Grano 438 (GR). Fresh sample mass 5 g in a volume of 50 mL. Pump flow rate 3.5 mL/min.

3.2.2 Comparison of the results obtained by the preparation procedures using FAAS and the FIAAS method

It was found that in general, the dry ash procedure and the FIAAS method had not significant differences when the Sheffe test was applied in the comparison of the results for all the elements (See table 8). The results are higher than those of the wet ash procedure and the direct aspiration of the leachates by FAAS, with the exception of Potassium. This trend is the consequence of the stronger matrix effect in these procedures. In the wet ash treatment the simple mineralization could not be total as occurs with the wet ash. In the case of the direct aspiration of the leachates by FAAS, the particle size is higher and the matrix was not totally eliminated. The nebulisation and atomization processes were affected in a different way as the aqueous standards used for calibration. The FIAAS procedure and the subsequent dispersion reduced the matrix effects. The addition of nitric acid allowed to the extraction of the analites to the aqueous phase with the increment in the nebulisation and atomization efficiency as compared to the procedure when the leachate is directly aspirated by FAAS.

	Preparation Procedure							
	Dry Ash		Wet Ash		Ultrasonic Leaching-FAAS		Ultrasonic Leaching-FIAAS	
Element	YG	G	YG	G	YG	G	YG	G
Ca %	0.400*(0.6)	0.310 (1)	0.38(0.6)	0.30 (3)	0.37 (0.9)	0.30 (1)	0.400 (1)	0.34 (2)
K %	1.03 (1)	1.38(0.5)	1.06(0.6)	1.80(0.7)	1.07 (0.2)	1.53 (0.5)	1.00 (2)	1.40 (2)
Fe µg/g	27 (3)	45 (3)	28.2 (4)	41 (5)	21 (3)	39 (3)	20.0 (3)	40 (3)
Zn µg/g	16.7 (3)	19.3 (3)	15.2 (2)	17.7 (3)	11.7 (3)	15.4 (3)	17.3 (4)	19.5 (3)
Mn µg/g	17.1 (2)	40 (4)	12.9 (5)	34 (3)	14.1 (5)	36(4)	15.2 (4)	39 (4)

Table 8. Comparison of elemental concentrations in onion bulb samples using different methods by Flame Atomic Absorption Spectrometry. N=4. In parenthesis Relative Standard Deviation. YG: Yellow granex cultivar and G: Texas Grano 438.

3.2.3 Comparison to the TXRF technique

A good agreement was found between the results obtained by FIAAS and TXRF (See Figure 6). The results obtained by FAAS when the leachates are directly aspired, were significantly lower (p=0.05) with the exception of the element potassium, in the same way as in the wet ash and dry ash procedures, as explained before. It is important to point that the TXRF technique has not the matrix effects as the FAAS technique. The agreement between FIAAS and TXRF demonstrates the effectiveness of the proposed procedure of ultrasonic leaching and FIAAS analysis for the reduction of the matrix effects and its reliability in the analysis of onion bulb samples.

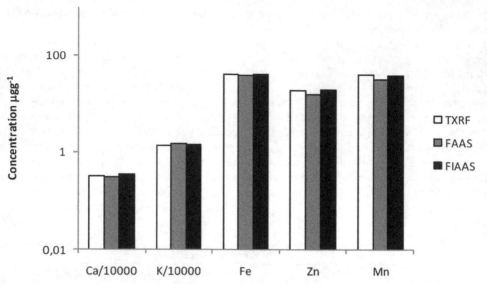

Fig. 6. Comparison of the results by FAAS and FIAAS to the TXRF technique for the leachates of onion bulb samples using ultrasonic aided extraction.

4. Conclusions

A simplified sample preparation procedure was evaluated for the analysis of human brain samples. The feasibility of the preparation of crude brain slurry samples for Atomic absorption spectrometry analysis, in terms of precision and accuracy was demonstrated. These slurry samples must be properly diluted in 5% V/V HNO_3 to aid the analyte extraction to the aqueous phase, to carry out the determination of copper and zinc by flame atomic absorption spectrometry. Optimal slurry concentration was in the range of 2.3% w/V to 12.5% w/V for zinc determination while for copper the slurry concentration must be higher than 20% w/V for an appropriate detection and less than 24% w/V to avoid matrix interferences. A good agreement between results by TXRF and FIAAS confirmed the accuracy of the proposed sample preparation procedure. The mean precision for the zinc and copper determination was less than 5% for most of the samples.

The determination of K, Ca, Mn, Fe, and Zn in fresh onion bulb samples using ultrasonic extraction is a reliable procedure when a FIAAS set up is used. The process must be aided with nitric acid at a concentration level of 5% v/V with five g of homogenized fresh sample in 50 mL. It was demonstrated the substantial reduction of matrix effects if a FIAAS method is applied. The procedure is fast, simple and has lower cost compared to the wet ash and dry ash procedures. The accuracy was verified by comparison to the independent technique TXRF and demonstrated by the good agreement found. The precision for the ultrasonic extraction and FIAAS set up is less than 5% of relative standard deviation for all analyzed elements.

5. Acknowledgments

The author thanks to the CDCHT-UCLA for the financial support of this research with the project AG-040-2000.

6. References

Abdullahi, M. S., Uzairu, A. and Okunola, O. J. Quantitative determination of heavy metal concentrations in onion leaves. *Int. J. Environ. Res.*, 3(2), pp 2009271-274,. ISSN: 1735-6865.

Abdullahi, M. S.*, Uzairu, A. and Okunola, O. J. Determination of some trace metal levels in onion leaves from irrigated farmlands on the bank of River Challawa, Nigeria. *African Journal of Biotechnology*, 7 (10), 2008, pp. 1526-1529. ISSN 1684-5315.

Akan J. C., Abdulrahman F.I., Sodipo O.A., Lange A. G.. Physicochemical parameters in soil and vegetable samples from Gongulon agricultural Site, Maiduguri, Borno State, Nigeria. Journal of American Science, 6(12). 2010, 78-87. ISSN 1545-1003.

Alvarez J. , Marcó L. M., Greaves E.D., Rivas R.. Determination of calcium, potassium, manganese, iron Koper and zinc levels in representative samples of two onion cultivars using TXRF and three preparation procedures. *Spectrochimica Acta B*, 58. 2003, pp 2183-2189. ISSN: 0584-8547.

Andrade Korn Maria das Gracas, Elane Santos da Boa Morte, Daniele Cristina Muniz Batista dos Santos, Jacira Teixeira Castro, Jose´ Tiago Pereira Barbosa, Alete Paixao Teixeira, Andrea Pires Fernandes, Bernhard Welz, Wagna Piler Carvalho dos Santos, Eduardo Batista Guimaraes Nunes dos Santos, and Mauro Korn. Sample preparation for the determination of metals in food samples using Spectroanalytical Methods— A Review. *Applied Spectroscopy Reviews*, 43, 2008, pp67-92,. ISSN 0570-4928 print/ 1520-569X online.

Andrási E., Orosz, L., Scheibler H., Bezur L., and Ernyei L., Concentrations of elements in brain tumours. *Mikrochim. Acta*, 118, 1995, pp 113-121,. ISSN 0026-3672.

Ariyama, K., Aoyama, Y., Mochizuki, A., Homura Yuji, Kadokura M., and Yasui, A.. Determination of the Geographic Origin of Onions between Three Main Production Areas in Japan and Other Countries by Mineral Composition. *J. Agric. Food Chem.*, 55 (2), 2007,pp 347–354. ISSN: 0021-8561.

Arroyo, J. , Marcó, L. , Malavé, R., Anzola E., Gómez L., Domínguez J.RF. and Alvarado, J.. Determination of Manganesum and Potassium in cement and gypsum samples by Slurry-FIAAS and alkaline fussion as analytical alternatives to the conventional procedure of leaching. *Rev. Téc. Ing. Univ. Zulia*, 25,(1), 2002, pp.42-48. ISSN 0254-0770.

Brandao G. C., Matos G., Ferreira S.L.C. Slurry sampling and high-resolution continuum source flame atomic absorption spectrometry using secondary lines for the determination of Ca and Mg in dairy products. *Microchemical Journal*, 98, 2011, pp 231–233. ISSN: 0026-265X.

Bugallo R., Alonso, Río Segade, S., Fernández Gómez, E.. Comparison of slurry sampling and microwave-assisted digestion for calcium, magnesium, iron, copper and zinc determination in fish tissue samples by flame atomic absorption spectrometry. *Talanta*, 72(1), 2007, pp 60-65. ISSN: 0039-9140.

Da Silva Erik G.P., do N. Santos A., Costa A. C.S., da N. Fortunato D. M., José N., Korn M. G.A., dos Santos W. N.L., Ferreira S. L.C.. Determination of manganese and zinc in powdered chocolate samples by slurry sampling using sequential multi-element flame atomic absorption spectrometry. *Microchemical Journal*, 82(2), 2006, pp 159-162. ISSN: 0026-265X.

Da Silva Erik G.P., Hatje, V., dos Santos W. N.L., Costa L. M., Nogueira A R.A., Ferreira S. L.C.. Fast method for the determination of copper, manganese and iron in seafood samples. *Journal of Food Composition and Analysis* 21, 2008, pp 259–263. ISSN: 0889-1575.

Fry, R.C., and Denton, M.B.; Anal. Chem., 49, No.9, 1413-1417, (1977). "High Solids Sample Introduction for Flame Atomic Absorption Analysis." , ISSN 0003-2700.

Hashmi D. R., Ismail S. and Shaikh G.H.. Assessment of the level of trace metals in commonly edible vegetables locally available in the markets of Karachi city. Pak. J. Bot., 39(3), 2007, pp 747-751. ISSN (printed): 0556-3321.

Horning M. S., Blakemore L. and Trombley, P.. Endogenous mechanisms of neuroprotection: role of zinc, copper, and carnosine. Brain Research, 852(1, 3), 2000, Pages 56-61 ISSN:0006-8993.

Januzzi, G., Krug, F.J., and Arruda, M.;. Application of the slurry technique to the determination of selenium in fish samples by Electrothermal Atomic Absorption Spectrometry. J. Anal. At. Spectrom., 12., 1997, pp375-378, ISSN: 0267-9477.

Koleva B. D. and Ivanova, E. H. Flow Injection Analysis with Atomic Spectrometric Detection. Eurasian Journal of Analytical Chemistry, 3(2), 2008,pp 183-211. ISSN: 1306-3057

Lech T., Lachowicz T. Application of ICP-OES to multielement analysis of biological material in forensic inorganic toxicology. Problems of Forensic. Sciences, 77, 2009, pp 64–78. ISSN 1230-7483.

Manutsewee Niwat, Aeungmaitrepirom, W., Varanusupakul, P., Imyim, A. Determination of Cd, Cu, and Zn in fish and mussel by AAS after ultrasound-assisted acid leaching extraction. Food Chemistry, 101(2), 2007, pp 817-824. ISSN 0308-8146.

Marcó, L., Graves, E., Alvarado, J., Analysis of human blood serum and human brain samples by total reflection X-ray fluorescence spectrometry applying Compton peak standardization. Spectrochimica Acta B, 54, 1999, pp 1469-1480. ISSN: 0584-8547.

Marcó P. Lué Merú, Hernández, E.A.., Pascusso, C and Alvarado, J.. Determination of manganese in brain samples by slurry sampling graphite furnace atomic absorption spectrometry. Talanta, 59(5), 2003, 897-904. ISSN: 0039-9140.

Marco, L. M., Hernández-Caraballo, E.A.. Direct analysis of biological samples by total reflection X-ray fluorescence, Spectrochimica Acta Part B, 59, 2004, pp 1077– 1090. ISSN: 0584-8547.

Mohamed, N., and Fry, R., Slurry Atomization Direct Spectrochemical Analysis of Animal Tissue. Anal. Chem., 53, 1981, pp 450-455. ISSN 0003-2700.

Mokgalaka N. S., Wondimu T. and McCrindle, R.. Slurry nebulization icp-oes for the determination of Cu, Fe, Mg, Mn and Zn in bovine liver. Bull. Chem. Soc. Ethiop., 22(3), 2008, 313-321. ISSN 1011-3924.

Religa D.; Strozyk, D.; Cherny R.A., ; Volitakis I., Haroutunian V., ; Winblad, B., Naslund, J. ; and Bush A.I., Elevated cortical zinc in Alzheimer disease. Neurology, 67, 2006, pp69–75. ISSN: 1526-632X.

Skip Kingston H.M.. Standardization of Sample Preparation for Trace Element Determination Through Microwave-Enhanced Chemistry. Atomic Spectroscopy, 19 (2), 1998, pp 27-30.ISSN 0195-5373.

Taylor A., Branch, S., Halls, D., Patriarca, M. and White, M.. Atomic spectrometry update. Clinical and biological materials, foods and beverages. J. Anal. At. Spectrom., 17, 2002, pp 414–455. ISSN 0267-9477.

Taylor A., Branch, S., Day, M.P., Patriarca, M., and White, M., Atomic spectrometry update. Clinical and biological materials, foods and beverages. J. Anal. At. Spectrom., 21, 2006, pp 439-491. ISSN 0267-9477.

Trojanowics Marek. Flow Injection Analysis. Instrumentation and Applications. Word Scientific Publishing. 2000. 481 pp ISBN 981-02-2710-8.

12

Activation of Bentonite and Talc by Acetic Acid as a Carbonation Feedstock for Mineral Storage of CO$_2$

Petr Ptáček, Magdaléna Nosková, František Šoukal, Tomáš Opravil,
Jaromír Havlica and Jiří Brandštetr
Brno University of Technology, Faculty of Chemistry,
Centre for Materials Research, Brno CZ-61200
Czech Republic

1. Introduction

The average global temperature has been slightly increasing by 0.76 °C over last 150 years. If the current state will continue, average Earth temperature will increase at the end of this century for about 1.1 – 6.3 °C according to applied emission scenario. Main reason of the observed global warming is the increasing contents of greenhouse gases (GHG), such as carbon dioxide (CO$_2$), methane (CH$_4$) a nitrous oxide (NO$_2$), in the Earth atmosphere. The most important greenhouse gas is CO$_2$. Carbon dioxide is considering as responsible for about two-third of the enhanced "greenhouse effect". Its atmospheric concentration has risen from the pre-industrial levels of 280 ppm to 380 ppm in 2005. Human emissions of greenhouse gases are very likely responsible for global warming of the planet surface (IPCC, 2007). The increasing carbon dioxide content in the atmosphere and its long-term effect on the climate has led to increasing interest and research of the possibilities of capture, utilization and long-term storage of carbon dioxide (Yang at al., 2008; Jiang, 2011)

Fossil fuels have been used as the world's primary source of energy upon over the 20th century and this trend is expected to continue throughout the 21st century (Yang at al., 2008; Maroto-Valer at al., 2005). There is a direct link between emissions of carbon dioxide (C$_e$), human population (P), economic development that is indicated by gross domestic product (GPD), production of energy (E), amount of carbon-based fuels used for production of energy (C) and CO$_2$ sinks (S$_{CO2}$):

$$C_e = P \frac{GDP}{P} \frac{E}{GDP} \frac{C}{E} - S_{CO_2} \qquad (1.1)$$

The emissions of anthropogenic carbon dioxide are increasing with population (P), standard of living (GPD/P); the energy intensity of economy (E/GPD) and the carbon intensity of the energy system (C/E). On the contrary C$_e$ is decreasing with S$_{CO2}$. Examination of the Eq.1.1, in principle, proposes that there are five ways to reduce atmospheric emissions of anthropogenic CO$_2$, of which the first two, i.e. reduction in population and/or decline in economic output are naturally unacceptable. Reducing the carbon intensity of the energy system can be achieved by using hydrogen-rich fuels and renewable energy sources. The last term indicates that emissions of carbon dioxide can be partially or totally covered by the artificial increase in the capacity and uptake rate of CO_2 sinks (S_{CO2}). Carbon sequestration includes terrestrial or marine photosynthetic fixation of CO_2 by plants and soils, and subsequent long-term storage of CO_2 as carbon-rich biomass or capture and long-term storage of CO_2 emissions at source prior to potential release. These techniques are collectively known as carbon capture and storage (CCS) (Bachu, 2008; Zhang at al., 2007; Piromon at al., 2007; Huesemann, 2006; Hoffert at al., 1998; Kaya, 1995).

The Carbon Capture and Storage technologies have been considered as suitable method for reduction of CO_2 emission, they are relatively abundant, cheap, available and globally distributed, thus enhancing the security and stability of energy systems (Bachu, 2008). CCS can be effectively integrated into various energy systems (Jiang, 2011). The CO_2 capture can be performed following three different technological concepts: post-combustion capture systems, pre-combustion capture systems and oxy-fuel capture systems (Damen at al., 2006; Pires at al., 2011).

The main options for CO_2 storage are:

1. Geological storage – CCGS (Carbon Capture and Geological Storage);
2. Ocean storage – CCOS (Carbon Capture and Ocean Storage);
3. Mineral storage – CCMS (Carbon Capture and Mineral Storage).

On the other hand there are fears that CCS technologies that offer the extension of the fossil-fuel era by perhaps a few 100 years are a double-edged sword. CCS technology is designed to limit emissions of CO_2 to the atmosphere, but it extends the period during which CO_2 is emitted (Spreng at al., 2007).

Carbon Capture and Geological Storage methods are using the geological media for storage of carbon dioxide at depths of more than one kilometer. Geological media suitable for CO_2 storage requires sufficient capacity, possibilities for CO_2 transport and preventing the CO_2 migration or escaping. Sedimentary basins may possess these requirements, because generally only sandstone and carbonate rock have needed to provide the porosity and then storage capacity and permeability. Confining low-permeability shales and evaporites such as salt beds and anhydrites provide primary physical barrier for CO_2 leakage (Gibbins & Chalmers, 2008; Bachu, 2008; Zhang at al., 2007; Pauwels at al., 2007; Friedmann at al., 2006; Gale, 2004; Soong at al., 2004; Bouchard & Delaytermoz, 2004; Torp & Gale, 2004; Xu at al., 2004).

Carbon dioxide may be stored in geological media by various means with various physical (Physical trapping) and chemical mechanisms (Chemical trapping) as a result of its properties at the pressure and temperature conditions found in Earth's subsurface. Physical trapping of CO_2 occurs when CO_2 is immobilized as a free gas or supercritical fluid. There are two types of physical trapping. Static trapping of mobile CO_2 in stratigraphic and structural traps or in man-made caverns is applied. The second possibility is represented by residual-gas trapping in the pore space at irreducible gas saturation. Chemical trapping occurs when CO_2 is absorbed into organic materials contained in coals and shales (adsorption trapping). Carbon dioxide may react directly or indirectly with mineral resulting to the geologic formation characterized by the precipitation of secondary carbonate minerals - mineral trapping. In direct carbonation process, gaseous CO_2 is in first stage dissolved during indirect (aqueous) process and reacts with solid mineral in following operation (Bachu, 2008; Alexander at al., 2007; Xu at al., 2004).

The dissolution of alkaline aluminosilicate minerals by CO_2 contributes to increasing of concentration of soluble carbonates and bicarbonates in solution, thereby enhancing "solubility trapping". The chemical reactions inducted by CO_2 injection are described by Eq.1.1 and 1.2 (Xu at al., 2004). The weathering of alkaline rocks, such as alkaline or alkaline earth silicates is thought to have played a great role in the historical reduction of the atmospheric CO_2 content in atmosphere of Earth (Kojima at al., 1997).

$$CO_2(g) + H_2O \rightarrow H_2CO_3 \rightarrow H^+ + HCO_3^- \qquad (1.2)$$

$$HCO_3^- + M^{2+} \rightarrow MCO_3(\downarrow) + H^+ \qquad (M^{2+} = Ca^{2+}, Mg^{2+}, Fe^{2+}...) \qquad (1.3)$$

Total estimated storage capacity of geological reservoirs is about 920 Gt CO_2 in depleted oil and gas fields, 400 – 10 000 Gt in deep saline reservoirs and 20 Gt in coal mine coal deposits. The cost for carbon dioxide capture and following storage in geologic formations is estimated about 4 – 48 EUR/t CO_2 (Friedmann at al., 2006; Gale, 2004). The research works concerning in risk assessment of CO_2 geological storage is mentioned in work (Gale, 2004).

Deep-sea storage of anthropogenic CO_2 is an attractive concept that offers large storage capacity comparing to other options. However, storing CO_2 in oceans is limited by its high cost, technology development, potentially high environmental impact, because the storage capacity of the ocean has not been defined. The oceanic processes are controlled long-term processes and large scale storage has been discussed only in general terms. Addition of anthropogenic CO_2 would change the CO_2 chemistry in the ocean by reducing pH at the site. The effects of long-term influence of low pH on planktonic ecosystem and oceanic biological processes are virtually unknown. Addition and CO_2 storage would probably dissolve carbonate deposits on the seafloor and suppress oxidation of organic matter (Wong & Matear, 1998; Bachu & Adams, 2003).

Mineral storage based on carbonation is a promising CCS method for long-term storage of CO_2 in continental inland utilization. The carbon dioxide is stored through mineral trapping mechanism that requires the participation of cations, including Ca^{2+}, Fe^{2+}, and Mg^{2+}, that can form stable solid carbonate phases (Giammar at al., 2005). This processing accelerates the natural weathering of silicate minerals, where these minerals react with CO_2 and form carbonate minerals and silica. Although the calcium silicate has been successfully carbonated at temperatures and pressures relevant for industrial processes, its natural resources are too small and expensive to be of practical interest. Therefore, current research activities focus mostly on carbonation of magnesium silicates (Teir at al., 2007). Overall course of carbonation process of wollastonite ($CaSiO_3$), olivine (Mg_2SiO_4) and serpentine ($Mg_3Si_2O_5(OH)_4$) may be described by Eq.1.4 – 1.6, respectively (Alexander, 2007; Wouter at al., 2007).

$$CaSiO_3(s) + CO_2(g) \rightarrow CaCO_3(\downarrow) + SiO_2 \qquad (1.4)$$

$$Mg_2SiO_4 + 2\,CO_2 \rightarrow 2\,MgCO_3(\downarrow) + SiO_2 \qquad (1.5)$$

$$Mg_3Si_2O_5(OH)_4 + 3\,CO_2 \rightarrow 3\,MgCO_3(\downarrow) + 2\,SiO_2 + 2\,H_2O \qquad (1.6)$$

The magnesium bearing minerals typically contain about 40 % of magnesium, whereas the content of calcium is approximately 10 – 15 %. Reactivity of olivine is higher than serpentine, but serpentine reactivity is strongly increasing by physical and chemical activation. Physical activation such as heat pre-treatment (calcination) at approximately

630 °C may remove water (dehydroxylation) from serpentine structure. The conversion to magnesite ($MgCO_3$) is higher (59.4 %) than the value (7.2 %) found for untreated samples (Alexander at al., 2007). Carbon dioxide sequestration capacity some of major rock forming minerals is listed in Table.1.1.

Mineral	Composition	Storage capacity [$kg_{CO2} \cdot m^{-3}$]
Plagioclase (anorthite)	$CaAl_2Si_2O_8$	436,4
Olivine (forsterite-fayalite)	$Mg_2SiO_4 - Fe_2SiO_4$	2014,7 - 1896,3
Pyroxene group - enstatite	$(Mg,Fe^{2+})Si_2O_6$	1404,2
Augite	$(Ca,Na)(Mg,Fe^{2+},Fe^{3+},Al,Ti)(Si,Al)_2O_6$	1306,3
Amphibole group – anthophy-llite - cummingtonite	$(Mg,Fe^{2+})_7Si_8O_{22}(OH)_2$	1169,5 - 1041,8
Common hornblende	$Ca_2(Mg,Fe^{2+},Al)_5(Si,Al)_8O_{22}(OH)_2$	1000,4
Calcinum amphiboles - tremolite	$Ca_2Mg_5Si_8O_{22}(OH)_2$	1119,3
Mica group - galuconite	$K_x(Fe^{3+},Al,Mg,Fe^{2+})_2(Si,Al)_4O_{10}(OH)_2; x < 1$	61,97
Mica group-phlogopite	$KMg^{2+}_3(Si_3Al)O_{10}(OH,F)_2$	881,8
Mica group-biotite	$K(Mg,Fe^{2+})_3(Si_3Al)O_{10}(OH,F)_2$	671,0
Serpentine	$Mg_3Si_2O_5(OH)_4$	1232,7
Chlorite group	$(Mg,Al,Fe^{2+})_{12}(Si,Al)_8O_{20}(OH)_{16}$	923,4
Clay minerals - illite	$(K,H_3O^+)Al_2(Si,Al)_4O_{10}(OH)_2$	78,42
Clay minerals - smectite	$(Ca_{0,5},Na)_{0,7}(Al,Mg,Fe)_4(Si,Al)_8O_{20}(OH)_4 \cdot nH_2O$	161,2

Table 1.1. Carbon dioxide sequestration potential of some major rock according to work (Xu at al., 2004).

If the rate-limiting step in the aqueous carbonation scheme is leaching of calcium or magnesium, then the production of carbonates may by accelerate via acceleration of dissolution stage. Inorganic (HCl, H_2SO_4, H_3PO_4) as well as organic acids (CH_3COOH), complexing agents and hydroxides ($NaOH$) were used for chemical activation of minerals. Hydrochloric acid enhances the magnesium ions liberation, however energy intensity production of $Mg(OH)_2$ has been increasing. Complexing agents were used to polarize and weaken the magnesium bonds within the serpentine structure. The most effective is treatment by H_2SO_4 which increases the surface area from 8 to 330 $m^2 \cdot g^{-1}$. Sulphuric acid pre-treatment enables aqueous carbonation of $Mg(OH)_2$ under milder condition. Temperature and pressure were reduced from 185 on 20 °C and 12.7 to 4.6 MPa. Process may by write as follows (Alexander at al., 2007; M.-Valer at al., 2005):

$$Mg_3Si_2O_5(OH)_4 + 3\,H_2SO_4 \rightarrow 3\,Mg^{2+} + 3\,SO_4^{2-} + 2\,SiO_2 + 5\,H_2O \tag{1.7}$$

$$Mg^{2+} + SO_4^{2-} + 2\,NaOH \rightarrow Mg(OH)_2 + Na_2SO_4 \tag{1.8}$$

$$Mg(OH)_2 + CO_2 \rightarrow MgCO_3(\downarrow) + 2H_2O \tag{1.9}$$

Industrial by-products, such as iron and steel slags and cement based material, may contain very height percentage of calcium and magnesium oxides and therefore they may be carbonated and exploited for CO_2 mineral storage. Calcium and magnesium can be leached out by acetic acid. Such a process consists of two main steps. The first one, where calcium ions are extracted from natural calcium silicate mineral:

$$CaSiO_3 + 2\,CH_3COOH \rightarrow Ca^{2+} + 2\,CH_3COO^- + SiO_2 + H_2O \tag{1.10}$$

$$Ca^{2+} + 2\,CH_3COO^- + CO_2 + H_2O \rightarrow CaCO_3(\downarrow) + 2\,CH_3COOH \tag{1.11}$$

And the second one, where carbon dioxide was introduced into the solution after removing of SiO_2 and calcite has been precipitated from the solution according to Eq.1.11. Acetic acid is recovered in this step and recycled for using of extraction in the first step (Eq.1.10). Similar reaction proceeds with magnesium silicates:

$$MgSiO_3 + 2\,CH_3COOH \rightarrow Mg^{2+} + 2\,CH_3COO^- + SiO_2 + H_2O \tag{1.12}$$

$$Mg^{2+} + 2\,CH_3COO^- + CO_2 + H_2O \rightarrow MgCO_3(\downarrow) + 2\,CH_3COOH \tag{1.13}$$

However, there are also small contents of many other compounds from iron and steel slags (such as heavy metals) which would be released by acetic acid (Teir at al., 2007).

1.1 Kinetics of silicate minerals, rocks, glass and raw materials dissolution

The main reasons for investigation of dissolution and precipitation reactions of silicate minerals and raw materials is in importance to understand the extent and environmental significance of the chemical weathering in nature (Cama at al., 1999;), study of its potential to utilization as the source of the divalent cations that is necessary for the sequestering of carbon dioxide into carbonates (Saldi at al., 2007), in order to improve their catalytic activity (Komadel & Madejová, 2006; Pushpaletha at al., 2005), study the puzzolanic activity in mortars and cements (Massazza, 1993) drug delivery (Viseras at al., 2010), synthesis of geopolymers (Buchwald at al., 2009), zeolites (Baccouche at al., 1998) and organic-clay composites (Yehia at al., 2012).

The kinetics of mineral dissolution is an area in geochemistry that has received considerable attention over the past several years (Knauss at al., 2003). Hence, numerous works dedicated to investigation of clay mineral dissolution kinetics can be found in the current literature (Table 1.2).

A basic concept in chemical kinetics is that reactions consist of a series of different physical and chemical processes that can be broken down into different "steps". For dissolution, these steps generally include at a minimum (Morse & Arvidson, 2002; Dorozhkin, 2002):

1. Diffusion of reactants through solution to the solid surface;
2. Adsorption of the reactants on the solid surface;
3. Migration of the reactants on the surface to an "active" site (e.g., a dislocation);
4. The chemical reaction between the adsorbed reactant and solid which may involve several intermediate steps where bonds are broken and formed, and hydration of ions occurs;

5. Migration of products away from the reaction site;
6. Desorption of the products to the solution;
7. Diffusion of products away from the surface to the "bulk" solution.

Mineral	Solution properties	E_A [kJ·mol⁻¹]	Rate limiting step of process	Reference
Wollastonite	pH 3 – 8	79.2	Diffusion [a]	Rimstidt & Dove, 1986
	Diluted acetic acid, pH 2 – 3.5	47.1	Two-dimension diffusion	Ptáček at al., 2011
Enstatite	pH 1 – 11; 28 – 168 °C	48.5	Reaction [b]	Oelkers & Schott, 2001
Forsterite	pH 2; 25 – 65 °C	63.8	Reaction [b]	Oelkers, 2001
Olivine	pH 2 – 5; 65 °C	125.6	Reaction	Chen & Brantley, 2000
	3 M H_2SO_4; 60 – 90 °C	66.5	Reaction	Jonckbloedt, 1998
Serpentine	2 M H_2SO_4; 30 – 70 °C	68	Diffusion	Teir at al., 2007
	2 M HCl; 30 – 70 °C	70		
	2 M HNO_3; 30 – 70 °C	74		
Talc	pH 1 – 10.6; 25 – 150 °C	45.0		Saldi at al., 1995
Anorthite	pH 2.4 – 3.2; 45 – 95 °C	18.4 [c]		Oelkers & Schott, 1995
Diopside	pH 2 – 12; 25 – 70 °C	40.6	Surface reaction	Knauss at al., 1993
Basaltic glass	pH 7.8 – 8.3; 90 °C	9.8	Diffusion	Daux at al., 1997

[a] Under low pH values. [b] Forming of rate-controlling precursor complex. [c] Under pH = 2.6.

Table 1.2. Dissolution kinetics of silicates. Table is extracted from the work (Ptáček at al., 2011).

A central concept in dissolution kinetics supposes that one of these steps is the slowest than other. The reaction cannot proceed faster than the rate limiting step. Above mentioned steps 1 and 7 involve the diffusive transport of reactants and products through the solution to and from the surface. When this process is rate-limiting, the reaction is said to be diffusion controlled. Steps 2– 6 occur on the surface of the solid and when one of them is rate controlling the reaction is said to be surface controlled (Morse & Arvidson, 2002; Dorozhkin, 2002).

The dissolution of solids in liquids (or melts) consists of a surface chemical reaction and transport of the reaction components to the reaction boundary (Šesták, 1984). Many multicomponent silicate minerals under acidic condition are dissolved incongruently. The Ca^{2+} ions were replaced by H_3O^+ ions and leached layer of silica was formed. This layer wasn't homogeneous and its structure was changing with time as a consequence of polymerization of silanol groups (Weissbart & Rimstidt, 2000):

$$2 \equiv Si - OH \rightarrow \equiv Si - O - Si \equiv +H_2O \qquad (1.14)$$

Monosilic acid may be liberated from silicates which contain SiO_4^{4-} ions separated by metal cations – nesosilicates. Besides the temperature the solubility of an amorphous silica layer depends on pH and shows the minimum at pH 7. The accurate data are still missing because there is an extreme variation in the forms in which the amorphous silica can occur. The rate of dissolution is proportional to the concentration of H_3O^+ and OH^- ions in the range from 0 to 2 and from 3 to 6, respectively. The rate of diffusion or desorption of the silicic acid from the surface limits the rate of dissolution if pH is higher than 6 (Iler, 1979).

The dissolution mechanism of each multioxide silicate can be deduced from its structure. Note that in some cases, not all metal–oxygen bonds present in the structure need to be broken to completely destroy a mineral. Dissolution proceeds via the sequential equilibration of metal–proton exchange reactions until no further viable structure remains. The last of these sequential exchange reactions destroys the structure and it is irreversible in most cases. Assuming that at acidic conditions, the sequence of metal–proton exchange reactions during the dissolution of a multioxide silicate follow the order prescribed by the relative reactions rates of the single oxide dissolution as illustrated in Fig.1.1 (Oelkers, 2001).

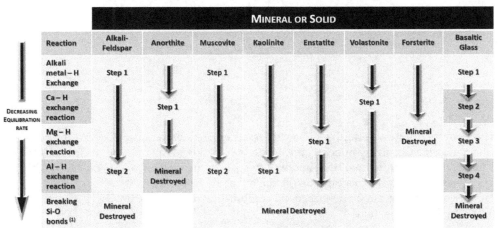

Fig. 1.1. Mechanism of dissolution of some minerals and basaltic glass at acidic condition according to Oelkers, 2001.

The dissolution rate of the clay minerals seems to be continuously decreasing with elapsed time due to the preferred dissolution of reactive edge surfaces. As edge surfaces are selectively dissolved, the percentage of these reactivity reactive sites decrease with time leading to a decrease in the average reactivity of the overall clay surface (Köhler at al., 2005). The derivation of rate law for congruent dissolution of silicate multioxides at close to equilibrium conditions will be derived using a general formula $M_{(1)}{}^{z1}{}_{n1}\, M_{(2)}{}^{z2}{}_{n2}\, M_{(3)}{}^{z3}{}_{n3}$ $O_{\Sigma(n(i)z(i))/2}$, which is representative for oxide composition of nesosilicates related to phenakite ($M_{(1)}M_{(3)}O_4$, where $M_{(1)}$ = Li, Be, Zn... and $M_{(3)}$ = Si), olivine ($M_{(1)}M_{(3)}O_4$, where $M_{(1)}$ = Ca, Mg, Fe^{2+}, Mn... and T = Si) and garnet ($M_{(1)}{}^{2+}{}_3\, M_{(2)}{}^{3+}{}_{(2)}(M_{(3)}O_4)_3$, where $M_{(1)}$ = Ca, Mg, Fe, Mn..., $M_{(2)}$ = Al, Fe^{3+}, Cr^{3+}, V^{3+}, Ti^{4+} ..., $M_{(3)}$ = Si, Fe^{3+}, Al...) structural groups:

$$\sum_{i=1}^{m}\left(M_{i,n_i}^{z_i^+}\right)O_{\sum_{i=1}^{m}\left(n_i z_i^+\right)/2} + \left(\sum_{i=1}^{m}\left(n_i z_i^+\right)\right)H^+ \leftrightarrow \sum_{i=1}^{m} n_i \; M_i^{z_i^+} + \sum_{i=1}^{m}(\frac{n_i z_i^+}{2})H_2O \tag{1.15}$$

For example, the members of olivine subgroup such as calcio-olivine, forsterite, fayalite, tephroite... for that $M_{(1)}$ = Ca, Mg, Fe^{2+}, Mn..., trivalent cations does not present and $M_{(3)}$ = Si, are then dissolved according to following reaction scheme:

$$M_2^{2+} SiO_4 + 8\,H^+ \leftrightarrow 2\,M^{2+} + Si^{4+} + 4\,H_2O \tag{1.16}$$

The kinetics of this reversible chemical reaction involving competition between two elementary – forward (+) and reverse (–) reactions, can be easily expressed by applying the Van't Hoff law such that:

$$r = r_+ - r_- = r_+ \left(1 - \frac{r_-}{r_+}\right) \tag{1.17}$$

The equilibrium constant K of forward and reverse Q reaction 1.15 can be then expressed as follows:

$$K = \frac{\prod_{i=1}^{m} a_{M_i^{z_i^+}}^{n_i}}{a_{H^+}^{\sum_{i=1}^{m}\left(n_i z_i^+\right)}} = \prod_{i=1}^{m}\left(\frac{a_{M_i^{z_i^+}}}{a_{H^+}^{z_i^+}}\right)^{n_i} \tag{1.18}$$

$$Q = \frac{1}{K} \tag{1.19}$$

The saturation state of a fluid is often expressed in terms of the ratio (Q/K); if by common convention the dissolving mineral appears on the left side of the reaction, values of (Q/K) < 1 indicate undersaturation of the fluid with respect to the mineral, and conversely, (Q/K) > 1 is representative of supersaturation (Hellmann at al., 2009).

The dissolution rate can be described via combination of Eq.1.17 with law 1.18 and 1.19 by following kinetic equation:

$$r = r_+ - r_- = k_+ \prod_{i=1}^{m} a_{M_i^{z_i^+}}^{n_i} - k_- \; a_{H^+}^{\sum_{i=1}^{m}\left(n_i z_i^+\right)} = k_+ \prod_{i=1}^{m} a_{M_i^{z_i^+}}^{n_i} \left(1 - \frac{Q}{K}\right) \tag{1.20}$$

where the k represents the reaction rate constant. The chemical affinity of described reaction is defined as follow (Hellmann at al., 2009; Gérard at al., 1998):

$$A_r = -RT \; ln\left(\frac{Q}{K}\right) \implies \Delta_r G = RT \; ln\left(\frac{Q}{K}\right) \tag{1.21}$$

so that can be derived that:

$$\frac{Q}{K} = exp\left(-\frac{A_r}{RT}\right) = exp\left(\frac{\Delta_r G}{RT}\right) \tag{1.22}$$

The R is the universal gas constant ($J \cdot mol^{-1} \cdot K^{-1}$) and $\Delta_r G = -A_r$ denotes Gibbs energy ($J \cdot mol^{-1}$) and chemical affinity ($J \cdot mol^{-1}$) of reaction. The dissolution rate at near to equilibrium conditions when $r_+ + r_- \approx 0$ requires that $Q \approx K$ and the ratio $k_+/k_- \approx K$:

$$exp\left(\frac{\Delta_r G}{RT}\right) = \frac{Q}{K} \tag{1.23}$$

The overall dissolution rate should through combination of Eq.1.22 with Eq. 1.23 expressed as:

$$r = k_+ \prod_{i=1}^{m} a_{M_i^{z+}}^{n_i} \left(1 - exp\frac{\Delta_r G}{RT}\right) \tag{1.24}$$

The temperature dependence of dissolution rate constant is given by Arrhenius law (Oelkers, 2001):

$$k_+(T) = A\ exp\left(-\frac{E_a}{RT}\right) \tag{1.25}$$

The combination of Eq.1.24 and Eq.1.25 leads to equation:

$$r = \prod_{i=1}^{m} a_{M_i^{z+}}^{n_i}\ A\ exp\left(-\frac{E_a}{RT}\right)\left(1 - exp\frac{\Delta_r G}{RT}\right) \tag{1.26}$$

where A is pre-exponential (frequency) factor and E_a is apparent activation energy. Under conditions that are not far from equilibrium conditions (please refer to Eq.1.23) where $exp\ (\Delta_r G/RT) \approx 1$ can be dissolution rate expressed as:

$$r \approx \prod_{i=1}^{m} a_{M_i^{z+}}^{n_i}\ A\ exp\left(-\frac{E_a}{RT}\right) \tag{1.27}$$

or

$$ln\ r \approx ln \prod_{i=1}^{m} a_{M_i^{z+}}^{n_i} + ln\ A - \frac{E_a}{R}\frac{1}{T} \tag{1.28}$$

The kinetic parameter of dissolution process can be then estimated from Arrhenius plot as the slope ($-E_a/R$) of the dependence of ln r on reciprocal temperature. Assuming information about ionic product of released cations ($\prod a_{M(i)}^{n(i)}$), the value of A can be calculated from the intercept with y-axis.

A general scheme for the dissolution of a mineral or glass can be written as follow (Wieland et al.,1988):

Reactants ± Aqueous Species ↔ Precursor Complex ↔ Activated complex → Products (1.29)

The precursor complex has the same chemical formula as the activated complex, but the activated complex has more energy. Within the context of transition-state theory (TST), the activated complex is in equilibrium with other species that precede it in the reaction sequence. It follows that a mineral dissolution rate can be considered to be proportional to the concentration of this "rate-controlling" precursor complex at the surface in accord with (Oelkers, 2001):

$$r_+ = k_+ X_p \tag{1.30}$$

where k_+ refers to a rate constant consistent with the P precursor complex and X_P stands for the mole fraction of the precursor complex at the surface.

The dissolution mechanism of this mineral or glass is often initiated by the formation of the precursor complex through one or more exchange reactions. The process leads to formation of the leached surface through the metal-proton exchange. The next part of the dissolution reaction is destruction of the leached surface (Oelkers, 2001), i.e. incongruent dissolution takes place. The overall mechanism then may consist of a series of "i" elementary steps:

$$r = r_+^i - r_-^i = \sigma_i r \tag{1.31}$$

The exponent σ is generally known as Temkin's average stoichiometric number, which is equal to the ratio of the rate of destruction of the activated or precursor complex relative to the overall dissolution rate. The σ value is related to the stoichiometric number of precursor complexes that can be formed from one mole of the commonly adopted chemical formula of a mineral or glass and it can have a value other than one (Aagaard & Hegelson, 1982). The average stoichiometric coefficient for the overall dissolution process that consists from i-steps can be defined as follows (Gin at al., 2008):

$$\sigma = \frac{\sum_{i=1}(\sigma_i \, \Delta_r G_i)}{\sum_{i=1} \Delta_r G_i} = \frac{\Delta_r G}{\sum_{i=1} \Delta_r G_i} \tag{1.32}$$

For reaction near to equilibrium we obtain:

$$\frac{r_+}{r_-} = \prod_{i=1} \frac{r_+^i}{r_-^i} = exp\left(-\frac{\Delta_r G}{RT}\right) \tag{1.33}$$

and

$$r = r_+\left(1 - exp\left(\frac{\Delta_r G}{\sigma RT}\right)\right) \tag{1.34}$$

From the general law of mineral dissolution proposed by Aagaard and Helgeson, 1982 it can be derived by the same way as before:

$$r = k_+ \prod_{i=1}^{m} a_{M_i^{z_i}}^{n_i} \left(1 - exp\frac{\Delta_r G}{\sigma RT}\right) \tag{1.35}$$

As it was pointed by Gin at al., 2008, the Eq.1.24 is often presented as direct application of transition state theory. In fact, this law may be derived using simple kinetic concepts (notably the Van't Hoff law) irrespective of any hypotheses concerning the reaction mechanisms. The notion of an activated complex associated with an elementary step is theoretically compatible with the kinetic law 1.24, assuming an equilibrium existing between the activated complex and reactants in the forward and reverse directions. However, this notion is not required to obtain Eq.1.24 and indeed leads to a paradox that lies in the fact that equilibrium was assumed between the activated complex and reactants in the forward direction, but that a second equilibrium was also assumed between the activated complex and the product in the reverse direction. This implies equilibrium between the products and the reactants, so the net rate should be zero. This paradox, of course, does not call into question the expression of the kinetic constants: the forward rate simply offsets the reverse rate. Postulating equilibrium between the reactants forming the activated complex in both directions and the activated complex therefore implies that Eq.1.24 is valid only at equilibrium.

1.2 Clay minerals

Human life and the existence of many organisms on this planet are connected with clays. Clay minerals are the basic constituents of clay raw materials and clay raw material has always played the substantial role in human life (Table 1.3) due to their wide-ranging properties, high resistance to atmospheric conditions, geochemical purity, easy access to their deposits near the earth's surface and low price. A majority of clays is known for its plasticity. However, many clay raw materials are not plastic, or they are semi-plastic such as clay stones, clay shales, talc, pyrophyllite, vermiculite and coarser mica. The properties of clay minerals also reflect the state and distribution of the electrostatic charge of the structural layers. The negative charge is a result of the ionic substitutions in the octahedral and tetrahedral sheets of clay minerals (Konta 1995; Murray; 2000).

Paper industry	Kaolinite,	Adsorbents	Bentonite, chlorites, palygorskite...	Bonding material	Kaolin, bentonite, bentonite...
Ceramics	Kaolinite, illite, talc, vermikulite...	Adhesives	Kaolinite,	Water purification	Vermiculite
Plastics and rubber	Kaolinite, pyrophyllite....	Pharmaceutic and cosmetics	Kaolinite, bentonite, pyrophyllite....	Waste treatment	Micas, bentonite...
Catalysts	Bentonite, palygorskyte...	Insulating material	Vermiculite, micas...	Agricultural and forestry	Palygorskyte, vermi-kulite, bentonite...
Dyes and paints	Kaolinite, micas, pyrophyllite....	Molecular sieves	Palygorskite, sepiolite...	Polishing materials	Bentonite

Table 1.3. Traditional application area of clay minerals (Konta 1995; Murray; 2000).

A significant role for clay minerals in the origin of life was postulated by Bernal, 1967. Clay surface could adsorb and concentrate organic substances and some hypothesis supposed that clay crystals could function as the earliest genetic information storing material (C.-Smith, 1966 and 1982) and iron-rich clay have significant importance in the origin of the photosynthetic organisms (Hartman, 1975). Clay minerals, the essential constituents of argillaceous rocks, can be classified in seven groups according to their crystal structure and crystal chemistry. These groups are listed together with their properties and the most important members in the Table 1.4.

Group	Layer type	Length d_{001} [Å]	Interlayer charge	Interlayer contains [1]	Octahedral layer type:	Example
Kaolinite and serpentine	1 : 1 t-o		zero	---	Dioctahedral	Kaolinite, dickite, nacrite...
					Trioctahedal	Serpetine
Talc and pyrophyllite	2 : 1 t-o-t		zero	---	Dioctahedral	Talc
					Trioctahedal	Pyrophyllite
Smectites	2 : 1 t-o-t	9.6 – 21.0 [2]	0.2 – 0.6	Na^+, Ca^{2+}, K^+, Li^+, H_3O^+ and H_2O	Dioctahedral	Montmorillonite
					Trioctahedal	Saponite
Vermiculites	2 : 1 t-o-t	~14.3	0.6 – 0.9	Mg^{2+}	Dioctahedral	Dioctahedral vermikulite
					Trioctahedal	Trioctahedal vermikulite
Micas	2 : 1 t-o-t	~10.0	0.9 – 1.0	K^+, Na^+, H_3O^+, Ca^{2+}, □ [3]...	Dioctahedral	Muscovite, illite...
					Trioctahe-dal	Biotite, flogopite...
Chlorites	2 : 1 + 1 t-o-t + o	~14.3	different	Di- or tri-octahedral layer	Dioctahedral	Donbassite
					Trioctahedral	Klinochlore
Palygorskite and sepiolite	other [4]	---	different	other [4]	other [4]	Sepiolite, palygorskite

[1] Interlayer ions that are present predominantly are marked by bold.
[2] For untreated smectites is typical $d_{001} \approx$ 15 Å.
[3] Vacation.

[4] Chanel containing water and exchangeable hydrated cations. Water can be withdrawn without structural lattice changes similar to zeolites.

Table 1.4. Classification of phyllosilicates (Martin et al., 1991; Konta, 1995).

Clay minerals represent a large family of alumino-silicate structures with a range of chemical composition, structure and surface properties. Their crystal structure with a few exceptions consists of sheets firmly arranged in structural layers. Hence are these minerals termed as sheet silicates or phyllosilicates. The individual layers consist of two, three or four sheets. The sheets are formed either by tetrahedrons $[SiO_4]^{4-}$ which are abbreviated as

"T" or by $[AlO(OH)]^{6-}$ octahedrons which are signed as "O". The interior of tetrahedrons and octahedrons contains smaller metal cations, their apices are occupied by oxygen's from which some are connected to protons (as OH). All these fundamental structural elements are arranged to form a hexagonal network in each sheet (Caglar at al., 2008; Konta 1995).

Numerous rock-forming silicates (feldspars, granites, syenites, gneisses, arkoses, phonolites, rhyolites...) alter into clay minerals such as kaolinite (Eq.1.36), illite (Eq.1.37) and montmorillonite (Eq.1.38) through an intense hydrolysis, supported by natural acids (Konta, 1995):

$$2\ KAlSi_3O_8 + 2\ CO_2 + 11\ H_2O \rightarrow Al_2Si_2O_2(OH)_4 + 2\ KHCO_3 + 4\ H_4SiO_4 \tag{1.36}$$

$$5\ KAlSi_3O_8 + 4\ CO_2 + 20\ H_2O \rightarrow KAl_4(Si_7Al)O_{20}(OH)_4 + 4\ KHCO_3 + 8\ H_4SiO_4 \tag{1.37}$$

$$Mg^{2+} + 3\ NaAlSi_3O_8 + 4\ H_2O \rightarrow NaAl_3MgSi_8O_{20}(OH)_4 + 2\ Na + H_4SiO_4 \tag{1.38}$$

Dissolution and precipitation of any feldspar can be described by the general formula (Hellmann at al., 2009):

$$\begin{aligned} Na_xK_yCa_zAl_{1+z}Si_{3-z}O_8 + 8\ H_2O &\rightarrow x\ Na^+ + y\ K^+ \\ +z\ Ca^{2+} + (1+z)\,[Al(OH)_4]^- &+ (3-z) + H_4SiO_4 \end{aligned} \tag{1.39}$$

where x + y + z = 1. The main factors affecting the rates and mechanisms of dissolution include the pH, temperature, composition of the liquid phase and feldspar, feldspar granulometry, the influence of atmospheric condition and vegetation (Chardon at al., 2006; Augusto at al., 2000).

1.3 Properties and mineralogy raw materials main minerals – montmorillonite and talc

Bentonite occurs in the form of lenses in other sediments mostly as a weathering product after igneous material settled in water. It also commonly occurs as a product of supergene or hydrothermal alteration of some volcanic rocks, e.g. rhyolites, porphyres, phonolites, dacites, andesites and basalts. Smectites are especially formed through the decomposition of volcanic glass. The chemical composition of smectite, the dominant mineral of bentonites, is variable. It varies between montmorillonite $(Al_{1.67}(Mg,Fe^{2+})_{0.33}Si_4O_{10}(OH)_2 0.5Ca_{0.33} \cdot nH_2O)$ and beidellite $(Na_{0.5}Al_2(Si_{3.5}Al_{0.5})O_{10}(OH)_2 \cdot nH_2O)$. In the interlayer space of both smectites different cations are adsorbed, especially alkalis and alkaline earths (Konta 1995).

Smectites are an important class of clay minerals; they are utilized in many industrial processes due to their high CEC, swelling ability, and high surface area (Madejová at al., 2006). Montmorillonite was the name given to a clay mineral found near Montmorillon in France as long ago as 1874 (Grimshaw, 1971). Montmorillonite is classified as a dioctahedral clay mineral with the 2:1 type of layer linkage that is related to the group of smectites (Caglar at al., 2008).

Dioctahedral layered structure of 2:1 type represents T-O-T sheet layered mineral with two tetrahedral and one octahedral layer where the centre of octahedron are predominantly occupied by trivalent cations such as Al^{3+}, Fe^{3+}, Cr^{3+}, V^{3+}, etc. The structure of montmorillonite is shown in Fig.1.2.

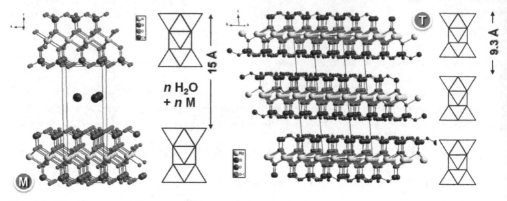

Fig. 1.2. The structure of montmorillonite (M) and talc (T).

Talc ($Mg_6Si_8O_{20}(OH)_4$) is a common 2:1 layer lattice silicate, the structure of which consists of two tetrahedral silicate sheets separated by an octahedral Mg-O(OH) sheet, i.e. it is the trioctahedral magnesian analogue of pyrophyllite ($Al_2Si_4O_{10}(OH)_2$). Among its many uses, talc is an important raw material for magnesium ceramics (steatites, cordierite, enstatite and forsterite products). As the ceramic raw material, its thermal decomposition behaviour is of considerable interest (MacKenzie & Meinhold, 1994). Talc and pyrophyllite crystallize during metamorphic or hydrothermal processes (Konta, 1995). The structure of talc is shown in Fig.1.2.

2. Leaching experiment

All experiments reported in this work were performed on bentonite from locality Obrnice (Czech Republic) produced by the company Keramost a.s., that was used as the source of Na, Ca - montmorillonite, and talc produced by Združena v.d. Spišká nová Ves, plant Gelnica from locality Gemerská poloma (Slovak Republic). The composition of montmorillonite and talc can be expressed by the empiric formula $Na_{0.2}Ca_{0.1}Al_2Si_4O_{10}(OH)_2(H_2O)_{10}$ and $Mg_3Si_4O_{10}(OH)_2$, respectively.

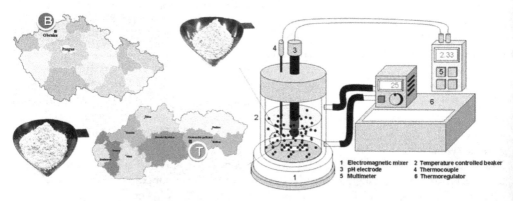

Fig. 2.1. Schematic illustration of the bentonite (B) and Talc (T) leaching experiment.

Leaching procedure was performed using the well stirred suspension of clay mineral in diluted solution of acetic acid (Lachema, p.a.) of concentration 3 $dm^3 \cdot mol^{-1}$. Temperature of leaching bath ranged from 22 to 50 °C. The temperature of double wall glass reactor was adjusted using external water flow of temperature controlled water bath (thermostat). Sample was poured on by solution of acetic acid that was preheated to the applied leaching temperature in water bath of thermostat. Hence, the stirring of system by magnetic stirrer was used. Suspension contained 12.5 g of wollastonite per dm^3 of leaching solution. The pH value of dispersing medium for 24 h leaching experiment was continuously measured by pH meter connected to PC (Fig.2.1).

Solid part of suspension was separated by filtration through dense filter paper (red strip) after leaching. Filter cake was washed three times by slightly acidified (acetic acid) distilled water. The quantities of ions in original sample and leachate were determined by the Inductively Coupled Plasma Atomic Emission Spectroscopy (ICP-OES; ICP IRIS Iterdip II XSP duo). Filter cake was dried at 110 °C; its properties and composition were subsequently investigated by simultaneous TG–DTA–EGA, FT-IR, BET and SEM.

Thermal analysis – simultaneous termogravimetry, differential thermal analysis and effluent gas analysis (TG–DTA and EGA) were performed with TG–DTA analyzer (Q600, Thermal Instruments) connected with FT-IR spectrometer (iS10, Thermo Scientific) through TGA/FT-IR interface (Thermo Scientific) heated to temperature 200 °C. That enables to study the composition of gas phase that was formed during processes which take place in heated sample. All experiments were performed with heating rate 20 $°C \cdot min^{-1}$ using argon with flowing rate 100 $cm^3 \cdot min^{-1}$ as the carrier gas, i.e. in the inert atmosphere.

Infrared spectra were collected upon mid-IR region via KBr pellets technique using FT-IR spectrometer iS10. Specimens were ground with dry spectroscopic grade KBr powder using the sample to KBr mass ratio of 1 : 100. The spectrum was obtained from 128 scans collected with resolution of 8 cm^{-1}. Scanning electron microscopy (SEM) was performed with a model BS 340 (Tesla). The X-ray diffractometer Siemens D500 with $CuK\alpha$ radiation at 40 kV and 40 mA was utilized for identification the phase composition of raw material and leached samples. Brunauer-Emmett-Teller (BET) analysis (Chembet 3000, Quantachrome Instruments) was used to determine of leached samples specific surface.

2.1 Evaluation of leaching test

The method applied for monitoring of the leaching process is the same as for study of leaching of calcium from wollastonite (Ptáček at al., 2010). The buffer system of weak acid (CH_3CO_2H) and its salt ($Ca(CH_3CO_2)_2$ or $Mg(CH_3CO_2)_2$) with a strong base, i.e. $Ca(OH)_2$ or $Mg(OH)_2$, was formed during dissolution of raw material. With respect to reaction stoichiometry, the amount of formed acetate ions was double to concentration of Ca^{2+} ions released from wollastonite. Hence following subform of well known Henderson buffer equation may be used for estimation of the course of leaching process:

$$pH(T) = pK_a(T) + log \frac{2[Ca^{2+}]}{[CH_3CO_2H]} \tag{2.1}$$

where pK_a denotes dissociation constant of acetic acid at given temperature. All variables in Eq.2.1 depend on the temperature.

2.2 Evaluation of leaching process kinetics

The monitoring of the progress of leaching experiment reflects the following facts and presumptions:

1. The amount of calcium and magnesium released into the solution is much higher than other elements extracted from raw material during leaching experiments, i.e. the amount of other metals in the solution is negligible;
2. Large excess of acetic acid in the system ensures its stable concentration level;
3. Henderson–Hasselbach buffer equation (Eq.2.1) can be applied for the reaction mixture;
4. Leached calcium was instantaneously transported out of surface by intensive stirring of the system.

The steady-state dissolution rate for applied temperature $r_+(T)$ $(mol \cdot m^{-2} \cdot s^{-1})$ can be calculated using following equation (Oelkers, 2001):

$$r_+(T) = \frac{([M^{2+}]_i - [M^{2+}]_t)V}{v_{M(Ac)_2} S(t-t_i)} = \frac{\Delta[M^{2+}]V}{S \Delta t} \tag{2.2}$$

where $[M^{2+}]_i$ and $[M^{2+}]_t$ are an initial t_i and general time t concentrations of $M^{2+} = Ca^{2+}$ and Mg^{2+} ions, respectively. The initial time of the process means the beginning of an induction period, so that the amount of Ca and Mg released during dissolution of calcite and dolomite can be excluded. The quantities V, $v_{M(Ac)2}$ and S are a volume of the system, stoichiometric number of $M(Ac)_2$ ($v_{M(Ac)\,2} \approx 0.3$ for the Ca-montmorillonite and $v_{M(Ac)\,2} \approx 0.3$ for talc) and total surface area of sample introduced into the reactor, respectively. The term $\Delta[M^{2+}]/\Delta t$ of Eq.2.2 can be determined as the slope of the linear part of the plot of concentration vs. time (Cama, 1999). This method of $r_+(T)$ value estimation is in particular favourable for the systems with very complicated stoichiometry of ongoing reactions such as in studied montmorillonite clay.

The reached stage of the system during the leaching process can be characterized by fractional conversion (degree of conversion) as follows:

$$y = \frac{[M^{2+}]_i - [M^{2+}]_t}{[M^{2+}]_i - [M^{2+}]_\infty} \tag{2.3}$$

where bottom index i, t and ∞ denotes the initial (beginning of the induction period), currently measured and final value of M^{2+} ions concentration. The degree of conversion can hold values from 0 to 1 and its time dependence enables to estimate mechanism and kinetics of leaching process by linearization procedure. The method is based on the formula:

$$g(y) = kt \quad [T = konst.] \tag{2.4}$$

where k is the rate constant of the process. If the kinetic function g(y) corresponding to the proper mechanism was chose, the dependence of g(y) on t should be straight line with the slope k on wide interval of y. The mathematic expression of the kinetic function can be found in published literature (Vlaev at al., 2008; Duan at al., 2008; Saikia at al., 2002; Šesták, 1984). The variation of mineral dissolution rates with temperature is commonly described using the empirical Arrhenius law - Eq.1.25 (Oelkers, 2001; Cama at al., 1999). The estimation of the apparent activation energy and the pre-exponential (frequency) factor (A) is based on the logarithmic form of the Arrhenius law:

$$\ln r_+(T) = \ln A - \frac{E_a}{R}\frac{1}{T} \tag{2.5}$$

using values of r_+ determined for several temperatures. The plot of ln k vs. T^{-1} (Arrhenius plot) should be straight line, where the slope $(-E_a/R)$ yields to the apparent activation energy of the process and y-axis intercept is then equal to the ln A. For the early stage of dissolution process, the concentration of M^{2+} ions in leaching solution is increasing with time almost linearly. It stands to the reason that the initial part of dissolution process enables to estimate the dissolution rate constant as:

$$k = \frac{dy}{dt} \quad [s^{-1}] \tag{2.6}$$

3. Results and discussion

There are many factors affecting the course of experiment such as pH of leaching solution, kind and solvent composition, temperature, pressure, particle size distribution and particle shape, concentration of solid in the suspension and stirring intensity. Hence, the initial state of raw material serving as the source of clay mineral should be characterized. The surrey of used raw materials composition and properties are listed in the Table 3.1.

Mineral	Montmorillonite			Talc		
Empirical formula	$Na_{0.2}Ca_{0.1}Al_2Si_4O_{10}(OH)_2(H_2O)_{10}$			$Mg_3Si_4O_{10}(OH)_2$		
Classification (Strunz)	VIII/H.19-20			VIII/H.09-40		
Colour	Light yellow		(14)	Light grey		(14)
Composition [%] Na_2O	1.13 (1)	1.76 (2)	1.38 (3)	--- (1)	--- (2)	1.93 (3)
K_2O	---	0.84	1.23	---	---	3.16
CaO	1.02	0.45	11.88	---	---	3.35
MgO	---	2.74	3.75	31.88	35.12	16.94
Al_2O_3	18.57	32.85	25.79	---	---	21.64
Fe_2O_3	---	13.72	11.29	---	---	1.68
SiO_2	43.77	41.50	36.63	63.37	58.43	51.29
TiO_2	---	---	1.65	---	---	---
H_2O	36.09	6.14	6.39	4.75	6.45	6.65
X_{50}/X_{90} [µm] (4)	9.12/ 37.67			29.33/ 71.65		
SH (5) [g·cm⁻³]	0.76			0.71		
SHS (6) [g·cm⁻³]	0.81			0.73		
Moisture (7) [%]	7.68			0.51		
ZŽ (8) [%]/Colour	19.54 / red			12.71 / beige		
OH (9) [g·cm⁻³]	2.23			2.80		
SS (10) [m²·g⁻¹]	95.55			2.10		
ζ(11) [mV]	-13.5			-13.9		

Admixtures [12]	Carbonates [13], illite (3), clinochlore (2), fluorapatite (6), barite (7) and rutile (5).	Dolomite (1), calcite (2), albit (3), pyrite (4) and quartz (5)

[1] Stoichiometric composition of mineral according to pertinent empirical formula.	[8] Loos on ignition (annealing at 1000°C to constant weight; according to standard ČSN 72
[2] Determined composition of clay mineral (dry state of sample).	0103).
[3] Analyzed composition of raw material (dry state sample).	[9] Bulk density (according to standard ČSN EN 993-17).
[4] Particle size analysis (Helos, Sympatec).	[10] Specific surface (BET, Chembet 3000, Quantachrome Instruments).
[5] Pour density (according to standard ČSN EN 725-8).	[11] Electrokinetic "zeta" potential (suspension of 0.1 g·dm⁻³).
[6] Bulk density in the shaken state (according to standard ČSN EN 725-8).	[12] Main admixture mineral found by following method: XRD, FT-IR, SEM and TA. The content of
[7] Determined by humidity analyzer Kern MLS 50-3 (sample was dried at 110 °C to constant weight.	crystalline phases was estimated by XRD in semi-quantitative mode.
	[13] Siderite (1), Ankerite (4) and Dolomite (7).
	[14] Monoclinic - prismatic class symmetry.

Table 3.1. The composition and properties of clay raw materials.

3.1 Thermal analysis

Results of thermal analysis allow identification of main mineral phases and estimate their content in the clay raw material. The typical TG-DTA and EGA patterns of clay raw materials that were used as the source of montmorillonite and talc are shown in Fig.3.1. The DTG curve is plotted in order to reach higher sensitivity to distinguish between individual steps of thermogravimetric analysis.

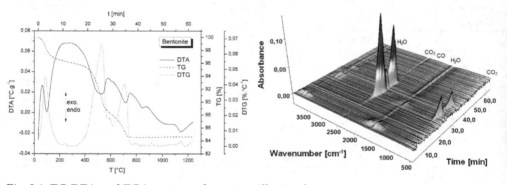

Fig. 3.1. TG-DTA and EGA pattern of montmorillonite clay.

TA of bentonite performed up to 1250 °C shows that mass of sample is decreasing for about 15.42 % within to the series of six endothermic steps. Evaporation of adsorbed water leads to the first endothermic peak of the maximum at temperature about 99.4 °C. The mass of sample was up to 165.2 °C (ousted point of DTG peak) reduced at about 3.71 %. The water vapour released from the sample is also well visible on EGA. The water can be also detected in the spectrum of gas phase upon the temperature interval ranged from 225 to 260 °C, where water molecules has been ousted from the interlayer space of montmorillonite and admixture of illite. The process shows maximum rate at temperature 245.0 °C.

The dehydroxylation of clay minerals, decomposition of carbonates and burning of organic admixtures are the main overlapping processes whose take place within temperature interval from 380 to 600 °C. The DTA shows broad endothermic peak having a composed structure at temperature 533.9 °C. The bands of carbon dioxide and water are well visible on EGA plot. These processes are affected together via partial pressure formed gas species. For example the water formed by dehydroxylation of montmorillonite slows down the diffusion of oxygen into burning organic material and shifts the organic matter process to the higher temperatures, while water vapour formed by combustion of organic admixtures leads to increasing of partial pressure of water vapours. That results into decreasing rate of dehydroxylation of clay minerals (Ptáček at al., 2010). Oxygen deficiency leading to reduction condition during TA is indicated by bands of carbon monoxide on the results of EGA.

The effects of carbonates on the above mentioned processes should be explained using the Richardson's diagrams (Richardson, 1974) as follows. The Bell-Boudoir's reaction (Eq.3.1) shows thermodynamic equilibrium at temperature 720 °C, so that the carbon monoxide is the more stable at higher temperature than carbon dioxide. That means that CO_2 formed by the thermal decomposition of carbonates at temperatures near to temperature of equilibrium or higher facilitates the residual carbon removing process.

$$CO_2(g) + C(s) \xleftrightarrow{T_{eq.} \approx 720\,°C} 2\,CO(g) \tag{3.1}$$

The two carbonates are identified in the analysis sample – siderite ($FeCO_3$) and dolomite ($CaMg(CO_3)_2$). Thermal decomposition of siderite that takes place at temperatures up to 410 °C are participate on the broad DTA endothermic effect at 533.9 °C. Annealing dolomite is decomposed within two steps that are represented by reactions 3.2 and 3.3. The first step takes place at 602.7 °C and second at 718.3 °C. The both processes are well visible on EGA.

$$CaMg(CO_3)_2(s) \rightarrow CaCO_3(s) + MgO(s) + CO_2(g) \tag{3.2}$$

$$CaCO_3(s) \rightarrow +CaO(s) + CO_2(g) \tag{3.3}$$

The formation of SO_2 was detected on EGA upon temperature interval from 800 to 870 °C due to presence of traces of pyrite. The endothermic peak at temperature 848.2 °C is related to the formation of cordierite that is connected with destruction of the phylosilicate structure of clay minerals. The eutectic melt was detected at temperature 1141.4 °C.

During thermal analysis of talc raw material performed up to temperature 1250 °C (Fig.3.2) is mass of the sample decreasing for about 13.12 %. The adsorbed water is removed up to 143 °C. The mass of sample was reduced for about 0.21 % during this process. The dehydroxylation of talc which takes place in temperature range from 720 to 970 °C and two steps of thermal decomposition of dolomite at 450 and 720 °C are the main occurring processes. The SO_2 bands in EGA plot indicate the presence of small amount of pyrite.

3.2 Infrared spectroscopy
The infrared spectrum of montmorillonite and talc clay is shown on Fig.3.3. The data published in literature (Eren & Afsin, 2008; Molina-Montes at al., 2008; Madejová at al., 2006; Tyagi at al., 2006; Kloprogge at al., 2005) were used for interpretation of raw material

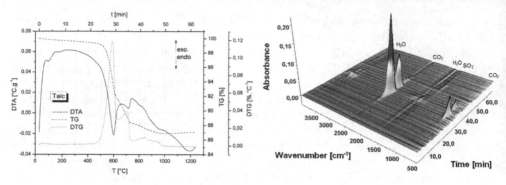

Fig. 3.2. TG-DTA plot and EGA pattern of talc raw material.

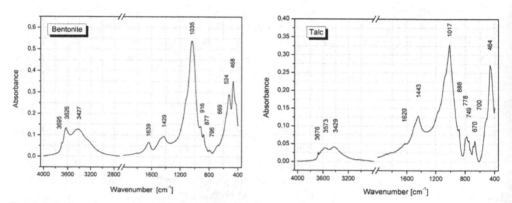

Fig. 3.3. Infrared spectrum of bentonite and talc raw material.

spectral features. The OH stretching bands are located at 3695 and 3626 cm^{-1}. The bending of AlAlO-H, AlFeO-H a AlMgO-H groups show bands at 916, 877 and 837 cm^{-1}. The stretching and bending band of physical adsorbed water are located at 3427 and 1639 cm^{-1}. The most intensive band at 1035 cm^{-1} is related to antisymmetric stretching of the ≡Si-O-Si≡ bridge. The deformation mode is placed at 524 cm^{-1}. The dolomite and quartz are identified by infrared spectroscopy as the main admixtures of clay raw material that was used as the source of montmorillonite.

The infrared spectrum of talc (Fig.3.3) shows stretching of MgO-H groups at wavenumber 3626 cm^{-1}. The deformation modes are located at 670 and 646 cm^{-1}. The band of antisymmetric stretching and bending mode of ≡Si-O bond shows maximum absorption intensity at 1017 and 453 cm^{-1}, respectively. The other bands belong to admixture minerals - clinochlore and dolomite.

3.3 Clay material particle size distribution and morphology

The SEM and particle size distribution analysis of clays is shown in Fig.3.4. Bentonite consists of massive aggregates. The most important admixture minerals of montmorillonite clay (Fig.3.5) are siderite ($FeCO_3$) and carbonates from dolomite group such as dolomite ($CaMg(CO_3)_2$) and ankerite ($CaFe(CO_3)_2$), phylosilicates illite (($K,H_3O^+)Al_2(Si,Al)_4O_{10}(OH)_2$)

and clinochlore $((Mg,Fe)_5Al(Si,Al)_4O_{10}(OH)_8)$. Further fluorapatite $(Ca_5(PO_4)_3F)$, barite $(BaSO_4)$ and rutile (TiO_2) are indentified in the clay. It's obvious that the carbonates serve as the source of Ca, Fe, Mg, Zn, Mn, etc. elements at the early stages of dissolution experiment. The particle size analysis of raw materials, i.e. bentonite clay and talc, used for leaching experiments are shown at Fig.3.4. The shape of particle size distribution curve of bentonite raw material reflects the complicate phase composition of sample that contains a significant amount of carbonates and other admixture minerals of different hardness compared to clay, i.e. minerals with different grindability. These admixtures are responsible for the right shoulders of the particle size distribution curve. The talc raw material with high content of clay phase shows almost ideal Gaussian profile of particle size distribution curve with median 29.33 µm (Table 3.1).

Fig. 3.4. SEM and particle size distribution analysis of clay raw material.

The layered structure of talc aggregates is shown at Fig. 3.6. The average size of (001) planes was via several measurements estimated on 200 µm. The calcite was identified as the main admixture mineral of talc raw material.

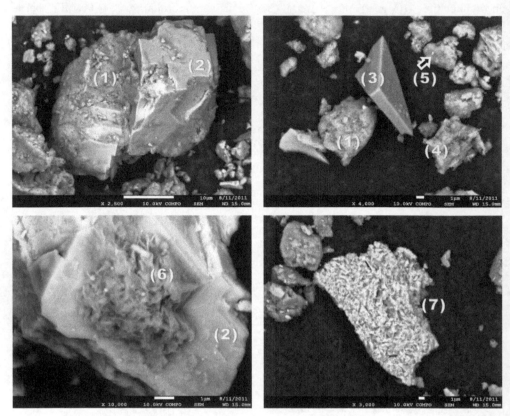

Fig. 3.5. The admixture mineral of montmorillonite clay (1): siderite (2), ankerite (3), illite (4), barite (5), clinochlore (6) and fluoroapatite (7).

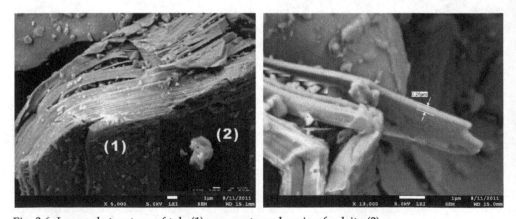

Fig. 3.6. Layered structure of talc (1) aggregate and grain of calcite (2).

3.4 Bentonite dissolution

The dissolution of main bentonite mineral, i.e. montmorillonite, in diluted solution of acetic acid should be expressed as following:

$$(Na,Ca)_{0.3}(Al,Mg)_2 Si_4 O_{10}(OH)_2 \cdot nH_2O +$$
$$y\,CH_3CO_2H + p\,H_3O^+ \rightarrow x\,CH_3CO_2Na + 0.3 - x\,Ca(CH_3CO_2)_2 \tag{3.4}$$
$$+z\,Mg(CH_3CO_2)_2 + 2 - z\left[Al(H_2O)_6\right]^{3+} + 4\,SiO_2 + q\,H_2O$$

where $y = 2z - x + 0{,}6$, $p = 5{,}4 - 2z + x$ and $q = 6z + n - 6$. On the other hand, with regard to the montmorillonite structure that is described in chapter 1.3, the release of cations from interlayer space is participating on the process. These ions are being exchanged by H_3O^+ according to Eq.3-2.

$$(Na,Ca)_{0.3}(Al,Mg)_2 Si_4 O_{10}(OH)_2 \cdot nH_2O +$$
$$(0.6 - x)\,CH_3CO_2H + (0.6 - x)\,H_3O^+ \rightarrow \tag{3.5}$$
$$(H_3O^+)_{0.6-x}(Al,Mg)_2 Si_4 O_{10}(OH)_2 \cdot nH_2O +$$
$$x\,Na(CH_3CO_2) + (0.3 - x)\,Ca(CH_3CO_2)_2$$

It was found by (Adams, 1987; Jovanovič and Janačkovič, 1991) that acid-activated (HCl or H_2SO_4 of different molar concentrations) bentonite leads to a dissolution or removal of the octahedral sheets and interlayer cations. Its resulting in an increase of the pore volume and pore diameter, an enrichment of residual amorphous SiO_2 and an increase of sorption properties.

The pH change of solvent during leaching process performed upon temperatures within range from 22 to 50 °C is shown in the Fig.3.7. The dependence of fractional conversion on the time was calculated according to formula 2.1 and 2.3 from measured pH on time dependence.

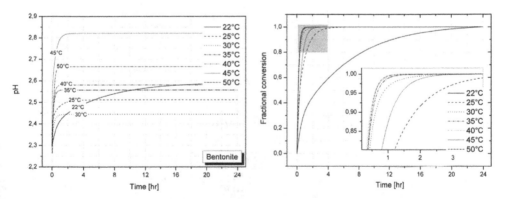

Fig. 3.7. The pH of leaching bath for experiment performed under different temperature and time dependence of fractional conversion.

The results of leaching experiment on montmorillonite clay show that mechanism of process is significantly affected by temperature. Linearization procedure leads to conclusion that the leaching process is handled by the stationary three-dimensional diffusion (D_4) at temperatures up to 25 °C, i.e. the course of leaching process can be characterized by Valensi-Ginstling-Brounstein (VGB) equation (Valensi, 1936; Ginstling and Brounstein, 1950):

$$g(y) = k\,t = 1 - \frac{2}{3}y - (1-y)^{2/3} \tag{3.6}$$

$$g(y) = k\,t = -\ln(1-y) \tag{3.7}$$

The Kolmogorov-Johnson-Mehl-Avrami (KJMA) equation shows the best results for experiments performed upon temperature interval from 30 to 40 °C. The kinetic function corresponding to the mechanism of random nucleation and subsequent growth of nuclei (F_1 or A_1) can be described by Eq.3.7.
At temperatures higher than 40 °C the leaching process is forced by chemical reaction of ¾th order ($F_{3/4}$), i.e. by mechanism non-invoking equation:

$$g(y) = k\,t = 1 - (1-y)^{1/4} \tag{3.8}$$

The Arrhenius plot is shown on Fig.3.8. The value of apparent activation energy that was determined upon the above mentioned temperature interval is listed in the Table 3.2.

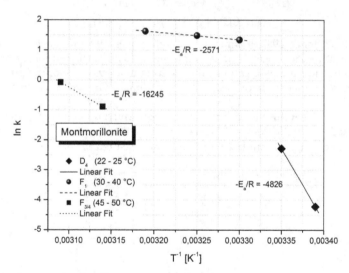

Fig. 3.8. The Arrhenius plot for the montmorillonite dissolved in diluted acetic acid.

T [°C]	Mechanism	k [s⁻¹]	D = R²	Eₐ [kJ·mol⁻¹]
22	D4: 1 - 2y/3 - (1 - y)^{2/3}	$1.47 \cdot 10^{-2}$	0.999	388.9
25		$1.02 \cdot 10^{-1}$	0.999	
30		3.82	0.999	
35	F1: -ln (1-y)	4.41	0.998	21.4
40		5.08	0.998	
45	F₃⁄₄: 1 - (1-y)^{1/4}	$4.10 \cdot 10^{-1}$	0.997	135.1
50		$9.23 \cdot 10^{-1}$	0.999	

Table 3.2. E_a of dissolution of montmorillonite clay. $D = R^2$ is the correlation coefficient of linear fit.

The results of ICP-OES analysis (Fig.3.9) of solvent after the leaching experiment show that predominantly extracted elements are Ca, Mg, Mg, Fe and Al. The amount of elements extracted per gram of clay raw material is listed in the Table 3.3.

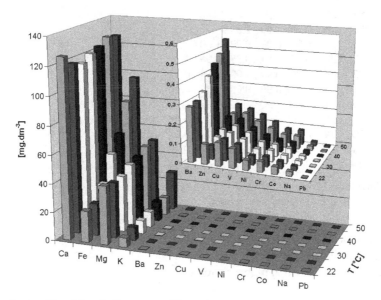

Fig. 3.9. Analysis of leaching bath composition.

With except of calcium where extracted amount is not correlated with temperature (Table 3.3), the amount of extracted elements is generally increasing with temperature. The higher temperature then enables to reach better activation of bentonite by acetic acid using higher temperatures due to increasing content of leached Fe and Mg. That behaviour results from the structure of mineral of smectite groups (Fig.3.2).

T [°C]	Element leached from the montmorillonite clay [µg·g⁻¹ of raw material]													
	Al	V	Cr	Co	Ni	Cu	Zn	Ba	Pb	Na	Mg	K	Ca	Fe
20	751.84	20.112	10.758	7.558	10.19	24.33	21.082	57.054	0.778	4.838	8238	1320	25220	4410
25	743.18	19.642	10.484	7.522	9.526	23.194	14.552	58.856	0.76	4.628	7850	1844	23780	4576
30	722.36	20.326	11.992	8.546	11.608	25.148	18.97	65.142	0.778	4.496	7946	1892	23250	6560
35	775.7	22.356	13.164	10.596	14.396	28.098	22.522	77.964	0.828	4.596	8820	2232	24160	10088
40	807.03	23.064	13.37	11.738	16.662	30.824	31.928	88.106	1.042	4.624	9148	2786	24670	12260
45	824.9	25.888	14.446	14.152	19.832	33.072	31.434	96.29	1.06	4.662	9932	2570	25560	16336
50	845.88	28.826	15.622	15.776	22.384	33.866	37.566	109.69	1.588	4.542	10160	5550	25350	19284

Table 3.3. Influence of temperature on the extraction process.

While calcium is placed in place interlayer space and should be then easily replaced by sodium by cation exchange process, magnesium is bonded in brucite sheet of T-O-T complex and it can be released only after its dissolution. That is also the reason for observed correlation of Mg on the amount of extracted Al and other cation (Fe^{3+}, Cr^{3+}, V^{3+}...) coordinated octahedrally in the "O" layer.

	T	Al	V	Cr	Co	Ni	Cu	Zn	Ba	Pb	Na	Mg	K	Ca	Fe
T	1.00														
Al	0.89	1.00													
V	0.94	0.94	1.00												
Cr	0.98	0.89	0.96	1.00											
Co	0.98	0.95	0.99	0.98	1.00										
Ni	0.97	0.95	0.98	0.98	1.00	1.00									
Cu	0.97	0.95	0.95	0.97	0.98	0.99	1.00								
Zn	0.89	0.95	0.93	0.92	0.94	0.95	0.96	1.00							
Ba	0.99	0.95	0.97	0.98	0.99	1.00	0.98	0.95	1.00						
Pb	0.85	0.87	0.94	0.86	0.90	0.90	0.84	0.90	0.91	1.00					
Na	-0.49	-0.08	-0.29	-0.42	-0.33	-0.32	-0.26	-0.15	-0.37	-0.31	1.00				
Mg	0.93	0.98	0.97	0.95	0.98	0.98	0.98	0.95	0.97	0.86	-0.14	1.00			
K	0.83	0.80	0.90	0.84	0.85	0.85	0.78	0.81	0.87	0.98	-0.46	0.79	1.00		
Ca	0.48	0.78	0.67	0.54	0.64	0.65	0.67	0.73	0.60	0.60	0.51	0.77	0.43	1.00	
Fe	0.98	0.95	0.98	0.98	1.00	1.00	0.98	0.94	1.00	0.90	-0.35	0.98	0.85	0.62	1.00

Table 3.4. Correlation table showing mutual relationships between temperature and amount of leached elements. The significant correlation is marked by bold.

The increasing efficiency of extraction process is shown in Table.3.5 as the calculated amount of carbon dioxide that may be captured by the extracted element in formed carbonate. The results indicate that extraction efficiency should be significantly improved by activation process performed at higher temperatures.

T [°C]		22	25	30	35	40	45	50
Ca	kg_{CO_2}/	27.7	26.1	25.5	26.5	27.1	28.1	27.9
Mg	1000 kg	14.9	14.2	14.4	16.0	16.6	18.0	18.4
Fe	raw	3.5	3.6	5.2	8.0	9.7	12.9	15.2
Σ	clay	46.1	43.9	45.9	50.5	53.3	58.9	61.4
$CaCO_3$		63.6	60.0	58.6	60.9	62.2	64.5	63.9
$MgCO_3$	kg	28.6	27.2	27.6	30.6	31.7	34.5	35.3
$FeCO_3$		9.1	9.5	13.6	20.9	25.4	33.9	40
Σ		101.3	96.7	99.8	112.5	119.4	132.8	139.2

Table 3.5. Bentonite clay activation efficiency.

The solid rest that is resulting from the leaching process was analysed by TA, IR and SEM to determine its properties for the usage in cements due to estimated puzzolanic activity or as absorption agents. Table 3.6 show that higher specific surface of leaching rest should be obtained for the sample prepared at temperature 35 °C.

T [°C]	22	25	30	35	40	45	50
SS $[m^2/g]$	91.4	96.8	96.0	106.6	102.5	97.1	97.8

Table 3.6. Influence the temperature of leaching bath on the specific surface of solid rest.

The results of infrared spectroscopy are shown in Fig.3.10. The results indicate that the raw material activation process is based mainly on the reaction 3.5. The increasing shift of the Si-O-Si stretching mode with temperature of leaching process indicate that the minerals are affected only by the formation of thin leached silica layer on the surface of aggregate, i.e. only the first step of incongruent dissolution process takes place.

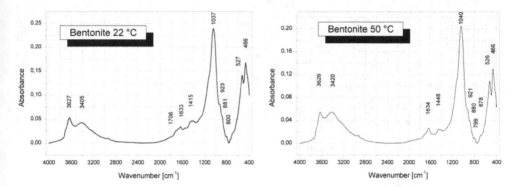

Fig. 3.10. Infrared spectrum of bentonite leached by acetic acid.

The results of thermal analysis (Fig.3.11) indicate introducing the salt of acetic acid into interlayer space of bentonite. The evaporation of acetic acid and burning of acetic salt are well visible on EGA. Thermal decomposition of acetates that is according to EGA connected with formation of acetone and carbon dioxide. Presence of carbon monoxide and dioxide bands upon the same temperature interval indicates the partially reduction condition of process that leads to formation of the calcium and magnesium carbonates:

$$M(CH_3COO)_2 + 5/2\,O_2 \rightarrow MCO_3 + 3\,H_2O + 3\,CO \qquad (M = Ca \; \boldsymbol{and} \; Mg) \qquad (3.9)$$

Thermal decomposition of formed carbonates that takes place upon temperature interval from 700 to 900 °C is well visible on DTA as well as EGA pattern.

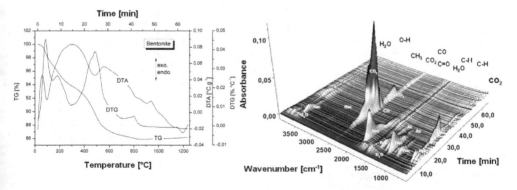

Fig. 3.11. Thermal analysis of solid rest after leaching process.

The SEM analysis of the clay after dissolution experiment is shown in Figure 3.12. The admixture of carbonate minerals (please see Fig.3.5) are dissolved at early stages of leaching process. The leached silica layer was formed on the surface of bentonite aggregates.

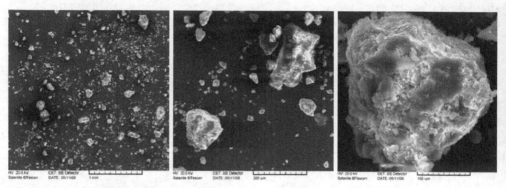

Fig. 3.12. The SEM of leached bentonite clay.

3.5 Dissolution of talc

The process of the dissolution of Talc in diluted solution of acetic acid should be described by following equation:

$$Mg_3Si_4O_{10}(OH)_2(s) + 6\,CH_3COOH(aq) \rightarrow$$
$$3\,(CH_3COOH)_2\,Mg(aq) + 4\,SiO_2(s) + 4\,H_2O(l)$$
$$(3.10)$$

The measured dependence of pH on the time of dissolution and fractional conversion time dependence calculated according to Eq.2.1 and 2.3 is shown on Fig.3.13. To compare with bentonite clay, the course of talc activation process seems to be less affected by the temperature of leaching bath. Hence only the limit temperatures are plotted in the Fig.3.13.

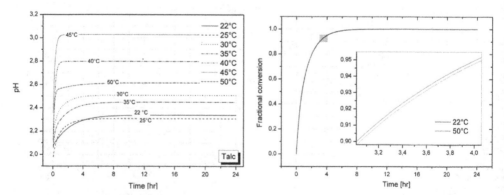

Fig. 3.13. The change of pH of leaching solution during activation of talc (a) and fractional conversion on time dependence (b).

The kinetic of leaching process should be described by the kinetic law:

$$y = 1 - \exp\left(-K t^{\frac{7}{6}}\right) \tag{3.11}$$

where kinetic exponent (Avrami's factors) has value of 1.2.

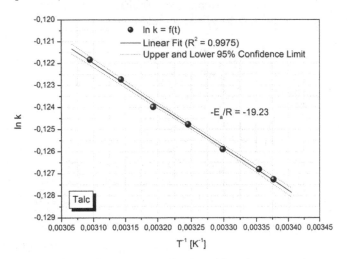

Fig. 3.14. The Arrhenius plot for the talc dissolved in diluted acetic acid.

The Arrhenius plot that is shown in Fig.3.14 was used for determination the apparent activation energy of the leaching test from the dependence of 160 ± 3 J·mol^{-1}.
The results of ICP-OES analysis of solvent after leaching tests performed within the temperature interval from 22 to 50 °C are plotted in Fig.3.15.

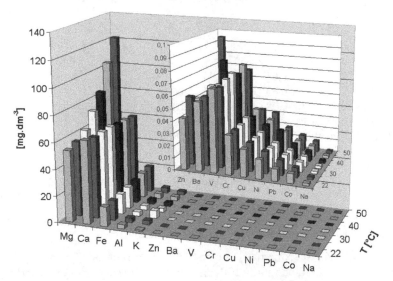

Fig. 3.15. The composition of leaching bath after leachig process.

There it is obvious that the amount of extracted magnesium and iron is strongly affected by the temperature while the calcium content is slightly decreasing with increasing temperature of solvent. It should be thus supposed that Ca come to solution in very short time after pouring the solvent during dissolution of the admixture of carbonates in the raw material. The negative temperature dependence is probably caused by absorption of calcium on leached layer that is formed on the surface of talc aggregates.

T [°C]	Element leached from the montmorillonite clay [$\mu g \cdot g^{-1}$ of raw material]													
	Al	V	Cr	Co	Ni	Cu	Zn	Ba	Pb	Na	Mg	K	Ca	Fe
20	612.55	13.53	6.44	1.6	3.17	4.51	8.48	11.29	1.96	0.4	10790	170	12580	2950
25	631.85	13.15	6.55	1.61	3.62	4.76	11.28	11.49	2.53	0.43	11330	620	12300	2970
30	661.65	13.93	6.67	1.62	3.38	4.79	10.38	12.63	2.53	0.44	12340	1190	12400	3100
35	696.86	14.39	7.16	1.71	3.4	4.94	9.6	12.21	2.49	0.44	14600	590	12450	3310
40	775.73	13.72	6.57	1.74	4.76	5.44	10.39	15.86	2.62	0.44	16870	540	12150	3460
45	804.73	14.82	7.28	1.85	3.75	5.63	10.79	12.2	2.48	0.46	20870	620	12100	3810
50	868.34	13.43	6.54	1.86	3.77	6.24	18.92	12.51	2.66	0.42	24300	570	11920	4050

Table. 3.7. Influence of temperature on the extraction process.

The increasing content of calcium in the leaching bath (please refer to Table 3.7) as well as the correlation between extracted amount of Al (Table 3.8) should be explained analogically with leaching test of montmorillonite clay.

	T	Al	V	Cr	Co	Ni	Cu	Zn	Ba	Pb	Na	Mg	K	Ca	Fe
T	1,00														
Al	0,98	1,00													
V	0,37	0,27	1,00												
Cr	0,38	0,26	0,93	1,00											
Co	0,96	0,97	0,43	0,44	1,00										
Ni	0,51	0,54	-0,05	-0,10	0,41	1,00									
Cu	0,96	0,99	0,14	0,16	0,94	0,50	1,00								
Zn	0,68	0,72	-0,31	-0,21	0,63	0,17	0,82	1,00							
Ba	0,42	0,43	0,04	-0,10	0,28	0,91	0,36	0,03	1,00						
Pb	0,69	0,61	0,05	0,19	0,51	0,57	0,62	0,55	0,50	1,00					
Na	0,43	0,30	0,70	0,71	0,34	0,41	0,21	-0,20	0,43	0,59	1,00				
Mg	0,97	0,98	0,28	0,29	0,98	0,40	0,98	0,76	0,27	0,54	0,23	1,00			
K	0,14	0,02	0,19	0,17	-0,06	-0,03	0,01	0,08	0,13	0,57	0,57	-0,02	1,00		
Ca	-0,89	-0,91	-0,01	-0,06	-0,85	-0,59	-0,95	-0,80	-0,38	-0,72	-0,27	-0,90	-0,11	1,00	
Fe	0,97	0,98	0,33	0,33	0,99	0,39	0,98	0,73	0,28	0,53	0,26	1,00	-0,01	-0,88	1,00

Table 3.8 Correlation table showing mutual relationships between temperature and amount of leached elements. The significant correlation is marked by bold.

The increasing efficiency of extraction process is shown in Table.3.9. The results indicate that extraction efficiency should be significantly improved by increasing of extracted Mg amount at higher temperatures.

T [°C]		22	25	30	35	40	45	50
Ca	kg$_{CO2}$/	13.8	13.5	13.6	13.7	13.4	13.3	13.1
Mg	1000 kg	19.6	20.5	22.4	26.4	30.6	37.8	44.1
Fe	raw	2.3	2.3	2.4	2.6	2.7	3.0	3.19
Σ	clay	35.7	36.4	38.4	42.7	46.6	54.1	60.3
CaCO$_3$		31.7	31.0	31.3	31.4	30.7	30.5	30.1
MgCO$_3$	kg	37.4	39.3	42.8	50.7	58.5	72.4	84.3
FeCO$_3$		6.1	6.2	6.4	6.9	7.2	7.9	8.4
Σ		75.3	76.5	80.5	88.9	96.4	110.8	122.8

Table 3.9. Bentonite clay activation efficiency.

Table 3.10 show that higher specific surface of leaching rest after dissolution experiment.

T [°C]	22	25	30	35	40	45	50
SS [m²/g]	3.2	3.3	2.9	3.1	3.1	3.1	3.6

Table 3.10. Influence the temperature of leaching bath on the specific surface of solid rest.

The infrared spectra of solid rest after leaching process are shown in Fig.3.16. The spectrum features indicate that the changes caused by leaching process are much lesser than in the case of bentonite.

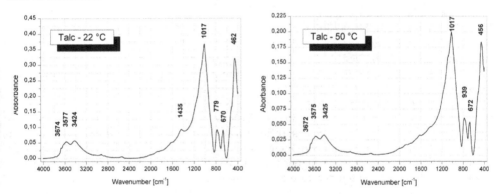

Fig. 3.16. Infrared spectroscopy of solid rest after activation of talc by acetic acid.

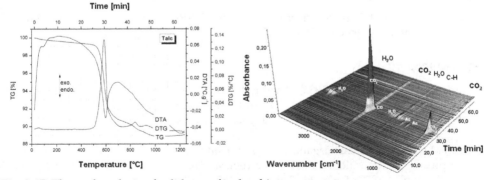

Fig. 3.17. Thermal analysis of solid rest after leaching process.

The typical results of thermal analysis shown on Fig.3.14. lead to the same conclusion. Thermal decomposition of acetates that is according to EGA connected with formation of carbon dioxide and acetone and dehydroxylation of talc are the main observed processes. With except of carbonate admixtures that were naturally dissolved, the results of SEM (Fig.3.18) do not show any significant changes in the activated talc material.

Fig. 3.18. Leached talc raw material.

4. Conclusion

The initial stage of bentonite leaching process is on exchange of Ca and K from the interlayer space of montmorillonite and illite. The dissolution of T-O-T complex that is promoted by higher temperature then leads to the release of Mg and other octahedrally coordinate ions. Storage capacity of bentonite clay for CCS should be then significantly improved by activation process performed at elevated temperature. Increasing temperature promotes the rate of incongruent leaching process. The process of activation of talc shows also significant influence of leaching bath on the process. While amount of extracted calcium remains constant or slightly decrease due to absorption phenomena, the amount of extracted calcium should be significantly improved with increasing temperature of leaching bath. The capacity for CO_2 caption is at about 35 % higher for the clay of montmorillonite. This difference is decreasing with increasing temperature of leaching bath.

5. Acknowledgment

This paper is supported by the research project of ERDF no. CZ.1.05/2.1.00/01.0012" Centres for Materials Research at FCH BUT".

6. References

Aagaard, P., Helgeson, H.C. (1982). Thermodynamic and kinetic constraints on reaction rates among minerals and aqueous solutions: I. Theoretical considerations. *American Journal of Science*, Volume 282, Issue 3 (March 1982), Pages 237-285, ISSN 0002-9599.

Adams, J.M. (1987). Synthetic organic chemistry using pillared cation exchanged and acid-treated montmorillonite catalysts - a review. *Applied Clay Science*, Volume 2, Issue 4 (September 1987), Pages 309-342, ISSN 0169-1317.

Alexander, G., Maroto-Valer, M. M., Gafarova-Aksoy, P. (2007). Evaluation of reaction variables in the dissolution of serpentine for mineral carbonation. *Fuel*, Volume 86, Issues 1-2 (January 2007), Pages 273-281, ISSN 0016-2361.

Augusto, L., Turpault, M.-P., Ranger, J. (2000). Impact of forest tree species on feldspar weathering rates. *Geoderma*, Volume 96, Issue 3 (June 2000), Pages 215-237, ISSN 0016-7061.

Baccouche, A., Srasra, E., Maaoui, M.E. (1998). Preparation of Na-P1 and sodalite octahydrate zeolites from interstratified illite–smectite. *Applied Clay Science*, Volume 13, Issue 4 (October 1998), Pages 255-273, ISSN 0169-1317.

Bachu, S. (2008). CO_2 storage in geological media: Role, means, status and barriers to deployment. *Progress in Energy and Combustion Science*, Volume 34, Issue 2 (April 2008) Pages 254-273, ISSN 0360-1285.

Bachu, S., Adams, J.J. (2003). Sequestration of CO_2 in geological media in response to climate change: capacity of deep saline aquifers to sequester CO_2 in solution. Energy Conversion and Management, Volume 44, Issue 20 (December 2003), Pages 3151-3175, ISSN 0196-8904.

Bemal, J.D. (1967). *The Origin of Life*, Universe Books, ISBN 0876631154, London.

Bouchard, R., Delaytermoz, A. (2004). Integrated path towards geological storage. *Energy*, Volume 29, Issues 9-10 (July-August 2004), Pages 1339-1346, ISSN 0360-5442.

Buchwald, A., Hohmann, M., Posern, K., Brendler, E. (2009). The suitability of thermally activated illite/smectite clay as raw material for geopolymer binders. *Applied Clay Science*, Volume 46, Issue 3 (November 2009), Pages 300-304, ISSN 0169-1317.

Caglar, B., Afsin, B., Tabak, A., Eren, E. (2009). Characterization of the cation-exchanged bentonites by XRPD, ATR, DTA/TG analyses and BET measurement. *Chemical Engineering Journal*, Volume 149, Issues 1-3 (July 2009), Pages 242-248, ISSN 1385-8947.

Cairns-Smith, A.G. (1982). *Genetic Take-over and the Mineral Origins of Life*, Cambridge University Press, ISBN 0-521-34682-7, New York.

Cairns-Smith. A.G. (1966). The origin of life and the nature of the primitive gene. *Journal of Theoretical Biology*, Volume 10, Issue 1 (January 1966), Pages 53-88, ISSN 0022-5193.

Cama. J., Ayora, C., Lasaga, A. (1999). The deviation-from-equilibrium effect on dissolution rate and on apparent variations in activation energy. *Geochimica et Cosmochimica Acta*, Volume 63, Issue 17 (October 1999), Pages 2481-2486, ISSN 0016-7037.

Chardon, E.S., Livens, F.R., Vaughan, D.J. (2006). Reactions of feldspar surfaces with aqueous solutions. *Earth-Science Reviews*, Volume 78, Issues 1-2 (September 2006), Pages 1-26, ISSN 0012-8252.

Chen, Y., Brantley, S.L. (2000). Dissolution of forsteritic olivine at 65 °C and 2 < pH < 5. *Chemical Geology*, Volume 165, Issues 3-4 (April 2000), Pages 267-281, ISSN 0009-2541.

Damen, K., Faaij, A., Turkenburg, W. (2006a). Health, safety and environmental risks of underground CO_2 storage – overview of mechanisms and current knowledge. *Climate Change*, Volume 74, Issues 1-3 (2006), Pages 289–318, ISSN ISSN: 0165-0009.

Daux, V., Guy, Ch., Advocat, T., Crovisier, J.-L., Stille, P. (1997). Kinetic aspects of basaltic glass dissolution at 90°C: role of aqueous silicon and aluminium. *Chemical Geology*, Volume 142, Issues 1-2 (October 1997), Pages 109-126, ISSN 0009-2541.

Dorozhkin S.V. (2002). A review on the dissolution models of calcium apatites, *Progress in Crystal Growth and Characterization of Materials*, Volume 44, Issue 1 (2002), Pages 45-61, ISSN 0960-8974.

Dove, P.M., Crerar, D.A. (1990). Kinetics of quartz dissolution in electrolyte solutions using a hydrothermal mixed flow reactor. *Geochimica at Cosmochimica Acta*, Volume 54, Issue 4 (April 1990), Pages 955-969, ISSN 0016-7037.

Duan, Y., Li, J., Yang. X., Hu, L., Wang, Z., Liu, Y., Wang, C. (2008). Kinetic analysis on the non-isothermal dehydration by integral master-plots method and TG–FTIR study of zinc acetate dihydrate. *Journal of Analytical and Applied Pyrolysis*, Volume 83, Issue 1 (September 2008), Pages 1-6, ISSN 0165-2370.

Eren, E., Afsin, B. (2008). An investigation of Cu(II) adsorption by raw and acid-activated bentonite: A combined potentiometric, thermodynamic, XRD, IR, DTA study. *Journal of Hazardous Materials*, Volume 151, Issues 2-3 (March 2008), Pages 682-691, ISSN 0304-3894.

Friedmann, S.J., Dooley, J.J., Held, H., Edenhofer, O. (2006). The low cost of geological assessment for underground CO_2 storage: Policy and economic implications. *Energy Conversion and Management*, Volume 47, Issues 13-14 (August 2006), Pages 1894-1901, ISSN 0196-8904.

Gale, J. (2004). Geological storage of CO_2: What do we know, where are the gaps and what more needs to be done? *Energy*, Volume 29, Issues 9-10 (July-August 2004), Pages 1329-1338, ISSN 0360-5442.

Gérard, F., Fritz, B., Clément, A., Crovisier, J.-L. (1998). General implications of aluminium speciation-dependent kinetic dissolution rate law in water–rock modelling. *Chemical Geology*, Volume 151, Issues 1-4 (October 1998), Pages 247-258, ISSN 0009-2541.

Giammar, D.E., Bruant, R.G.Jr., Peters, C.A. (2005). Forsterite dissolution and magnesite precipitation at conditions relevant for deep saline aquifer storage and sequestration of carbon dioxide. *Chemical Geology*, Chemical Geology, Volume 217, Issues 3-4 (April 2005), Pages 257-276, ISSN 0009-2541.

Gibbins, J., Chalmers. H. (2008). Carbon capture and storage. *Energy Policy*, Volume 36, Issue 12 (December 2008), Pages 4317-4322, ISSN 0301-4215.

Gin, S., Jégou, Ch., Frugier, P., Minet, Y., Theoretical consideration on the application of the Aagaard–Helgeson rate law to the dissolution of silicate minerals and glasses. *Chemical Geology*, Volume 255, Issues 1-2 (September 2008), Pages 14-24, ISSN 0009-2541.

Ginstling, A.M., Brounstein, B.I. (1950). Journal of Applied Chemistry of the USSR (English translation) 23 (1950) 1327–1338.

Grimshaw, R.W. (1971). *The Chemistry and Physics of Clays and Allied Ceramic Materials*, Techbooks, ISBN 1878907441, London.

Hartman, H. (1975). *Journal of Molecular Evolution*, Volume 4, Issue 4 (1975), Pages 359-370, ISSN 0022-2844.

Haydn H.H. (2000). Traditional and new applications for kaolin, smectite, and palygorskite: a general overview. *Applied Clay Science*, Volume 17, Issues 5-6 (November 2000), Pages 207-221, ISSN 0169-1317.

Hellmann, R., Daval, D., Tisserand, D. (2009). The dependence of albite feldspar dissolution kinetics on fluid saturation state at acid and basic pH: Progress towards a universal relation. *Comptes Rendus Geoscience*, Volume 342, Issues 7-8 (July-August 2010), Pages 676-684, ISSN 1631-0713.

Hoffert, M.I., Caldeira, K., Jain, A.K., Haites, E.F., Harvey, L.D.D., Potter, S.D., Schlesinger, M.E., Schneider, S.H., Watts, R.G., Wigley, T.M.L., Wuebbles, D. J. (1998). Energy implications of future stabilization of atmospheric CO_2 content. *Nature*, Volume 395 (October 1998), Pages 881-884, ISSN: 0028-0836.

Huesemann, M.H. (2006). Can advances in science and technology prevent global warming? *Mitigation and Adaptation Strategies for Global Change*, Volume 3, Issue 11 (2006), Pages 539 – 577, ISSN 1381-2386.

Huijgen, W.J. J. , Comans, R.N. J., Witkamp, G.-J. (2007). Cost evaluation of CO_2 sequestration by aqueous mineral carbonation. *Energy Conversion and Management*, Volume 48, Issue 7 (July 2007), Pages 1923-1935, ISSN 0196-8904.

Iler, R.K. (1979). *The Chemistry of Silica: Solubility, Polymerization, Colloid and Surface Properties and Biochemistry*. A Wiley-Interscience Publication, ISBN 0-471-02404-X.

IPCC (Intergovernmental panel on climate change). Climate change 2007: The physical science basis. Fourth assessment report, IPCC Secretariat, Geneva, Switzerland, 2007.

Jiang, X. (2011). A review of physical modelling and numerical simulation of long-term geological storage of CO_2. *Applied Energy*, Volume 88, Issue 11 (November 2011), Pages 3557-3566, ISSN 0306-2619.

Jonckbloedt, R.C.L. (1989). Olivine dissolution in sulphuric acid at elevated temperatures — implications for the olivine process, an alternative waste acid neutralizing process. *Journal of Geochemical Exploration*, Volume 62, Issues 1-3 (June 1998), Pages 337-346, ISSN 0375-6742.

Jovanović, N., Janaćković, E.J. (1991). Pore structure and adsorption properties of an acid-activated bentonite. *Applied Clay Science*, Volume 6, Issue 1, (May 1991), Pages 59-68, ISSN 0169-1317.

Kaya, Y. (1995). The role of CO_2 removal and disposal. *Energy Conversion and Management*, Volume 36, Issues 6-9 (June-September 1995), Pages 375-380, ISSN 0196-8904.

Kloprogge, J.T., Mahmutagic, E., Frost, R.L., Mid-infrared and infrared emission spectroscopy of Cu-exchanged montmorillonite. *Journal of Colloid and Interface Science*, Volume 296, Issue 2 (April 2006) Pages 640-646, ISSN 0021-9797.

Knauss, K.G., Nguyen, S.N., Weed, H.C. (1993). Diopside dissolution kinetics as a function of pH, CO_2, temperature, and time. *Geochimica et Cosmochimica Acta*, Volume 57, Issue 2 (January 1993), Pages 285-294, ISSN 0016-7037.

Köhler, S.J., Bosbach, D., Oelkers E.H. (2005). Do clay mineral dissolution rates reach steady state? *Geochimica et Cosmochimica Acta*, Volume 69, Issue 8 (April 2005), Pages 1997-2006, ISSN 0016-7037.

Komadel, P., Madejová, J. (2006). Chapter 7.1 Acid Activation of Clay Minerals. *Developments in Clay Science*, Volume 1 (2006), Pages 263-287.

Konta, J. (1995). Clay and man: Clay raw materials in the service of man. *Applied Clay Science*, Volume 10, Issue 4 (November 1995), Pages 275-335, ISSN 0169-1317.

MacKenzie, K.J.D., Meinhold, R.H. (1994). The thermal reactions of talc studied by ^{29}Si and [25]Mg MAS NMR. *Thermochimica Acta*, Volume 244, Issue 3 (October 1994), Pages 195-203, ISSN 0040-6031.

Madejová, J., Pálková, H., Komadel, P. (2006). Behaviour of Li^+ and Cu^{2+} in heated montmorillonite: Evidence from far-, mid-, and near-IR regions. *Vibrational Spectroscopy*, Volume 40, Issue 1 (January 2006), Pages 80-88, ISSN 0924-2031.

Maroto-Valer, M.M., Fauth, D.J., Kuchta, M.E., Zhang, Y., Andrésen, J.M. (2005). Activation of magnesium rich minerals as carbonation feedstock materials for CO_2 sequestration. *Fuel Processing Technology*, Volume 86, Issues 14-15 (October 2005), Pages 1627-1645, ISSN 0378-3820.

Martin, R.T., Bailey, S.W., Eberl, D.D., Fanning, D.S., Guggenheim., S., Kodama, H., Pevear, D.R., Srodon, J., Wicks, F.J. (1991). Report of the clay minerals society nomenclature committee: revised classification of clay materials. *Clays and Clay Minerals*, Volume 39, Issue 3 (1991), Pages 333–335, ISSN 0009-8604.

Massazza, F. (1993). Pozzolanic cements. *Cement and Concrete Composites*, Volume 15, Issue 4 (1993), Pages 185-214, ISSN 0958-9465.

Molina-Montes, E., Timón, V., Hernández-laguna, A., Sainz-díaz, C.I. (2008). Dehydroxylation mechanisms in Al^{3+}/Fe^{3+} dioctahedral phyllosilicates by quantum mechanical methods with cluster models. *Geochimica et Cosmochimica Acta*, Volume 72, Issue 16 (August 2008), Pages 3929-3938, ISSN 0016-7037.

Morse J.W., Arvidson R.S. (2002). The dissolution kinetics of major sedimentary carbonate minerals. *Earth-Science Reviews*, Volume 58, Issues 1-2 (July 2002), Pages 51-84, ISSN 0012-8252.

Oelkers, E.H. (2001). General kinetic description of multioxide silicate mineral and glass dissolution. *Geochimica et Cosmochimica Acta*, Volume 65, Issues 21 (November 2001), Pages 3703-3719, ISSN 0016-7037.

Oelkers, E.H., Schott, J., An experimental study of enstatite dissolution rates as a function of pH, temperature, and aqueous Mg and Si concentration, and the mechanism of pyroxene/pyroxenoid dissolution, *Geochimica et Cosmochimica Acta*, Volume 65, Issue 8 (April 2001), Pages 1219-1231, ISSN 0016-7037.

Oelkers, E.H., Schott. J., Experimental study of anorthite dissolution and the relative mechanism of feldspar hydrolysis. *Geochimica et Cosmochimica Acta*, Volume 59, Issue 24 (December 1995), Pages 5039-5053, ISSN 0016-7037.

Pauwels, H., Gaus I., Michel le Nindre, Y., Pearce, J., Czernichowski-Lauriol, I. (2007). Chemistry of fluids from a natural analogue for a geological CO_2 storage site (Montmiral, France): Lessons for CO_2–water–rock interaction assessment and monitoring. *Applied Geochemistry*, Volume 22, Issue 12 (December 2007), Pages 2817-2833, ISSN 0883-2927.

Pires, J.C.M., Martins, F.G., Alvim-Ferraz, M.C.M., Simões, M. (2011). Recent developments on carbon capture and storage: An overview. *Chemical Engineering Research and Design*, Volume 89, Issue 9 (September 2011), Pages 1446-1460, ISSN 0263-8762.

Pironon, J., Jacquemet, N., Lhomme T., Teinturier S. (2007). Fluid inclusions as micro-samplers in bath experiments: A study of the system C-O-H-S-cement for the potential geological storage of industrial acid gas. *Chemical Geology*, Volume 237, Issues 3-4 (March 2007), Pages 264-273, ISSN 0009-2541.

Ptáček, P., Nosková, M., Brandštetr, J., Šoukal, F., Opravil, T. (2010). Dissolving behaviour and calcium release from fibrous wollastonite in acetic acid solution. *Thermochimica Acta*, Volume 498, Issues 1-2 (January 2010), Pages 54-60, ISSN 0040-6031.

Ptáček, P., Nosková, M., Brandštetr, J., Šoukal, F., Opravil, T. (2011). Mechanism and kinetics of wollastonite fibre dissolution in the aqueous solution of acetic acid. *Powder Technology*, Volume 206, Issue 3 (January 2011), Pages 338-344, ISSN 0032-5910.

Ptáček, P., Šoukal, F., Opravil, T., Havlica, J., Brandštetr, J. (2011). The kinetic analysis of the thermal decomposition of kaolinite by DTG technique. *Powder Technology*, Volume 208, Issue 1 (March 2011), Pages 20-25, ISSN 0032-5910.

Pushpaletha, P., Rugmini, S., Lalithambika, M. (2005). Correlation between surface properties and catalytic activity of clay catalysts. *Applied Clay Science*, Volume 30, Issues 3-4 (November 2005) Pages 141-153, ISSN 0169-1317.

Richardson, F.D. (1974). *Physical Chemistry of Melts in Metallurgy, Volume 2*, Academic Press Inc, ISBN 0125879024, New York.

Rimstidt, J.D., Dove, P.M. (1986). Mineral/solution reaction rates in a mixed flow reactor: wollastonite hydrolysis. *Geochimica et Cosmochimica Acta*, Volume 50, Issue 11 (November 1986), Pages 2509-2516, ISSN 0016-7037.

Saikia, N., Sengupta. P., Gogoi, P.K., Borthakur, P.Ch. (2002). Kinetics of dehydroxylation of kaolin in presence of oil field effluent treatment plant sludge. *Applied Clay Science*, Volume 22, Issue 3 (December 2002), Pages 93-102, ISSN 0169-1317.

Saldi, D.G., Köhler, J.S., Marty, N., Oelkers, H.E. (2007). Dissolution rates of talc as a function of solution composition, pH and temperature. *Geochimica et Cosmochimica Acta*, Volume 71, Issue 14 (July 2007), Pages 3446-3457, ISSN 0016-7037.

Šesták, J. (1984). *Thermal analysis. Part D: Thermophysical properties of solids – Their measurements and theoretical thermal analysis*, Elsevier, ISBN 0-444-99653-2, Amsterdam.

Soong, Y., Goodman, A.L., McCarthy-Jones, J.R., Baltrus, J.P. (2004). Experimental and simulation studies on mineral trapping of CO_2 with brine. *Energy Conversion and Management*, Volume 45, Issues 11-12 (July 2004), Pages 1845-1859, ISSN 0196-8904.

Spreng, D., Marland, G., Weinberg, A. M. (2007). CO_2 capture and storage: Another Faustian Bargain? *Energy Policy*, Volume 35, Issue 2 (February 2007), Pages 850-854, ISSN 0301-4215.

Stumm, W. (1997). Reactivity at the mineral-water interface: dissolution and inhibition. *Colloids and Surfaces A: Physicochemical and Engineering Aspects*, Volume 120, Issues 1-3 (February 1997), Pages 143-166, ISSN 0927-7757.

Teir, S., Eloneva, S., Fogelholm, C.-J., Zevenhoven, R. (2007). Dissolution of steelmaking slags in acetic acid for precipitated calcium carbonate production. *Energy*, Volume 32, Issue 4 (April 2007), Pages 528-539, ISSN 0360-5442.

Torp, T.A., Gale, J. (2004). Demonstrating storage of CO_2 in geological reservoirs: The Sleipner and SACS projects. *Energy*, Volume 29, Issues 9-10 (July-August 2004), Pages 1361-1369, ISSN 0360-5442.

Tyagi, B., Chudasama, Ch.D., Jasra R.V. (2006). Determination of structural modification in acid activated montmorillonite clay by FT-IR spectroscopy. *Spectrochimica Acta Part A: Molecular and Biomolecular Spectroscopy*, Volume 64, Issue 2 (May 2006), Pages 273-278, ISSN 1386-1425.

Ullman W.J., Kirchman D.L., Welch S.A., Vandevivere P. (1996). Laboratory evidence for microbially mediated silicate mineral dissolution in nature, *Chemical Geology*, Volume 132, Issues 1-4 (October 1996), Pages 11-17, ISSN 0009-2541.

Valensi, G. (1936). Kinetics of oxidation of metallic spherules and powders. *Comptes Rendus*, 202 (1936) 309–312.

Viseras, C., Cerezo, P., Sanchez, R., Salcedo, I., Aguzzi, C. (2010). Current challenges in clay minerals for drug delivery. *Applied Clay Science*, Volume 48, Issue 3 (April 2010), Pages 291-295, ISSN 0169-1317.

Vlaev, L., Nedelchev, N., Gyurova, K., Zagorcheva, M. (2008). A comparative study of non-isothermal kinetics of decomposition of calcium oxalate monohydrate. *Journal of Analytical and Applied Pyrolysis*, Volume 81, Issue 2 (March 2008), Pages 253-262, ISSN 0165-2370.

Weissbart, E.J., Rimstidt, J.D. (2000). Wollastonite: incongruent dissolution and leached layer formation. *Geochimica et Cosmochimica Acta*, Volume 64, Issue 23 (December 2000), Pages 4007-4016, ISSN 0016-7037.

Wong, C.S., Matear, R.J. (1989). Ocean disposal of CO_2 in the North Pacific Ocean: Assessment of CO_2 chemistry and circulation on storage and return to the atmosphere. *Waste Management*, Volume 17, Issues 5-6 (1998), Pages 329-335, ISSN 0956-053X.

Xu, T., Apps, J.A., Pruess, K. (2004). Numerical simulation of CO_2 disposal by mineral trapping in deep aquifers. *Applied Geochemistry*, Volume 19, Issue 6 (June 2004), Pages 917-936, ISSN 0883-2927.

Yang, H., Xu, Z., Fan, M., Gupta, R., Slimane, R.B., Bland A.E., Wright, I. (2008). Progress in carbon dioxide separation and capture: a review. *Journal of Environmental Sciences*, Volume 20, Issue 1 (2008), Pages 14-27, ISSN 1001-0742.

Yehia, A.A., Akelah, A.M., Rehab, A., El-Sabbagh, S.H., Nashar D.E. El., Koriem, A.A. (2012). Evaluation of clay hybrid nanocomposites of different chain length as reinforcing agent for natural and synthetic rubbers. *Materials & Design*, Volume 33 (January 2012), Pages 11-19, ISSN 0261-3069.

Zhang, Y., Oldenburg, C.M., Finsterle S., Bodvarsson, G.S. (2007). System-level modelling for economic evaluation of geological CO_2 storage in gas reservoirs. *Energy Conversion and Management*, Volume 48, Issue 6 (June 2007), Pages 1827-1833, ISSN 0196-8904.

Permissions

The contributors of this book come from diverse backgrounds, making this book a truly international effort. This book will bring forth new frontiers with its revolutionizing research information and detailed analysis of the nascent developments around the world.

We would like to thank Muhammad Akhyar Farrukh, for lending his expertise to make the book truly unique. He has played a crucial role in the development of this book. Without his invaluable contribution this book wouldn't have been possible. He has made vital efforts to compile up to date information on the varied aspects of this subject to make this book a valuable addition to the collection of many professionals and students.

This book was conceptualized with the vision of imparting up-to-date information and advanced data in this field. To ensure the same, a matchless editorial board was set up. Every individual on the board went through rigorous rounds of assessment to prove their worth. After which they invested a large part of their time researching and compiling the most relevant data for our readers. Conferences and sessions were held from time to time between the editorial board and the contributing authors to present the data in the most comprehensible form. The editorial team has worked tirelessly to provide valuable and valid information to help people across the globe.

Every chapter published in this book has been scrutinized by our experts. Their significance has been extensively debated. The topics covered herein carry significant findings which will fuel the growth of the discipline. They may even be implemented as practical applications or may be referred to as a beginning point for another development. Chapters in this book were first published by InTech; hereby published with permission under the Creative Commons Attribution License or equivalent.

The editorial board has been involved in producing this book since its inception. They have spent rigorous hours researching and exploring the diverse topics which have resulted in the successful publishing of this book. They have passed on their knowledge of decades through this book. To expedite this challenging task, the publisher supported the team at every step. A small team of assistant editors was also appointed to further simplify the editing procedure and attain best results for the readers.

Our editorial team has been hand-picked from every corner of the world. Their multi-ethnicity adds dynamic inputs to the discussions which result in innovative outcomes. These outcomes are then further discussed with the researchers and contributors who give their valuable feedback and opinion regarding the same. The feedback is then collaborated with the researches and they are edited in a comprehensive manner to aid the understanding of the subject.

Apart from the editorial board, the designing team has also invested a significant amount of their time in understanding the subject and creating the most relevant covers. They scrutinized every image to scout for the most suitable representation of the subject and create an appropriate cover for the book.

The publishing team has been involved in this book since its early stages. They were actively engaged in every process, be it collecting the data, connecting with the contributors or procuring relevant information. The team has been an ardent support to the editorial, designing and production team. Their endless efforts to recruit the best for this project, has resulted in the accomplishment of this book. They are a veteran in the field of academics and their pool of knowledge is as vast as their experience in printing. Their expertise and guidance has proved useful at every step. Their uncompromising quality standards have made this book an exceptional effort. Their encouragement from time to time has been an inspiration for everyone.

The publisher and the editorial board hope that this book will prove to be a valuable piece of knowledge for researchers, students, practitioners and scholars across the globe.

List of Contributors

R. García and A. P. Báez
Centro de Ciencias de la Atmósfera, Universidad Nacional Autónoma de México, Ciudad Universitaria, Mexico City, Mexico

Pourya Biparva
Department of Nanotechnology, Islamic Azad University, Langaroud Branch, Langaroud, Iran

Amir Abbas Matin
Research Department of Analytical Chemistry, Iranian Academic Center for Education, Culture & Research (ACECR), Urmia, Iran

Hélcio José Izário Filho, Maria da Rosa Capri, Ângelo Capri Neto, Marco Aurélio Kondracki de Alcântara and André Luís de Castro Peixoto
Universidade de São Paulo, Brazil

Rodrigo Fernando dos Santos Salazar
Universidade Federal de São Carlos, Brazil

Ajai Prakash Gupta
Patent Cell Division, India

Suphla Gupta
Plant Biotechnology Department, Indian Institute of Integrative Medicine, Jammu-180001, Jammu & Kashmir, India

T.V. Sheina and K.N. Belikov
State Scientific Institution "Institute for Single Crystals" of National Academy of Sciences, Ukraine

Mark F. Zaranyika and Albert T. Chirenje
Chemistry Department, University of Zimbabwe, Mt Pleasant, Harare, Zimbabwe

Courtie Mahamadi
Chemistry Department, Bindura University of Science Education, Bindura, Zimbabwe

Roozbeh Javad Kalbasi and Neda Mosaddegh
Department of Chemistry, Shahreza Branch, Islamic Azad University, 311-86145 Shahreza, Isfahan, Iran Razi Chemistry Research Center, Shahreza Branch, Islamic Azad University, Shahreza, Isfahan, Iran

Lué-Merú Marcó Parra
Universidad Centro-Occidental Lisandro Alvarado, Decanato de Agronomía, Dpto. Química y Suelos, Núcleo Tarabana, Cabudare, Edo. Lara,Venezuela

Petr Ptáček, Magdaléna Nosková, František Šoukal, Tomáš Opravil, Jaromír Havlica and Jiří Brandštetr
Brno University of Technology, Faculty of Chemistry, Centre for Materials Research, Brno CZ-61200 ,Czech Republic

Printed in the USA
CPSIA information can be obtained
at www.ICGtesting.com
JSHW011445221024
72173JS00004B/953